W0042739

Ontogeny of Olfaction

Principles of Olfactory Maturation in Vertebrates

Edited by Winrich Breipohl

With 78 Figures

Springer-Verlag
Berlin Heidelberg New York
London Paris Tokyo

Editor:
Professor Dr. WINRICH BREIPOHL
Department of Anatomy, University of Queensland
St. Lucia/Brisbane, Australia 4067

Co-Editor:
Professor Dr. RAIMUND APFELBACH
Institut für Biologie III, Universität Tübingen
7400 Tübingen, FRG

Editorial Assistants:
MARIAN KREMER, Institut für Anatomie
Universitätsklinikum Essen, 4300 Essen, FRG

BRIAN KEY, Department of Anatomy
University of Queensland, St. Lucia/Brisbane
Australia 4067

MICHAEL RAUWOLF, Institut für Anatomie
Universitätsklinikum Essen, 4300 Essen, FRG

ISBN-13: 978-3-642-71578-5   e-ISBN-13: 978-3-642-71576-1
DOI: 10.1007/978-3-642-71576-1

Library of Congress Cataloging in Publication Data. Ontogeny of olfaction. Includes index.
1. Smell. 2. Ontogeny. I. Breipohl, W. [DNLM: 1. Central Nervous System–growth & development. 2. Olfactory Bulb–growth & development. 3. Olfactory Mucosa–growth & development.
WV 301 059] QP458.058   1986   596',01'826   86-22113

2131/3130-543210

This book is dedicated to my former colleagues, coworkers, and friends in the Medical Faculty at the University of Essen (FRG) and to my new colleagues and coworkers at the University of Queensland (Australia).

PREFACE

The impacts of specifically experienced external and internal environments upon phylogenetically established pathways of ontogenetic development seem to be responsible for the intra-species variation of organisms. Therefore normal function, as well as disorders of sensory systems, can often be better understood by considering the principles of ontogenetic maturation and the time schedule of environmental influences during that period.

Sensory organs and systems have long been considered to be hereditally determined for their analysis of the environment. However, it is becoming more and more clear that their specificity depends also upon the sequence and nature of environmental impacts, impacts during highly sensitive periods of postnatal life being most effective.

The present book, using an interdisciplinary approach, has brought together various descriptions of developmental processes for one of the phylogenetically oldest sensory analytical circuits, the olfactory system. Although investigations of a wide range of vertebrates are included, from marsupials to man, the main attention was given to rodents, the most frequently used experimental animal in this fields of research. Authors have been asked for reviews and original considerations on the rationales of developmental principles in the olfactory system, and wherever appropriate to highlight its uniqueness or similarities with other parts of the nervous system. Authors were also asked to consider future needs of research in their fields, to describe their own approaches in this context, and to outline general horizons of neurodevelopmental research. As a result the book contains many new sophisticated explanations of the maturation and function of the olfactory system, with several articles presenting data which highlight urgent needs to modify some present hypotheses.

Although being aware of the Sysiphos-like type of all scientific investigations and the relativity of knowledge, I hope that this collection of articles may lead to better understanding of the role of inborn and acquired features for the normal structure and function as well as disorders of the vertebrate nervous system in general and the olfactory system in particular. This book, hopefully, may also stimulate the reader's scientific enjoyment and encourage his own future contributions in this field of neurobiology.

Winrich Breipohl

CONTENTS

E. MATURATION OF TERTIARY OLFACTORY CENTERS

F. POSTNATAL DEVELOPMENT OF OLFACTORY-GUIDED BEHAVIOR

LIST OF CONTRIBUTORS

A. INTRODUCTION

Ontogeny of Human Olfactory Function

RL Doty

Smell and Taste Center, Department of Otorhinolaryngology and Human Communication, and Department of Physiology, School of Medicine, University of Pennsylvania, Philadelphia, PA 19104, USA

1 Introduction

Until recently, little was known about the onset of olfactory function in humans or about the nature of alterations in smell function across the lifespan. Unequivocal answers to such basic questions as to whether neonates can detect odorants (independent of trigeminal stimulation) or to what degree smell ability declines in old age were not available. During the last decade a number of behavioral studies have provided definitive answers to such questions, ushering in a new era of human behavioral research. This research, complemented by anatomical studies, indicates that the human olfactory system is operative at birth and that massive loss of smell function occurs in many older individuals. However, as will be addressed in this chapter, numerous other questions concerning the nature of age-related changes in human olfactory function remains unanswered.

2 Olfaction in the Neonate

2.1 Anatomical Considerations

The olfactory neuroepithelium is well developed in the human fetus, with cilia appearing from olfactory knobs by 9 weeks. By 11 weeks completely differentiated olfactory cells are observed (Pyatkina 1982). Adult-like lamination of the olfactory bulb is seen at 18.5 weeks (Humphrey 1940). Both periglomerular and inter-glomerular cells are seen at this early stage, as is a well-defined granule cell layer[1]. However, olfactory marker protein is not yet present in the 22-week fetus, even though it is present in the adult (Nakashima et al. 1985). This

---

[1]The appearance of the granule cells at this early stage is unlike the situation in rats and mice, where a large proportion of these cells make their appearance after birth, originating as late as several weeks postnatally (Altman 1969; Altman and Das 1966; Hinds 1968 a,b). Apparently the full compliment of granule and periglumerular cells is not needed for many early olfactory functions, since newborn rats use odor cues in suckling and can be readily conditioned to approach or avoid novel odors (Blass et al. 1977; Rudy and Cheatle 1977). Furthermore fetal rats (gestational day 20) that have received in utero injections of apple juice and lithium chloride show postpartum evidence of having developed a conditioned aversion to the odor of apple juice (Smotherman 1982) (for review, see Doty 1985b).

protein first appears in rat receptor neurons on the day they evidence selectivity to odorants (Gesteland et al. 1982) and, thus, may be a marker for functional olfactory receptor cell activity.

The trigeminal nerve (CN V), which mediates intranasal and intraoral irritative responses to volatile and non-volatile chemicals and skin sensations (e.g., coolness, fullness, sharpness, warmth; cf. Doty et al. 1978), is well formed in utero (Gasser and Hendrickx 1969; Hogg 1941) and is functional at birth. Indeed, the perioral areas supplied by the mandibular and maxillary divisions of this nerve is the first region of the embryo to be sensitive to cutaneous stimulation (circa 7.5 weeks). The ophtalmic division, which distributes free nerve endings throughout the nasal mucosa, has been observed in month-old embryos and is presumably functional by 10.5 weeks of gestational age (Brown 1974, Humphrey 1966; Streeter 1908).

As other mammals, humans possess a well-developed vomeronasal organ in utero (cf. Bossy 1980, Humphrey 1940, Nakashima et al. 1984, Pyatkina 1982). However, unlike the case of most other forms, a functioning vomeronasal organ is apparently not present postnatally, and the accessory olfactory formation undergoes degenerative rather than generative changes during the second third of gestation (Humphrey 1940).

2.2 Behavioral Studies of Neonatal Olfactory Function

Responses to General Odors. A number of early researchers observed in the movements and facial expressions of infants presented with various odorants, including asafoetida, bone oil, and orange extract (e.g., Genzmer 1873, Kroner 1881, Kussmaul 1896, Peterson and Rainey 1910-1911). Although most of these workers concluded that odorants elicit behavioral responses in newborns (and, in some cases, premature infants as young as 7 months of gestational age), their results were not definitive, since blind scoring procedures and blank stimulus controls were not used (see Disher 1934, for a review). Nonetheless, the general phenomena observed by these investigators have been confirmed by more recent studies in which such controls were incorporated.

The most extensive of these early studies was performed by Peterson and Rainey (1910-1911) at the New York Lying-In Hospital. Two hundred and seven normal term babies (as well as several premature ones) were tested. The odorants used were asafoetida, compound spirts of orange, oil of rose geranium, tincture of gentian, and mother's milk. No details are provided as to how the stimuli were presented, only that most of them induced behavioral responses. A sample of the reactions elicited in individual subjects under the postpartum age of 6 hours is presented in Table 1. It is of interest that sucking was the most common behavior elicited by the more pleasant (to adults) odorants (orange extract, oil of geranium), whereas grimaces and head movements occurred most commonly in response to the unpleasant odorant (asafoetida).

To eliminate subjectivity inherent in observations such as those noted above, Engen et al. (1963) used electronic transducers to monitor leg withdrawal, general body activity, respiration, and heart rate of 20 neonates (32 to 68 h old) before, during and following the presentation of odorants on cotton Q-tips. In the first of two experiments, a higher percentage of responses was observed for acetic acid than for phenyl ethyl alcohol, suggesting that unpleasant or irritating odorants may be particularly effective in producing the changes. In the second experiment,

Table 1 Reactions of newborn infants (6 h old or younger) to odorants noted by Peterson and Rainey (1910-1911)

Subject #	Sex	h Postpartum	Odorant	Reaction
1	F	1st	O	Removed fist from mouth but continued to suck
2	F	1st	A	After 30 s started, moved head, expression of disgust or grimice followed by sucking
3	F	1st	O	Slight grimice, sucking movement
6	M	15 min	A	Grimice
7	M	1st	O	Opened eyes, sucking movement
11	M	2nd	O	Dilated nostrils
12	M	2nd	O	Vigorous sucking
16	F	3rd	G	Stopped crying, sucking
20	M	3rd-4th	A	Grimice - upward movement of hands and feet
22	M	3rd-5th	O	Respiration quickened - squirmed
23	F	4th	G	After several seconds stopped crying, sucking movement
32	M	5th	O	Sucking movement
33	M	5th	O	Grimace and restlessness followed by sucking
34	F	5th	O	Sucking movement, squirmed
37	F	5th-7th	O	Sucking, restlessness, turned head, upward movement of hands
40	M	6th	A	Accelerated breathing
42	M	6th	A	Grimice, sucking, turned head away
46	F	6th	O	Sucking movement
48	M	6th-10th	G	Opened and shut eyes

A = asafoetida; G = oil of geranium; O = compound spirits of orange.

anise and asafoetida were found to elicit increases in the dependent measures, with a greater percentage of responses occurring to the more unpleasant of the stimuli (asafoetida). However, unlike the case with acetic acid and phenyl ethyl alcohol, there was a significant decrement in responses across ten successive stimulus trials. Recovery of responses occurred following the presentation of the other odorant. Engen and Lipsitt (1965) argue from a subsequent finding (that greater recovery to a two-component odor mixture occurs following the presentation of the component least similar in odor to the mixture) that such response decrements are due to habituation, rather than to sensory fatigue.

In another series of related experiments, Lipsitt et al. (1963) presented successively higher concentrations of asafoetida until a response was elicited. In general, sensitivity appeared to increase over the first 4 days of life. To what extent this increase in reactivity reflects temporal changes in the olfactory systems, the motor response systems, learning, sensitization, or such factors as airway patency (e.g., from mucus clearing) is not known.

Because many of the aforementioned studies used odorants with probable trigeminal activity, Self et al. (1972) chose to test 32 infants on each of the first four postpartum days with compounds they believed to be free from such activity (oil of anise, tincture of asafoetida, oil of lavender, tincture of valerian, and water). Although responses were noted in most of the infants in at least one of the test days, large individual differences were present, and a relatively high rate of responding occurred for water alone. Furthermore, the same infant rarely gave uniform responses across the test days, and the two dependent measures were often not in agreement. Nonetheless, responses to the odorants were observed regardless of the infant's level of sleep (irregular or deep), and indivuals seemed to fall into three levels of responding: low, moderate, and high. No sex differences were observed in the dependent measures.

More recently, Steiner (1977, 1979) performed cinematographic studies of the facial reactions of neonates (less than 12 h old) to odorants presented on cotton swabs. The stimuli included banana extract, vanilla extract, artificial shrimp flavor of a "fishy" character, artificial butter flavor, and the artificial odor of rotten eggs. Compared to the control condition, in which a swab dipped in propylene glycol was presented, apparently all the subjects displayed facial responses to the odorants. These responses fell into two general classes, one resembling a "smiling" expression accompanied by sucking movements (interpreted as a reaction of acceptance, satisfaction or liking) and the other typified by the depression of the mouth angles or a pursing of the lips (interpreted as dislike or rejection). The banana, vanilla, and butter odors elicited the first type of reaction, whereas the shrimp and rotten egg odors elicited the second type.

Taken together, these studies suggest that newborn infants respond to odorous volatiles. It is not clear in all cases, however, whether the observed responses reflect stimulation of olfactory receptors or free nerve endings from the trigeminal nerve, since few odorants are "pure" olfactory stimulants (Doty et al. 1978). Nonetheless, the differential responses noted to relativily weak and likely nonirritating stimuli (e.g., anise, butter, vanilla) strongly suggest that infants can detect odorants independent of trigeminal activity.

Responses to Maternal Odors. Perhaps the most compelling evidence that neonates can detect odorous volatiles via olfactory, rather than trigeminal, cues comes from recent studies showing that infants preferentially orient towards or respond to body odors of their own mothers compared to those from unfamiliar mothers (cf. Russell 1976). For example, Macfarlane (1975) found, in a pioneering experiment, that infants spent more time orienting towards a breast pad lowered into its crib which was previously worn by its lactating mother than a juxtaposed breast pad which was clean or worn by an unfamiliar lactating mother. The differential reaction to the breast pads increased across the first few days of life, suggesting that either learning or development was associated with the response.

Evidence that learning is prepotent comes from a recent series of similar studies by Cernoch and Porter (1985) in which axillary odors were used. These authors demonstrated that breast fed neonates of approximately 2 weeks of age preferentially oriented towards axillary odors from their mother relative to axillary odors from (a) another lactating mother or (b) a nonlactating unfamiliar mother. However, bottle-fed infants failed to show this phenomenon, suggesting that the breast-fed infants learned to respond to the maternal odors. No preference was observed when the axillary odors of the father were juxtaposed to those of an unfamiliar male.

Responses to Irritating Vapors. In addition to the observation of Engen et al. (1963) that the presentation of acetic acid vapors changes bodily movements and cardiovascular responses of neonates, there is evidence that neonates have the capability of responding directionally to the presence of irritating vapors. Rieser et al. (1976) videotaped movements of infants aged from 16 to 131 h in bassinets following the brief positioning of two open-ended glass sleeves in front of their nostrils one containing cotton saturated with a low concentration of ammonium hydroxide (6 M) and the other cotton alone. On 64% of 304 trials the newborns turned away from the side of the ammonia stimulus, and 30% they turned towards that side.

3 Olfaction in Children and Teenagers

Compared to research done on infants, little work has been done to evaluate the olfactory function of children or teenagers. However, it is clear that children recognize and respond to a wide range of odorants and, in some cases, evidence preferences for odorants which seem different from those of adults (cf. Moncrieff 1966).

3.1 Behavioral Studies

Odor Identification Ability. Most children are familiar, at a relative young age, with a broad range of odors. For example, by 6 years of age American girls, on the average, correctly identify over 75% of the odorants on the University of Pennsylvania Smell Identification Test (UPSIT), a standardized 40-odorant test (cf. Doty et al. 1984 a,b; 1985 a,b). By 10 years of age they are performing, for all practical purposes, at adult levels on this test. Boys, while not performing at the level of girls, still correctly identify approximately two-thirds of the odors by the age of 7, and achieve adult levels of performance before the teenage years (Fig. 1).

The correlation between age and UPSIT scores across the 4 to 12 year age span is relatively strong (Pearson $r = 0.74$, $df = 293$, $p < 0.001$). To determine if the age-related changes in odor identification scores were mainly a reflection of knowledge of the visual sources of the odors of the odorant names used in the UPSIT, we tested a subset of 103 youngsters between the ages of 4 and 11,

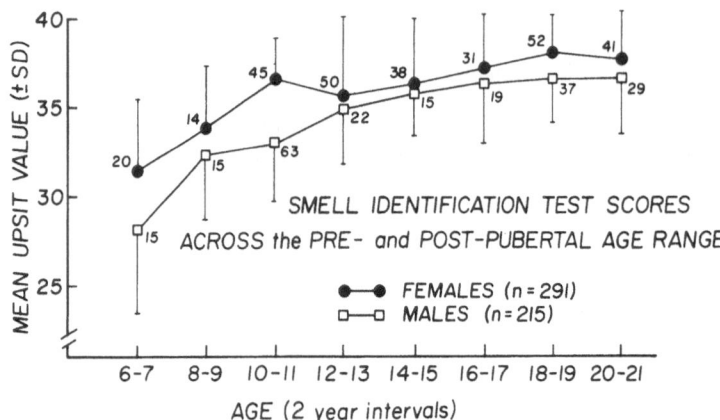

Fig. 1 Average scores on the University of Pennsylvania Smell Identification Test (UPSIT) as a function of subject gender across the prepubertal, adolescent, and early adult years (Doty 1985a)

inclusive, on both the UPSIT and the Picture Identification Test (PIT)[2]. It is apparent from Table 2 that even 4 to 5 year olds are familiar with the objects and name concepts used in the UPSIT, as evidenced by their high scores on the PIT. Although a 0.53 correlation was present between the UPSIT and PIT scores, removal of the age variable by means of a partial correlation resulted in a much lower r value (0.37, p < 0.001) (Table 2). These findings suggest that familiarity with the odorant source of its name unlikely to be the major cause of the changes in odor identification ability across the early years. If familiarity plays a role, it is presumable familiarity with the odor, per se, or its relationship to the odorant source.

Olfactory Threshold Sensitivity. Within the limits of the testing procedures used in their time, a number of early investigators concluded that children were not less sensitive to odorants than adults (for reviews, see McCartney, 1968, and Peto, 1936). Indeed, Toulouse and Vaschide (1899) tested three groups of 163 school children (3-5 years, 6 years, and 12 years of age) for their sensitivity to camphor and concluded that sensitivity was slightly greater in younger chidren, with girls being more sensitive than boys.

---

[2]The PIT is a test identical in format and form to the UPSIT except that drawings of the odor source, rather than the microencapsulated odorants, are presented. For example, the first item of the PIT reads "This picture looks most like: a) gasoline, b) pizza, c) peanuts, or d) lilac" instead of "This odor smells most like: a) gasoline, b) pizza, c) peanuts, or d) lilac." The pictures are located on the pages of the test booklets in such a manner that they can be turned over to reveal the question, resulting in the intervention of an equivalent amount of time as occurs with the UPSIT between stimulus exposure and generation of the forced-choice response.

Table 2 Mean scores on the University of Pennsylvania Smell Identification Test (UPSIT) and the Picture Identification Test (PIT) in four young age groups. Numbers in brackets are standard deviations

	Age group (Years)			
Test	4-5	6-7	8-9	10-11
UPSIT	20.11	28.00	33.81	35.45
	(6.24)	(5.74)	(3.24)	(3.03)
PIT	35.21	36.94	38.91	39.20
	(4.83)	(4.39)	(1.32)	(1.56)

In more recent work, Strauss (1970) tested olfactory sensitivity to phenyl ethyl alcohol in 300 persons falling into three groups: 8-10 years, 16-18 years, and 21-39 years. Although the threshold values were found to be lowest in the 8-10 year old group, these results are misleading since the Elsberg blast-injection procedure was used (see Wenzel 1948, for a critique). Aside from the problem of confounding olfactory and tactile responses, this procedure can result in lower thresholds in children simply because less volume is needed to stimulate the receptor region of smaller noses. The lack of a forsed-choise procedure further confounded this study.

In an experiment designed mainly to explore sex differences in olfactory thresholds, Koelega and Koster (1974) determined, using a forced-choise method of constant stimuli procedure, thresholds for amyl acetate in subjects ranging in average age from 9 to 20 years. On the average, the threshold values of the youngsters were of the same order of magnitude as those of the adults. However, in another experiment of this series, these authors noted that a number of the prepubescent youngsters were unable to detect the musk odors of pentadecanolide or oxahexadecanolide at the concentrations detected by adults, suggesting they may less sensitive to them.

Suprathreshold Odor Intensity Perception. Le Magnen (1952) asked 25 men, 25 women, 22 prepubescent boys, and 25 prepubescent girls to rate the intensity of crystalline Exaltolide (pentadecanolide) odor as either (a) absent or very weak, (b) weak, (c) strong, or (d) extremely strong. Significantly more of the boys, girls, and men rated the odor absent or very weak, as compared to the adult women, who typically rated the odor as strong or very strong. Assuming that this differences are nor due to differential response biases, they would support the notion that youngsters perceive this particular odorant as less strong than adult females.

Odor Preferences. There is some question as to when odor preferences arise in children. Assuming that facial expressions and movements toward breast pads reflect preferences, it would appear that clear-cut preferences are present at birth. However, the suggestion that 3- to 4-year-old children are more tolerant of unpleasant odors than adults (Stein et al. 1958) can be questioned, since this research did not control for the tendency of young children to respond agreeably

Table 3 Average rankings given to ten odorants by five age groups. Numbers in brackets at the bottom of each column indicate range of rankings. (After Moncrieff 1966)

Age group (years)				
0-7 (n=98)	8-14 (n=162)	15-19 (n=150)	20-40 (n=100)	>40 (n=49)
1.50 Strawberry	1,58 Strawberry	2.13 Strawberry	3.22 Strawberry	3.86 Musk
3.99 Vanillin	3.62 Vanillin	4.13 Vanillin	4.10 Lavender	4.05 Strawberry
4.94 Spearmint	4.28 Musk	4.90 Spearmint	4.43 Vanillin	4.36 Vanillin
5.35 Naphthalene	4.70 Naphthalene	4.90 Lavender	4.60 Musk	4.43 Lavender
5.56 Musk	5.19 Spearmint	4.96 Almond	5.16 Naphthalene	4.93 Spearmint
5.79 Rape oil	5.59 Almond	5.04 Musk	5.25 Almond	5.09 Neroli
5.81 Almond	6.60 Lavender	5.18 Naphthalene	5.30 Spearmint	5.29 Almond
6.01 Lavender	7.10 Neroli	7.09 Neroli	5.87 Neroli	5.77 Naphthalene
7.51 Neroli	7.42 Rape oil	7.46 Rape oil	7.80 Rape oil	7.97 Rape oil
8.54 Chlorophyll	8.88 Chlorophyll	9.20 Chlorophyll	9.26 Chlorophyll	9.26 Chlorophyll
(7.04)	(7.30)	(7.07)	(6.04)	(5.40)

in situations where either "yes" or "no" is required as a response. For example, Engen (1974) has shown that when a four year old child is presented with an odor that is unpleasant to adults and asked the question, "Does it smell pretty?", the child typically answers in the affirmative. However, when presented with the same odor and asked, "Does it smell ugly?", he is also more prone to also say yes, such that the number of positive responses to "Does it smell pretty?" is much higher than the number of negative responses to "Does it smell ugly?". Despite such response biases, however, Kneip et al. (1931) found the preference rankings of 7 to 13 year old children given to 14 diverse odorants to be similar to those obtained from 18 to 24 year olds (Spearman rs > 0.91), suggesting that "This high degree of consistency may be attributed to physiological mechanisms which are relatively mature at the age of seven years" (p. 416).

The possibility that age-related preferences exist for odors was also suggested by Moncrieff (1966), who found that persons under the age of 15 tended to rank, relative to adults, flower smells lower and fruity smells higher in preference. This phenomenon is seen in the data of Tabe 3, in which the preference for strawberry odor appears to dissipate with age. Since all 10 ranks were used by all groups, the differences observed in Tabel 3 cannot be attributed to artifacts such as the use of scale values. However, they may reflect an increase in individual differences among the adults.

Engen and Corbit (1970) present data from indirect scaling procedures that suggest children do not discriminate as well as adults between odorants on a hedonic or preference scale. Although their data agrees with that of Kneip et al. (1931) in demonstrating that the average preferences of children are similarly ordered to those of adults, the difference in preference values for the odorants decreased as the age of the children decreased, implying that the younger the child, the less fine the hedonic distinction.

More recently, Laing and Clark (1983) examined the preferences of 302 males aged 8-9, 14, and 16 years for 10 food-related odors: meat (roast beef gravy), fish, chocolate, onion, Vegemite, peanut butter, spearmint, chicken, expresso coffee, and parmesan cheese. Although these authors concluded "that no major change occurs in preferences for food related odors between the ages of 8 and 16 years and that puberty has no major effect of preferences", they did find small statistically significant differences between 8-9 year old subjects and the others for meat and chicken odors, and between the 14 year old subjects and the others for peanut butter odor. In addition, the preference ratings for the odor of coffee changed from dislike to like as the age of the subjects increased.

4 Conclusions

As evidenced in this review, preference measures have served a major role in the measurement of human olfactory function. Although this may reflect, in part, the ease with which such measures can be obtained (compared to measures requiring more accurate stimulus quantification, such as thresholds), multidimensional odor scaling studies of adults typically extract a strong hedonic dimension (e.g., Schiffman 1974). Furthermore, pleasantness appears in a number of odor classification schemes (Harper et al. 1968). Thus, hedonicity may well be a fundamental component of olfactory perception. Such a focus on hedonics is also observed in studies of taste function (see Pfaffmann, 1960), and may have a neurological basis. For example, Scott and Perrotto (1980) have collected neural data from the brainstem gustatory relay nuclei which appear to support the notion that a neural correlate for hedonics exists at that level, coexistent with neural correlates for quality and intensity.

In light of the relation between olfaction and hedonic mechanisms, it is of interest to understand the genesis of odor preferences. Specifically, to what extent are they inborn or altered by experience? Although to my knowledge no studies have directly examined the inheritance of human odor preferences, the heritability coefficients of taste preferences for different concentrations of sucrose are uniformly low, suggesting that the variability observed in these measures may be largely due to experience (Green et al. 1975); see also Beauchamp and Moran

1982, 1984)[3]. However, the failure to find a genetic basis for differences among individuals in preferences for various concentrations of sugar water is less profound than the fact that the sugar preference itself is likely innate. The same would appear to apply to odors. Thus, the observations of Steiner (1977, 1979) and others suggest that neonates have innate general odor preferences. Clearly, a major question is the extent to which such general odor preferences can be altered by learning. Presumably preferences to some odorants are highly malable and preferences to others less so. The situation is a complex one, however, since liking for a relatively neutral flavor can be enhanced by simply exposing a subject to it (Cain and Johnson 1978; Pliner 1982). Furthermore, even initially aversive tastants (and presumable odorants) such as chili pepper can become preferred within the context of a social situation (Rozin and Schiller 1980), even in chimpanzees (Rozin and Kennel 1983). The importance of subject variables (e.g., gender, age, education) likewise play a role in determining such preferences. For example, in a study of first- and second-generation Chinese adolescent immigrant boys to Canada, Hrboticky and Krondly (1984) found that the second generation boys and those with more accultured patterns of language use gave higher hedonic flavor and prestige ratings to desert, snack, and fast foods, and discriminated better among nutrient rich and poor foods.

Despite the recent progress that has been made in understanding the development of olfactory function, numerous questions remain to be answered. For example: To what range of odorants can the newborn respond? Can the odor-related facial responses of neonates be quantified in manner to provide a preference continuum? If so, does this continuum correlate with the preferences of adults? To what degree, if any, does postnatal morphogenesis occur in the human olfactory system? Are children more sensitive than adults to odorants? Are some odor qualities more susceptible to age-related alterations than others? At what age do strong odor preferences and aversions develop? Hopefully, answers to most of these and related questions will be forthcoming during the next decade of research in this fascinating area of study.

Summary

The present chapter reviews the available data on the development of the human olfactory function - data which lead to several basic conclusions: first, the human olfactory system is functional at birth; second, distinct facial responses resembling those of pleasure and displeasure are seen in human neonates during and following the presentation of odorants; third, the human neonate is responsive to air-borne irritants, such as ammonia, and has the capacity to localize

---

[3]Although the ability to detect most, if not all, odorants is dependent upon inborn mechanisms, evidence for heritability of olfactory sensitivity from twin studies is found for only some stimuli (e.g., androstenone; cf Hubert et al. 1980; Wysocki and Beauchamp 1984). In addition, it is noteworthy that both humans and rats can detect substances to which they have never been exposed during the course of ontogeny or evolution. For example, the detection thresholds of rats for perfluorcarbons are low and within the range of organic odorants used in olfactory threshold research (Marshall et al. 1981), suggesting that the olfactory system is responsive to an extremely wide array of stimulants for which direct selection pressures could not have occurred.

the direction from which they are presented; fourth, human infants can learn to distinguish, within the first few days of life, odors of their own mothers from those of strange mothers; and fifth, children and youngsters have similar general preferences to odorants as adults, although exceptions occur, likely as a function of experience and enculturation.

Acknowledgement. Supported by Grant NS 16365 from the National Institute of Neurological and Communicative Disorders and Stroke.

References

Altman J (1969) Autoradiographic and histological studies of postnatal neurogenesis. IV. Cell proliferation and migration in the anterior forebrain, with special reference to persistent neurogenesis in the olfactory bulb. J. Comp Neurol 137: 433-458

Altman J, Das GD (1966) Autoradiographic and histological studies of postnatal neurogenesis. I. A longitudinal investigation of the kinetics, migration, and transformation of cells incorporating tritiated thymidine in neonate rats, with special reference to postnatal neurogenesis in some brain regions. J Comp Neurol 126: 337-389

Baron G, Frahm HD, Bhatnagar KP, Stephan H (1983) Comparison of brain structure volumes in Insectivora and Primates. III. Main olfactory bulb (MOB). J Hirnforsch 24: 551-68

Beauchamp GK, Moran M (1982) Dietary experience and sweet taste preference in human infants. Appetite 3: 139-152

Beauchamp GK, Moran M (1984) Acceptance of sweet and salty tastes in 2-year-old children. Appetite 5: 291-303

Blass EM, Teicher HH, Cramer CP, Bruno JP, Hall WG (1977) Olfactory, thermal and tactile control of suckling in preauditory and previsual rats. J Comp Physiol Psychol 91: 1248-1260

Bossy J (1980) Development of olfactory and related structures in staged human embryos. Anat Embryol (Berlin) 161: 225-236

Brown JW (1974) Prenatal development of the human chief sensory trigeminal nucleus. J Comp Neurol 156: 307-335

Cain WS, Johnson F Jr (1979) Lability of odor pleasantness: Influence of mere exposure. Perception 7: 459-465

Cernoch JM, Porter RH (1985) Recognition of maternal axillary odors by infants. Child Develop, in press

Disher DR (1934) the reactions of newborn infants to chemical stimuli administered nasally. In: Dockeray FC (ed) Studies of infant behavior. Ohio State Univ Press, Columbus, pp 1-52

Doty RL (1985a) Gender and endocrine-related influences upon olfactory sensitivity. In: Meiselman HL, Rivlin RS (eds) Clinical measurement of taste and smell. MacMillan, New York, in press

Doty RL (1985b) Odor-guided behavior in mammals. Experientia, in press

Doty RL, Brugger WE, Jurs PC, Onndorff MA, Snyder PJ, Lowry LD (1978) Intranasal trigeminal stimulation from odorous volatiles: Psychometric responses from anosmic and normal humans. Physiol Behav 20: 175-185

Doty RL, Shaman P, Dann M (1984a) Development of the University of Pennsylvania Smell Identification Test: A standardized microencapsulated test of olfactoyr function. Physiol Behav (Monograph) 32: 489-502

Doty RL, Shaman P, Applebaum SL, Giberson R, Sikorski L, Rosensberg L (1984b) Smell identification ability: Changes with age. Science 226: 1441-1443

Doty RL, Newhouse MG, Azzalina JD (1985a) Internal consistency and short-term test-retest reability of the University of Pennsylvania Smell Identification Test. Chem Senses 10: 297-300

Doty RL, Applebaum SL, Zusho H, Settle RG (1985b) A cross-cultural study of sex differences in odor identification ability. Neuropsychologia 23: 667-672

Engen T (1974) Method and theory in the study of odor preference. In: Johnston JW, Moulton DG, Turk A (eds) Human responses to environmental odors. Academic Press, New York, pp 121-141

Engen T, Corbit TE (1970) Feasibility of olfactory coding of noxious substances to insure aversive responses in young children. Publication ICRL-RR 69-6, Injury Control Research Laboratory, U.S. Department of Health, Education and Welfare, Public Health Service, Environmental Health Service

Engen T, Lipsitt LP (1965) Decrement and recovery of responses to olfactory stimuli in the human neonate. J Comp Physiol 59: 312-316

Engen T, Lipsitt LP, Kay H (1963) Olfactory responses and adaptation in the human neonate. J Comp Physiol Psychol 56: 73-77

Gasser RF, Hendrickx AG (1969) The development of the trigeminal nerve in baboon embryos (Papio sp.) J Comp Neurol 136: 159-182

Genzmer A (1873) Untersuchungen über die Sinneswahrnehmungen des neugeborenen Menschen. Inaugural Disertation, Halle

Gesteland RC, Yancy RA, Farbman AI (1982) Development of olfactory receptor neuron selectivity in the rat fetus. Neuroscience 7: 3127-3136

Greene LS, Desor JA, Maller O (1975) Heredity and experience: Their relative importance in the development of taste preference in man. J Comp Physiol Psychol 89: 279-284

Harper R, Bate Smith EC, Land DG (1968) Odour description and odour classification. American Elsevier, New York, p 191

Hinds JW (1968a) Autoradiographic study of histogenesis in the mouse olfactory bulb - 1. Time of origin of neurons and neuroglia. J Comp Neurol 134: 287-304

Hinds JW (1968b) Autoradiographic study of histogenesis in the mouse olfactory bulb - 2. Cell proliferation and migration. J Comp Neurol 134: 305-322

Hogg ID (1941) Sensory nerves and associated structures in the skin of human fetuses of 8 to 14 weeks of menstrual age correlated with functional capability. J Comp Neurol 75: 371-410

Hrboticky N, Korndly M (1984) Acculturation to Canadian foods by Chinese immigrant boys: Changes in the perceived flavor, health value and prestige of foods. Appetite 5: 117-126

Hubert HB, Fabsitz RR, Feinleib M (1980) Olfactory sensitivity in humans: Genetic versus environmentel control. Scienc 208: 607-609

Humphrey T (1949) The development of the olfactory and accessory olfactory formation in human embryos and fetuses. J Comp Neurol 73: 431-468

Humphrey T (1966) The development of trigeminal nerve fibers to the oral mucosa, compared with their development to cutaneous surfaces. J Comp Neurol 126: 91-108

Kneip HH, Morgan WL, Young PT (1931) Studies in affective psychology. XI. Individual differences in affective reactions to odors. Amer J Psychol 43: 89-95

Koelega HS, Koster EP (1974) Some experiments on sex differences in odor perception. Ann New York Acad Sci 237: 234-246

Kroner T (1881) Über die Sinnesempfindungen der Neugeborenen. Breslauer Ärzliche Zeitschrift. Cited in Peterson and Rainey, 1910-1911

Kussmaul A (1859) Untersuchungen über das Seelenleben des neugeborenen Menschen. CF Winter'sche Verlagsbuchhandlung, Leibzig und Heidelberg

Laing DG, Clark PJ (1983) Puberty and olfactory preferences of males. Physiol Behav 30: 591-597

Le Magnen J (1952) Les phenomenes olfacto-sexuels chez l'homme. Arch Sci Physiol 6: 125-160

Lipsitt LP, Engen T, Kaye H (1963) Developmental changes in the olfactory threshold of the neonate. Child Develop 34: 371-376

Macfarlane A (1975) Olfaction in the development of social preferences in the human neonate. In: Macfarlane A (ed), CIBA Foundation Symposium 33: 103-117

Marshall DA, Doty RI, Lucero DP, Slotnick BM (1981) Odor detection thresholds in the rat for the vapors of three related perfluorcarbons and ethylene glycol dinitrate. Chem Senses 6: 421-433

Masolella V (1934) L olfatto nella diverse eta. Arch Ital Otol Rinol Laringol 46: 43-62

McCartney W (1968) Olfaction and odours. Springer-Verlag, Berlin.

Moncrieff RW (1966) Odour preferences. Wiley, New York

Nakashima T, Kimmelman CP, Snow JB, Jr (1984) Structure of human fetal and adult olfactory neuroepithelium. Arch Otolaryngol 110: 641-646

Nakashima T, Kimmelman CP, Snow JB, Jr (1985) Immunohistopathology of human olfactory epithelium, nerve and bulb. Laryngoscope 95: 391-396

Peto E (1936) Contribution to the development of smell feeling. Brit J Med Psychol 15: 314-320

Peterson F, Rainey LH (1910-1911) The beginnings of mind in the new born. Bull Lying Hosp 7: 99-122

Pfaffman C (1960) The pleasures of sensation. Psychol Rev 67: 253-268

Pliner P (1982) The effects of mere exposure on liking for edible substances. Appetite 3: 283-290

Pyatkina GA (1982) Development of the olfactory epithelium in man. Z Mikrosk Anat Forsch 96: 361-372

Rieser J, Yonas A, Wikner K, (1976) Radial localization of odors by newborns. Child Develop 47: 856-859

Rozin P, Kennel K (1983) Acquired preferences for piquant foods by chimpanzees. Appetite 4: 69-77

Rozin P, Schiller D (1980) The nature and aquisition of a preference for chili pepper by humans. Motiv Emot 4: 77-101

Rudy JW, Cheatle MD (1977) Odor-aversion learning in neonatal rats. Science 198: 845-846

Russell MJ (1976) Human olfactory communication. Nature 260: 520-522

Schiffman S (1974) Physicochemical correlates of olfactory quality. Science 185: 112-117

Scott TR, Perrotto RS (1980) Intensity coding in pontine taste area: guestatory information is processed similarly throughout rat's brain stem. J Neurophysiol 44: 739-750

Self PA, Horowitz FD, Paden LY (1972) Olfaction in newborn infants. Develop Psychol 7: 349-363

Smotherman WP (1982) Odor aversion learning by the rat fetus. Physiol Behav 29: 769-777

Stein M, Ottenberg P, Roulet N (1958) A study of the development of olfactory preferences. AMA Archiv Neurolog Psychiat 80: 264-266

Steiner JE (1977) Facial expressions of the neonate indicating the hedonics of food-related chemical stimuli. In: Weiffenbach JM (ed) Tast and development. DHEW Publication No. (NIH) 77-1068, pp 173-189

Steiner JE (1979) Human facial expressions in response to taste and smell stimulation. Adv Child Dev Behav 13: 257-295

Strauss EL (1970) A study on olfactory acuity. Ann Otol Rhinol Laryngol 79: 94-104

Streeter GL (1908) The peripheral nervous system in the human embryo at the end of the first month (10 mm). Amer J Anat 8: 285-302

Toulouse E, Vaschide N (1899) Mesure de l odorat chez les enfants. C R Soc Biol 51: 487-489

Wenzel BM (1948) Techniques in olfactometry: A critical review of the last one hundred years. Psychol Bull 45: 231-247

Wilson DB (1980) Embryonic development of the head and neck: part 4, organs of special sense. Head Neck Surg 2: 237-247

Wysocki CJ, Beauchamp GK (1984) Ability to smell androstenone is genetically determined. Proc Nat Acad Sci 81: 4899-4902

B. PRINCIPLES OF NEURONAL CELL PROLIFERATION

Neurogenesis in the Vertebrate Main Olfactory Epithelium

W Breipohl, A Mackay-Sim*, D Grandt**, B Rehn** and C Darrelmann**

Department of Anatomy, University of Queensland, St.Lucia 4067, Australia

*Department of Physiology, University of Adelaide, Australia

**Institut für Anatomie, Universitätsklinikum Essen, 4300 Essen, FRG

1 Introduction

It is generally accepted that the olfactory system is unique among the senses since the olfactory receptor cells, which are primary neurons, are thought to have a limited life span. Implicit in the current literature is the notion that these receptor cells are born, live and die by some "inherent clock" which, in mammals, lead to neuronal death after about 30 days (Moulton 1974; Graziadei and Monti Graziadei 1979). The turnover of receptor cells and their "programmed" replacement from a basally located progenitor cell compartment is thought to continue throughout the life of an animal and at a constant rate. This concept has gradually developed since Nagahara (1940) first reported the appearance of mitotic activity in the olfactory epithelium proprium of adult mice. Mitosis of basally located progenitor cells and repair of experimentally destroyed sensory cells was subsequently confirmed in a variety of other species (Lams 1940; Schultz 1941, 1960; Smith 1951; Bimes and Planel 1952; Westerman and von Baumgarten 1964; Andres 1965; Thornhill 1970; Graziadei and Metcalf 1971; Graziadei and Dehan 1973; Moulton 1974; Graziadei 1973; Breipohl and Ohyama 1981; Simmons et al. 1981; Matulionis 1982). Recent investigations on cell proliferation, cell death and repair by Hinds et al. (1984), Balboni and Vannelli (1982), Matulionis (1982) and our own group (Breipohl 1982b; Breipohl et al. 1985 a,b; 1986) lead to some doubt about a constant turnover of olfactory receptor cells throughout life.

The present article presents new results which question the idea that olfactory receptor cells are "automatically" replaced at a constant rate throughout life. A model is proposed which provides new directions from which to examine neuronal plasticity in this part of the nervous system.

2 Age

If receptor cells are constantly and continually replaced and if mitotic activity of the progenitor cells is appropriately matched to serve this replacement, one could assume a constant ratio of progenitor cells versus sensory cells at different ages. As can be shown in mice, however, the ratio of progenitor versus sensory cells undergoes considerable variation with age (Breipohl et al. 1985a). In pre- and young postnatal mice, the epithelial thickness mainly results from the number of progenitor cells while in adult animals the thickness is mainly caused by the receptor cells staggered in five to nine layers. In other words, in older animals more sensory cells would have to be replaced by fewer numbers of progenitor cells. Assuming the original theory is correct, then the number of mitotic figures in the progenitor cell compartment in relation to the sensory cells should be constant. However, the opposite is true (Breipohl et al., in preparation). On the basis of the theory of a constant and continual receptor cell replacement throughout life one must then postulate that a given number of progenitor cells could cope with the replacement of greater numbers of sensory cells just by shortening their generation cycle, i.e., speeding up their mitotic activity. This explanation seems rather unlikely, since the slowing down of the generation cycle of neurogenic matrix cells is a general characteristic of normal development in the nervous system (Korr 1979; Schultze and Korr 1981) and in fact the ratio of mitoses per progenitor cells declines with age (Breipohl et al. in preparation).

Other evidence against the hypothesis of a constant olfactory receptor cell replacement throughout life comes from investigation of cell death ratios. If receptor cells are constantly and continually replaced, then the ratio of dead or dying cells to the number of receptor cells should be constant among differently aged animals. Our own unpublished observations in mice and tiger salamander, however, indicate that this ratio declines with age. Furthermore, a general retardation in the maturation of recently divided progenitor cells probably occurs, because there seems to be an age-related increase in the ratio of immature receptor cells to mature receptor cells (Breipohl et al. unpublished observations). In other words, receptor cells in older animals spend longer periods in the differentiation phase of their life cycles. The only explanation that remains is that older olfactory sensory cell populations have less need for mitotic activity of the neurogenic matrix cells than younger populations. If so, it follows that the higher rate of mitotic activity in early postnatal life is due to the demand for newly produced sensory cells when the surface area to be covered by olfactory epithelium is growing. Thus, mitosis may serve to enlarge both the area of the olfactory epithelium and its thickness even in older animals, rather than merely serve to replace dead olfactory receptor cells.

Within the basal cell layer of adult vertebrates in which the "progenitor cells" lie, are two cell types: light and dark basal cells. Dark basal cells are missing from the olfactory epithelium pre-natally in rodents and in early pouch life of marsupials (Breipohl unpublished observations; Kratzing this Vol). With age, dark basal cells increase in number relative to light basal cells (Breipohl unpublished observations). The numbers of dark basal cells do not only depend on age but are also influenced by sex hormonal status (Balboni and Vannelli 1982). To date, an involvement of dark basal cells in the origin of new sensory cells has not been proven and therefore the term "progenitor cell" should be restricted to describe the light, or globose, basal cell. Balboni and Vannelli (1982) propose that dark basal cells are merely supportive tissue elements. Rehn et al. (1981) discuss the possibility that dark basal cells may be quiescent progenitor cells. If so, then age-related increase in dark basal cells could represent an increasing transition from active progenitor cells (light basal cells) into quiescent progenitor cells (dark basal cells).

Other evidence for age-related changes in olfactory neurogenesis comes from experiments in which the olfactory receptor cells have been chemically damaged by methyl-formimino-methylester. After destruction of olfactory receptor cells, (Rehn et al. 1981) and olfactory function (Brederek 1965; Schmidt et al. 1984) with this chemical, morphological repair is slower in adults compared to young mice (Breipohl et al. unpublished observations). Similarly, there was no replacement of receptor cells after 1% zinc sulfate treatment of old adult mice, whereas young adult mice retained the ability to replace these neurons (Matulionis 1982).

Thus, the ability to regenerate the receptor neurons does not appear to be constant throughout the life of an animal. With increasing age, there seems to be a gradual slowing down of the neurogenic process. It is quite possible that the decline of neurogenic activity and the decline in the rate of transformation of matrix daughter cells to mature olfactory receptor cells can help to explain both the reduction of olfactory knobs per unit area of the olfactory epithelium (Hinds et al. 1984) and the loss of olfactory ability with senescence (Doty this Vol).

In addition to these age-related changes in olfactory neurogenesis still another feature has to be investigated in more detail: cell death in the progenitor cell compartment. Breipohl et al. (1985a,b) gave evidence that not all the offsprings of the progenitor cells will migrate peripherally to develop into olfactory sensory cells. Thus death of progenitor cells, prior to leaving the basal cell layer, has to be considered to avoid overestimation of receptor cell replacement on the basis of ^{3}H-thymidine labeling counts. Apart from this, calculations of receptor cell replacement with this technique will have to include the possibility of age-related changes in the ratio of cell death versus mitosis in the progenitor cell compartment (Breipohl et al. in preparation).

3 Regional Differences

The most likely explanation for the high rates of mitosis in the olfactory epithelium of young animals is that the epithelium is undergoing a rapid period of expansion (Hinds et al. 1984). In rats, the epithelium even continues to expand at its rostral edge, well into adulthood (Hinds et al. 1984). Similar results have been observed in the other olfactory neuroepithelia as well (Organ of Masera: Breipohl and Naguro unpublished observations; Vomeronasal Organ: Segovia and Guillamon, 1982). During intra-uterine development, in the period of most rapid expansion of the olfactory epithelium, mitotic activity is seen throughout the thickness of the epithelium. When it differentiates, mitosis gradually becomes confined to the progenitor cell layer (Breipohl et al. 1973; Smart 1971; Kratzing this Vol). During this process in mouse and chicken epithelia, the most caudal regions of the epithelium differentiate first, with ciliated mature neurons observed here before being seen more rostrally (Breipohl et al. 1974b; Breipohl and Fernandez 1977; Breipohl and Ohyama 1981; Farbman and Menco this Vol).

Regional differences in the rate of cell genesis have also been observed in adult tiger salamander (Mackay-Sim and Patel 1984). In this case, the caudal epithelium is thinner than the rostral epithelium and undergoes much more rapid mitosis in the progenitor cell layer. These results may also reflect a continuing expansion of the adult olfactory epithelium as observed in the mouse. However, since the olfactory epithelium already occupies virtually the whole interior of the nasal cavity of salamanders, "expansion of the olfactory epithelium" implies an increase in the size of the nasal cavity. Expansion of the area covered by olfactory receptor cells by growing into areas formerly covered by ridges of ciliated and microvillous cells is a further likely explanation (Breipohl et al. 1984; Getchell and Getchell this Vol). Although salamanders do continue to grow throughout adulthood, the rostro-caudal differences in cell genesis seem to be too great to be explained by growth alone. Recent morphometric analysis of the olfactory epithelium in tiger salamander supports this conjecture. Counts of dying receptor cell nuclei, mature and immature receptor cell nuclei, and mitotic figures suggest that the rostral (thicker) epithelium undergoes the neurogenic process more slowly than the caudal (thinner) epithelium (Mackay-Sim et al. unpublished observations). Thus, the rate of neurogenesis may not be constant, even within the same animal, regardless of age.

The differentiation between young and older epithelial areas within the same animal raises the question of how regional patterns of sensitivity originate (Mozell DG 1966; Mozell MM 1970; Moulton 1976; Breipohl et al. 1976; Hornung and Mozell 1977; Leon et al. 1984; Mackay-Sim et al. 1982; Mackay-Sim and Patel 1984). Physiological differences may coincide with cytochemical differences illustrated, for example, by regional variations in receptor cell antigenicity for certain monoclonal antibodies (Mori et al. 1985). These differences may arise both during development through dynamic interactions with odour stimuli present during a "critical period" and from different stem cells. From other sensory systems it is known that neurons pass through critical periods during which they obtain their individual specifity (Marler 1970; Pettigrew and Freeman 1973; Blakemore and van Sluyters 1975). Olfactory neurons may also pass through critical periods (Leon et al. 1984; Rehn et al. 1986; Coopersmith and Leon this Vol; Panhuber this Vol; Schmidt this Vol). As a consequence, differences in receptor cell maturation combined with periodic differences in odour stimulation may lead to regional differences in odour sensitivity. Thus, regional differences in sensitivity may ultimately result from variations in the development of receptor neurons as well as regional variation in environmental destruction of receptor neurons.

According to our criteria, young epithelial areas are located most orally in mice and most caudally in tiger salamander. Previous investigations on the principles of the development of the olfactory epithelium in chicken and mouse (Breipohl et al. 1973, 1974a, b; Breipohl and Fernandez 1977) revealed the first signs of regional differences in maturation. During ontogeny, in the most caudal parts of the mouse olfactory epithelium, there is an enhanced maturity in comparison to the most orally located epithelial areas. In the latter there are fewer groups of olfactory terminals at the free surface of the epithelium and less advanced outgrowing of olfactory cilia can be observed. This indicates a caudo-oral shift of differentiation (and perhaps consecutively in imprinting?). Additionally, even in the more immature regions, there are individual cells of advanced maturity, as indicated by well-developed olfactory vesicle with long cilia (Breipohl et al. 1974b). Therefore the development of regional differences in sensitivity imprinting could well be modified by local insults and newly originated receptor cells and/or by the existence of different types of olfactory progenitor cell (Fujita et al. 1985; Morgan this Vol).

4 Environmental Factors

Certain "extrinsic factors" may help regulate the rate of neurogenesis by dictating the period of survival of mature receptor neurons. Several authors have commented that destruction of olfactory receptor cells by various treatments stimulates mitotic activity in the basal cell layer (Graziadei 1973; Rehn et al. 1981; Samanen and Forbes 1981), although this was not seen in all species (Simmons and Getchell 1981). It is possible, therefore, that the rate of progenitor cell mitosis is not constant, but may be regulated in some way by the presence or absence of mature receptor neurons and/or death of them. This regulation may be under local control, and it may explain why mitotic activity in normal epithelium seems to occur in small "patches" rather than evenly throughout the basal cell layer (Graziadei and Metcalf 1971; Breipohl et al. 1985 a,b).

In a study of mice in which the housing environment was not specified, Moulton et al. (1970) demonstrated that the total number of labeled nuclei remained constant for 8 days after tritiated thymidine injections, and remained constant at about 30% of this number for 90 days. Of those labeled nuclei remaining for 44-90 days, 80% were located in the receptor cell layer. Hinds et al. (1984) also demonstrated that mice receptor neurons may life for periods much longer than 30 days. They postulated that previous studies failed to show this, since older receptor cells may have died from infection. Accordingly, these authors housed their mice in a filterd-air low-germ environment. After injection with ^{3}H-thymidine, the number of labeled nuclei in the olfactory epithelium remained constant for 30 days, then dropped to about 30% of the original number by 2 months, remaining at this number until about 6 months. These results of Moulton et al. (1970) and Hinds et al. (1984) thus indicate that 30% of the new olfactory sensory cells remained in the epithelium for at least 3 to 6 months and that the "average" life-span of these cells was well over 30 days. In contrast, Graziadei and Monti Graziadei (1979) observed no labeled nuclei after 30 days in the olfactory epithelium of healthy mice "free of nasal infection or other respiratory diseases".

These conflicting results could mean that rhinitic infection may be only one factor in determining the life-span of the new olfactory receptor cells and the regenerative capacity of the sensory epithelium. Strain differences (Matulionis 1982) and other housing differences may also contribute. Age may also be a factor, although the ages of mice in the studies overlapped (Moulton et al. 1970: 5 months; Graziadei and Monti Graziadei 1979: 2-3 months; Hinds et al. 1984: 2-4 months). The housing conditions of these studies may have varied in the degree of environmental odour from urine, feces and other laboratory odours as well. Presumably the filtered-air environment presented in the study of Hinds' group provided fewer odours than the others. Although these conditions were not deemed odour-free, the mice were probably exposed to a reduced olfactory environment. The extra longevity of receptor neurones may therefore also be a consequence of this.

5 Hormonal Factors

Function of the olfactory system is closely related with endocrine pathways (cf. Doty 1976; Breipohl 1982a). Disturbances in the sense of smell are observed in humans with imbalances in thyroid, gonadal and adrenal hormones, and in glucose metabolism (Kallman et al. 1944; Jorgensen and Buch 1961; Henkin 1967, 1975; Marshall and Henkin 1971; McConnel et al. 1975). Although the mechanisms for these disturbances are poorly understood, in most instances it is possible that some of them interfere with the proliferation and maturation of olfactory neurons. Altman (1969) described a lower rate in mature neuronal offsprings of cerebellar matrix cells in hypothyroid rat. Hypothyroidism has also been implicated with human anosmia (McConnel et al. 1975). Our own recent experiments indicate that in hypothyroid adult mice, anosmia is associated with a failure in differentiation of recently divided progenitor cells. In hypothyroid mice, recently divided cells die before differentiating into mature neurons. This leads to a reduction in the

receptor cell population under hypothyroidism (Beard and Mackay-Sim 1986).
Thus, neurogenesis of the olfactory epithelium in adults is dependent on
thyroxine, similar to neurogenesis in the developing central nervous system.

6 Conclusions

In recent years olfactory neurogenesis has come to be regarded as a process
whose time course is dictated by the pre-determined life-span of the olfactory
receptor cells (Fig. 1) (for review see Barber 1982). It was implicitly assumed
that there is one point of entry into this cycle (mitosis of the progenitor cell) and
one point of exit (death of the mature receptor cell). It was also assumed that
there is minimal regulation of this process, the exception being that there may be
accaleration of the rate of progenitor cell mitosis after extensive receptor cell
death (Samanen and Forbes 1984). However, it is likely that individual receptor
cell death varies throughout the nasal cavity and throughout the life of an animal
and thus the model of neurogenic and maturational processes must be adapted to
account for these probabilities.

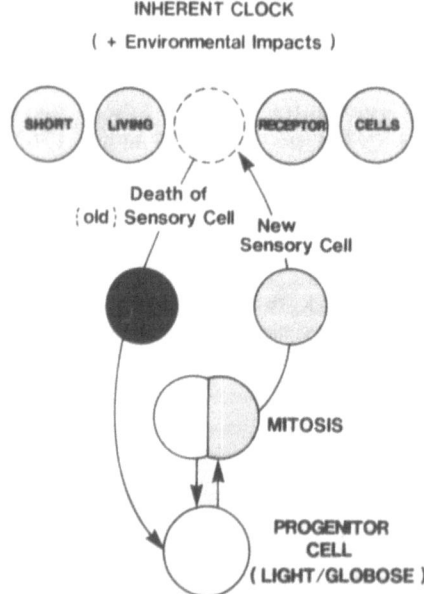

Fig. 1. The usually accepted
model of olfactory neurogenesis.
The lifespan of synaptic connected receptor cells is determined by an "inherent clock"
except when disease or environmental insults intervene. Massive
loss of sensory cells may
stimulate mitosis of progenitor
cells

The accepted model of the neurogenic process does not account for different
probabilities of receptor cell death and maturation, and it replaced all receptor
cells regularly, thereby creating two major disadvantages. First, the synaptic
connections between the receptor cells and the mitral cells would have to undergo
continual and frequent change, on average once every 30 days. Second, if the
only regulation of neurogenesis is through stimulation of progenitor cell mitosis by
(extensive) receptor cell death, then this would impart some latency in the
replacement process (Miragall and Monti Graziadei 1982). Thus, if disease or
inhalation of toxins led to widespread destruction of receptor neurons, there
would always be an extended period of "impairment".

In the light of the information discussed above, we propose a new model of neurogenesis and plasticity in the olfactory epithelium (Fig. 2). The underlying assumption at the heart of this model is that individual receptor cells may not have a predetermined life-span: they may die instead through other internal factors, environmental insult, disease, or perhaps even through "saturation and exhaustion" of their odour-receptor mechanisms.

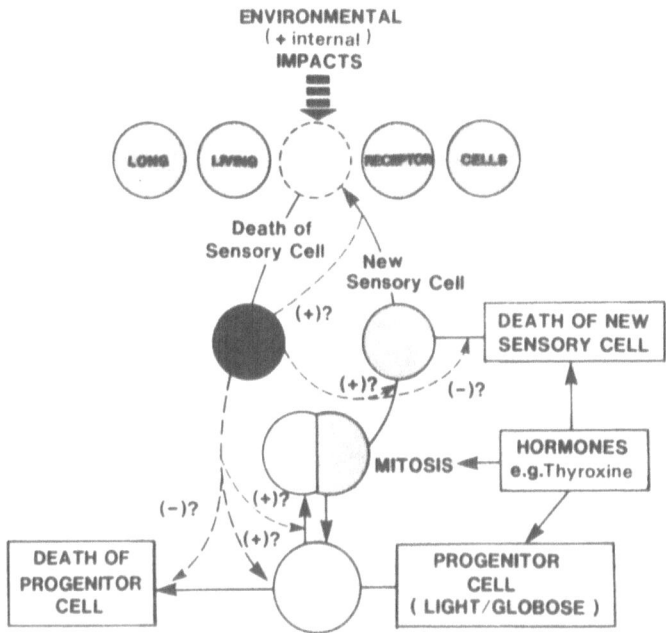

Fig. 2. New hypothesis on olfactory neurogenesis. Olfactory sensory cells syn-aptically connected with the olfactory bulb may tend to live as long as not being killed by environmental insults and disease. Cell death figures occurring in the receptor cell compartment may mainly be due to cell death of sensory cells not yet synaptically connected with the olfactory bulb. Proliferation of progenitor cells may be influenced by various parameters: death of sensory cells, death of pro-genitor cells, hormonal influences, e.g., thyroxine. There may be feedback mechanisms between all these parameters. The mechanisms for the origination of different types of sensory cells and eventually even progenitor cells is not under-stood so far, as discussed in this chapter, but may easily be incorporated in this model.

According to the new model neurogenetic proliferation in the PCC acts to provide a continuous supply of "almost mature" neurons should any "synaptically connected" neurons die. The distinction between "almost mature" and "synaptically connected" neurons is important. The assumption of almost mature neuronal cells in a state of apparently suspended development for at least several days and waiting to establish functional contact, remains on similar waiting phenomena in other sensory pathways (Rowe 1982). We assume that the death of a functioning, "synaptically connected" neuron would leave available post synaptic space in the olfactory bulb. This space could then be occupied immediately by an "almost mature" neuron from the same local epithelial region, whose axon has grown to the same region of the bulb by following other axons from that region. Regional specificity in the outgrowth of new axons towards the liberated postsynaptic places may be facilitated by the bundling of sensory cells in groups by the olfactory Schwann cells and specific membrane proteins on the outgrowing olfactory axons (Allen and Akeson 1985; Key and Giorgi 1986). Regional specificity may also be guaranteed by the special arrangement of peripheral and central olfactory glial cells that may protect the liberated postsynaptic places in the olfactory bulb against local axonal sprouting and "reserve" those spaces as "no man's land" until new sensory axons from the appropriate bundle have arrived (cf. Raisman 1985).

In this model it is assumed that the "almost mature" neuron would die if it does not find synaptic space. The neurogenic matrix activity then acts to keep up the supply of "almost mature" neurons, which may or may not eventually replace a "synaptically connected" neuron. This model overcomes the disadvantages of the model in the literature. First, specific regional synaptic connectivity with the bulb is maintained unless external factors dictate otherwise (Graziadei and Samanen 1980). Second, there would be much shorter latency in replacement after widespread destruction of receptor cells, thus decreasing the period of impairment.

In the new model, the point of entry into the neurogenic cycle is the same as before (mitosis of progenitor cells) but there now exist multiple points of exit since the model suggests a supply of "almost mature" neurons whose death is dictated by the availability of synaptic space in the bulb and whose development is dependent upon hormonal and nutritional state. In this regard the loss of radioactive nuclei after ^{3}H-thymidine injection may reflect not death of "synaptically connected" receptor neurons, but death at some stage before synaptic connections are made, as is clearly the case in hypothyroidism (Beard and Mackay-Sim 1986). Another exit in our model is cell death in the progenitor cell compartment (Breipohl et al. 1985b), which together with the rate of mitosis and differentiation of progenitor cells may well control the plasticity of the olfactory system. The possibility of the existence of subclasses of olfactory progenitor cells with different lifespans of their daughter cells could easily be incorporated in the new model, but the reason for a development of immunocytologically demonstrated differences in receptor cells is still unclear (Fujita et al. 1985).

The data presented highlights the uniqueness of the olfactory matrix in comparison with other neurogenic matrices. While the latter exhaust during intra-uterine life or shortly after birth the olfactory matrix remains active far into adulthood. Since it is so accessible for experimentation, the olfactory matrix provides a useful model for investigations on the developing nervous system with regard to cell proliferation, cell maturation, and cell death.

The new model for the olfactory system provides multiple sites for the regulation of the neurogenic cycle rather than a "clock-driven" cycle in which the rate of progenitor cell mitosis is controlled by the rate of receptor cell death. Probably rates of mitosis and development of new receptor cells and their survival can be altered in addition through local feedback mechanisms as well as through hormonal, nutritional, disease and age factors, although these pathways are still far from being understood appropriately.

In comparison with the established model, this new model considers a greater complexity in the neurogenesis and maturation of olfactory neurones. The outcome however, is a system that responds more quickly and more effectively to environmental assault whilst maintaining greater "synaptic stability".

Summary

Investigations and considerations of the regulation of olfactory sensory receptor cell replacement in vertebrates are presented. The data gives new insights into the dynamics of cell proliferation, cell death and cell differentiation in the olfactory epithelium and challenges the generally accepted idea of a limited lifespan of olfactory receptor cells. A new approach towards olfactory sensory cell replacement is proposed which may lead to a different overall concept of plasticity and postnatal differentiation in this part of the vertebrate nervous system.

Acknowledgments. The authors gratefully acknowledge the careful assistance of M.Kremer and financial support by the following grants: Br 358/5-2, BRF 52, NH&MRC 860587: 85/3452 to W.B.; N.H. & M.R.C. 850400: 84/4993, A.B.F. 8285 to A.M.S.

References

Allen WK, Akeson R (1985) Identification of a cell surface glycoprotein family of olfactory receptor neurons with a monoclonal antibody. J Neurosci 5: 284-296

Altman J (1969) Autoradiographic and histological studies of postnatal neurogenesis. III. Dating the time of production and onset of differentiation of cerebellar microneurons in rats. J Comp Neurol 136: 269-294

Andres KH (1965) Der Feinbau der Regio olfactoria von Makrosmatikern. Z Zellforsch 69: 140-154

Balboni GC, Vannelli GB (1982) Morphological features of the olfactory epithelium in prepubertal and postpubertal rats. In: Breipohl W (ed) Olfaction and endocrine regulation. IRL Press, London, pp 285-297

Barber PC (1982) Neurogenesis and regeneration in the primary olfactory pathway of mammals. Bibl Anat 23: 12-26

Beard MD, Mackay-Sim A (1986) Thyroid control of olfactory function in adult mice. Neurosci Lett (Suppl) 23: 532

Bimes C, Planel H (1952) Signification de l'assise basale du neuroepithelium olfactif. CR Assoc Anat 70: 199-204

Blakemore C, Sluyters RC van (1975) Innate and environmental factors in the development of the kitten visual cortex. J Physiol (London) 248: 663-716

Brederek H (1965) Mitt Nachr Chem Tech 13: 66

Breipohl W (ed) (1982a) Olfaction and endocrine regulation. IRL Press, London

Breipohl W (1982b) Discussions in olfaction and endocrine regulation. In: Breipohl W (ed) IRL Press, London, pp 20, 21, 308

Breipohl W, Fernandez M (1977) Scanning electron microscopic investigations of olfactory epithelium in the chick embryo. Cell Tissue Res 183: 105-114

Breipohl W, Ohyama M (1981) Comparative and developmental SEM studies on olfactory epithelia in vertebrates. Biomedical aspects and speculations. Biomed Res (Suppl) 2: 437-448

Breipohl W, Mestres P, Meller K (1973) Licht- und elektronen-mikroskopische Befunde zur Differenzierung des Riechepithels des weissen Maus. Verh Anat Ges 67: 443-449

Breipohl W, Laugwitz HJ, Bornfeld N (1974a) Topological relations between the dendrites of olfactory sensory cells and sustentacular cells in different vertebrates. An ultrastructural study. J Anat 117: 89-94

Breipohl W, Bijvank GJ, Pfefferkorn GE (1974b) Scanning electron microscopy of various sensory receptor cells in different vertebrates. Scanning Electr Microsc Proc 2: 557-564

Breipohl W, Zippel HP, Rückert K, Oggolter H (1976) Morphologische und elektrophysiologische Studien zur Struktur und Funktion des olfaktorischen Systems beim Goldfisch unter normalen und experimentellen Bedingungen. Beitr Elektronenmikrosk Direktabb Oberfl 9: 561-584

Breipohl W, Mackay-Sim A, Ummels M, Rehn B, Bhatnagar K (1984) Morphology of the olfactory epithelium in the larval (aquatic) and adult (terrestrial) tiger salamander. Symposium "Ontogeny of olfaction in vertebrates". September 9th – 12th, Tübingen

Breipohl W, Grandt D, Rehn B, Mackay-Sim A, Hierche H (1985a) Investigations of cell replacement in the olfactory epithelium. Neurosci Lett (Suppl) 19: 7

Breipohl W, Rehn B, Molyneux GS, Grandt D (1985b) Plasticity of neuronal cell replacement in the main olfactory epithelium of mouse. Abstr XII Int Anat Congr, London

Breipohl W, Darrelmann C, Rehn B, Tran-Dinh H (1986) Cell death and cell origin in the olfactory epithelium of mice. 21th Annu Conf Anat Soc Aust N Z. J Anat (in press)

Doty RL (ed) (1976) Mammalian olfaction, reproductive processes and behaviour. Academic Press, London New York

Fujita SC, Mori K, Imamura K, Obata K (1985) Subclasses of olfactory receptor cells and their segregated central projections demonstrated by a monoclonal antibody. Brain Res 326: 192-196

Graziadei PPC (1973) Cell dynamics in the olfactory mucosa. Tissue and Cell 1: 113-131

Graziadei PPC, Metcalf JF (1971) Autoradiographic and ultrastructural observations on the frog's olfactory mucosa. Z Zellforsch Mikrosk Anat 116: 305-318

Graziadei PPC, Dehan RS, (1973) Neuronal regeneration in frog olfactory system. J Cell Biol 59: 525-530

Graziadei PPC, Monti Graziadei GA (1978a) Continuous nerve cell renewal in the olfactory system. In: Jacobson M (ed) Handbook of Sensory Physiology, vol IX. Springer, Berlin Heidelberg New York, pp 55-82

Graziadei PPC, Monti Graziadei GA (1978b) The olfactory system. A model for the study of neurogenesis and axon regeneration in mammals. In: Cotman CW (ed) Neuronal plasticity. Raven Press, New York, pp 131-151

Graziadei PPC, Monti Graziadei GA (1979) Neurogenesis and neuron regeneration in the olfactory system of mammals. I. Morphological aspects of differentiation and structural organization of the olfactory neurons. J Neurocytol 8: 1-18

Graziadei PPC, Samanen DW (1980) Ectopic glomerular stuctures in the olfactory bulb of neonatal and adult mice. Brain Res 187: 467-472

Henkin RI (1967) Abnormalities of taste and olfaction in patients with chromatin negative gonadal dysgenesis. J Clin Endocrinol Metab 27: 1435-1440

Henkin RI (1975) The role of adrenal corticosteroids in sensory processes. In: Blanschko H, Smith AD, Seyers G (eds) Handbook of physiology, Sect 7, vol VI. Am Physiol Soc, Washington DC, pp 209-230

Hinds JW, Hinds PL, McNelly NA (1984) An autoradiographic study of the mouse olfactory epithelium: evidence for long-lived receptors. Anat Rec 210: 375-383

Hornung DE, Mozell MM (1977) Factors influencing the differential sorption of odorant molecules acrosss the olfactory mucosa. J Gen Physiol 69: 343-361

Hornung DE, Mozell MM (1980) Tritiated odorants to monitor retention in the olfactory and vomeronasal organs. Brain Res. 181: 488-492

Jorgensen MB, Buch NH (1961) Studies on the sense of smell and taste in diabetics. Acta Otolaryngol (Stockh) 53: 539-545

Kallman FJ, Schoenfeld WA, Barrara SE (1944) The genetic aspects of primary eunuchoidism. Am J Ment Defic 48: 203-36

Key B, Giogi P (1986) Selective binding of soy bean agglutinin to the olfactory system of Xenopus. Neuroscience (in press).

Korr H (1979) Autoradiographische Untersuchungen zur Proliferation verschiedener Zellelemente im Gehirn von Ratte und Maus: Ein Überblick. Verh Anat Ges 73: 1047-1049

Lams MH (1949) Recherches sur la vascularite de certaines epithelium. Le epithelium olfactif chez les mammiferes. Bull Acad Belg C Sci 5: 110-135

Leon M, Coopersmith R, Ulibarri C, Porter RH, Powers JB (1984) Development of olfactory bulb organization in precocial and altricial rodents. Dev Brain Res 12: 45-53

Mackay-Sim A. Shaman P, Moulton DG (1982) Topographic coding of olfactory quality: odorant-specific patterns of epithelial responsivity in the salamander. J Neurosurg 48: 584-596

Mackay-Sim A, Patel U (1984) Regional differences in cell density and cell genesis in the olfactory epithelium of the salamander Ambystomas tigrinum. Exp Brain Res 57: 99-106

Marler PA (1970) A comparative approach to vocal learning: song development in white crowned sparrows. J Comp Phys Psychol 71(2): 1-25

Marshall JR, Henkin RI (1971) Olfactory acuity, menstrual abnormalities and oocyte status. Ann Intern Med 75: 207-211

Matulionis DH (1982) Effects of the aging process on olfactory neuron plasticity. In: Breipohl W (ed) Olfactory and endocrine regulation. IRL Press, London, pp 299-308

McConnel RJ, Menendez CE, Smith FR, Henkin RI, Rirlin RS (1975) Defects of taste and smell in patients with hypothyroidism. Am J Med 59: 354-364

Miragall F, Monti Graziadei GA (1982) Experimental studies on the olfactory marker protein. II. Appearance of the olfactory marker protein during differentiation of the olfactory sensory neurons of the mouse: an immunohistochemical and autoradiographic study. Brain Res 239: 245-250

Mori K, Fujita SC, Imamura K, Obata K (1985) Immunohistochemical study of subclasses of olfactory nerve fibres and their projections to the olfactory bulb in the rabbit. J Comp Neurol 242: 214-229

Moulton DG (1974) Dynamics of cell populations in the olfactory epithelium. Ann N.Y.Acad Sci 237: 52-61

Moulton DG (1976) Spatial patterning of response to odors in the peripheral olfactory system. Physiol Rev 56: 578-593

Moulton DG, Celebi G, Fink RP (1970) Olfaction in mammals - two aspects: proliferation of cells in the olfactory epithelium and sensitivity to odours. In: Wostenholme GEW, Knight J (eds) Taste and smell in vertebrates. Churchill, London, pp 227-250

Mozell DG (1966) Spatial patterning of response to odors in the peripheral olfactory system. Physiol Rev 56: 578-593

Mozell MM (1970) Evidence for a chromatographic model of olfaction. J Gen Physiol 56: 46-63

Nagahara Y (1940) Experimentelle Studien über die histologischen Veränderungen des Geruchsorgans nach der Olfactorius durchschneidung. Beiträge zur Kenntnis des feineren Baus des Geruchsorgans. Jap J Med Sci V, Pathol 5: 165-199

Pettigrew J, Freeman R (1973) Visual experience without lines: effect on developing cortical neurons. Science 182: 599-601

Raisman G (1985) Specialized neurological arrangement may explain the capacity of vomeronasal axons to reinnervate central neurons. Neuroscience 14: 237-254

Rehn B, Breipohl W, Schmidt C, Schmidt U, Effenberger F (1981) Chemical blockade of olfactory perception by N-methyl-formimino-methylester in albino mice. Chem Senses 6: 317-328

Rehn B, Breipohl W, Mendoza AS, Apfelbach R (1986) Changes in granule cells of the ferret olfactory bulb associated with imprinting on prey odours. Brain Res 373: 114-125

Rowe MJ (1982) Development of mammalian somatosensory pathways. TINS 5: 408-411

Samanen DW, Forbes WB (1984) Replication and differentiation of olfactory receptor neurons following axotomy in the adult hamster. A morphometric analysis of postnatal neurogenesis. J Comp Neurol 225: 201-211

Schmidt C, Schmidt U, Breipohl W, Effenberger F (1984) The effect of N-methyl-formimino-methylester on the neural olfactory threshold in albino mice. Arch Otorhinolaryngol 239: 25-29

Schultz EW (1941) Regeneration of olfactory cells. Proc Soc Exp Biol 46: 41-43

Schultz EW (1960) Repair of the olfactory mucosa. Am J Pathol 37: 1-19

Schultze B, Korr H (1981) Cell kinetic studies of different cell types in the developing and adult brain of the rat and the mouse: a review. Cell Tissue Kinet 14: 309-325

Segovia S, Guillamon A (1982) Effect of sex steroids on the development of vomeronasal organ in the rat. Dev Brain Res 5: 209-212

Simmons P, Getchell TV (1981) Physiological activity of newly differentiated olfactory receptor neurons correlated with morphological recovery from olfactory nerve section in the salamander. J Neurophysiol 45: 529-549

Simmons P, Rafols JA, Getchell TV (1981) Ultrastuctural changes in olfactory receptor neurons following olfactory nerve section. J Comp Neurol 197: 237-257

Smart I (1971) Location and orientation of mitotic figures in the developing mouse olfactory epithelium. J Anat 109: 243-251

Smith CG (1951) Regeneration of sensory olfactory epithelium and nerves in adult frogs. Anat Rec 109: 661-671

Thornhill RA (1970) Cell division in the olfactory epithelium of the lamprey, Lampetra fluviatilis. Z Zellforsch 109: 147-157

Westerman RA, Baumgarten von (1964) Regeneration of olfactory paths in carp. Experientia 20: 519-520

Effects of Sex Steroids on the Development of the Vomeronasal System in the Rat

S Segovia and A Guillamón

Departamento de Psicobiología, Universidad Nacional de Educación a Distancia, Ciudad Universitaria, 28040 Madrid, Spain

1 Introduction

Abbreviations used:
AOB accessory olfactory bulb
VNO vomeronasal organ
TP   testosterone propionate
CNS central nervous system

The existence of morphological, physiological and biochemical sex differences in the central nervous system are well know (Arnold and Gorski 1984). These differences are normally established during a perinatal period of maximal susceptibility to sex steroids (Goy and McEwen 1980). Various nuclei that, either directly or indirectly, receive input from the vomeronasal organ (VNO) and the accessory olfactory bulb (AOB) (Scalia and Winans 1976) are targets for sex steroid hormones (Stumpf and Sar 1982) and present sexual dimorphism. For example: the medial amygdala (Staudt and Dörner 1976), the medial preoptic area (Dörner and Staudt 1968; Gorski et al. 1978, 1980) the ventro medial nucleus of the hypothalamus (Dörner and Staudt 1969) and the ventral premammillary nucleus (Dörner 1976). Moreover, many of these structures are also involved in reproductive behavior (Dörner 1976; Larsson 1979; Pfaff 1980).

With these facts in mind, we hypothesized that both the VNO and the AOB should present sexual dimorphism, which would be governed by gonadal steroids during the perinatal period.

2 Sex Differences in the Vomeronasal Organ

The rat VNO is a tubular structure situated in the anteroventral portion of the nasal cavity, adopting a symmetric position with respect to the nasal septum. Histologically, it presents two types of epithelia: respiratory, in the lateral part of the organ; and neurosensorial, occupying the medial position next to the nasal septum. In mammals, the neurosensorial epithelium of the VNO, which resembles the olfactory epithelium of the nasal mucosa, has bipolar neurons as receptor cells (Allison 1953; Graziadei 1971).

The possible existence of sexual dimorphism and its control by sex steroids in the VNO were investigated using the structure, volume, neuronal cell number, and neuronal nuclear size as morphological measures (Segovia and Guillamón 1982).

Four groups of rats (Wistar strain) were studied: (a) orchidectomized males (b) females androgenized with a single injection of 1.25 mg of testosterone propionate (TP) (c) sham-operated males and (d) females treated with TP vehicle. All these treatments were carried out on the day of birth (D1). At the age of 6 months, the animals were killed and the VNO was dissected and embedded in paraffin. The morphological parameters were evaluated as described in detail elsewhere (Segovia and Guillamón 1982).

Sex differences were observed in the volume of the organ, the volume of the neurosensorial epithelium, neuronal population, and neuronal nuclear size (Table 1).

Table 1 Effect of sex steroids on the development of the VNO

	VNO volume mm^3	Epithelium volume mm^3	Number of receptor cells	Nuclear size mm^2
Control males	1.7236+- 0.071^{aaa}	0.1838+- 0.013^{aa}	37961+- 1438^{aa}	22.88+- 0.007
Orchid- ectomized males	1.0779+- 0.036^{bbb}	0.1141+- 0.011^{bb}	31560+- 720^{bb}	23.73+- 0.01^{bbb}
Control females	1.0811+- 0.068	0.1134+- 0.012	30175+- 1154	23.52+- 0.009^{aaa}
Androgen- ized females	1.6631+- 0.125^{bb}	0.2023+- 0.011^{bbb}	38819+- 1576^{bb}	22.94+- 0.006^{bbb}

Data were submitted to analysis of variance; t-test were based on their appropriate error variance term. MEAN +/- SEM. [a]comparisons between control groups. [b]comparisons with respect to the control group of the same sex. [aa]$p < .005$; [aaa]$p < .001$; [bb]$p < .005$; [bbb]$p < .001$.

The VNO volume in the control males was significantly greater than that in the control females. If we consider the VNO volume of the control males as 100%, the control females show a volume of 62.7%. Castration of males on D1 was followed by a 37.47% decrease in VNO volume, a value close to the one observed in untreated females. Androgenization of females on D1 produced a considerable increase in VNO volume, which rose to only 3.52 less than that in the control males. Similar trends were observed with respect to VNO neurosensorial epithelium volume and the neuronal population (see Table 1). The treatment of newly born females with testosterone leads to an average neurosensorial epithelium heigth 10.06% above that of the male controls.

The sex differences observed in the size of the neuronal nuclei are the opposite to those described above. The nuclei of the bipolar neurons of the control female rats were larger than those of the control males, while the orchidectomy of the males on D1 or androgenization of the females abolished these differences (Table 1).

Sexual steroids appear to effect not only the organization and postnatal development of some nervous system structures, but their effects continue to adulthood. For example, Morishita et al. (1978) observed that prepubertal ovariectomy diminished the nuclear size of neurons in the hypothalamic ventromedial nucleus which receives vomeronasal input from the medial amygdala (Scalia and Winans 1976; Kevetter and Winans 1981). Segovia et al. (1984b) checked whether sex steroids also control normal VNO morphology in postpuberal animals. They found that gonadectomy produced a decrease in nuclear size of the bipolar neurons in the VNO in 90-day-old rats both. Gonadectomy also causes a 20% reduction in neuroepithelial height.

In conclusion, our results indicate that sexual dimorphism is present in the rat VNO. This dimorphism is established perinatally and can be abolished by experimental treatments on the day of birth. Furthermore, the maintenance of normal neurosensorial morphology in adulthood depends upon action of sex steroids.

3 Sex Differences in the Accessory Olfactory Bulb

The AOB is situated dorsocaudally in the olfactory bulb and its histology has been meticulously described by Ramón y Cajal (1901-1902). In 1935, Smith carried out a detailed study on the development of the AOB in the white rat. Although this author used both sexes in his experimental design he did not mention any sex related morphological differences. Yanay (1979) did not find any differences either in his study of the entire olfactory bulb. However, applying different analytical techniques, sex differences in the olfactory bulb were described by Orensanz et al. (1982). These authors found that the male presents higher levels of ^{3}H-dihydroergocryptine binding than the diestrous female. In relation to the protein content of membranes isolated from the olfactory bulb, Montero et al. (1983) found sex and intra-sex (estrous vs. diestrous females) differences.

To study the effects of sexual steroids during the early postnatal period on the development of the AOB (Segovia et al. 1984a) the volume of the AOB and the volume of the distinct layers that can be observed in this structure (VNO nerve-glomerular, mitral cell, plexiform and granule cell layers) were studied.

As can be seen in Table 2, sex differences exist in the AOB. The volume of the AOB is significantly greater in control males as compared to the control females. The castration of the males on D1 significantly reduces AOB volume, almost down to the typical value for control females. TP androgenization of females on D1, caused a significant increase in the volume of this structure, which rose to a value resembling that of the control males. With the exception of the VNO nerve-glomerular layer, each of the remaining layers presented sex differences on their volume (Table 2).

In a study of similar design, sex differences in the mitral cell population were investigated (Valencia et al. 1985). Control males showed more mitral cells than control females. The orchidectomy of the males (D1) caused mitral cell-number decrease. Moreover, administration of TP to the newborn females produced an increase in the mitral cell-number.

Table 2 Effects of sex steroids on the development of the AOB

	AOB volume mm^3	VO nerve-glomerular layer volume mm^3	Mitral cell layer volume mm^3	Plexiform layer volume mm^3	Granular cells layer volume mm^3
Control males	$0.211+- 0.05^a$	$0.036+- 0.01$	$0.065+- 0.01^{aaa}$	$0.032+- 0.006$	$0.073+- 0.02^{aaa}$
Orchidectomized males	$0.113+- 0.03^{bbb}$	$0.0017+- 0.008^{bb}$	$0.034+- 0.01^{bbb}$	$0.020+- 0.006^b$	$0.039+- 0.01^{bbb}$
Control females	$0.137+- 0.04$	$0.034+- 0.02$	$0.038+- 0.003$	$0.020+- 0.006$	$0.041+- 0.007$
Androgenized females	$0.231+- 0.09^b$	$0.052+- 0.02$	$0.072+- 0.03^b$	$0.036+- 0.013^b$	$0.063+- 0.02^b$

MEAN +/- SD. [a]comparisons between control groups. [b]comparisons with respect to the control group of the same sex. [a]$p < .05$; [aa]$p < .01$; [aaa]$p < .001$; [b]$p < .05$; [bb]$p < .01$; [bbb]$p < .001$.

These observations indicate that AOB sexual dimorphism is established by the action of gonadal steroids in the early postnatal period. In conclusion: these results confirm our initial hypothesis that both the VNO and AOB present sexual dimorphism.

4 Comments on Sexual Dimorphism in the Vomeronasal System

The hypothalamic structures that receive vomeronasal input are implicated in reproductive behavior and present morphological sex differences. We suggest that any concept concerning the functional anatomy of the vomeronasal system should include these structures (Fig. 1).

The sexual dimorphism, development and differentiation of, at least, the VNO, does not solely depend on the appropriate action of gonadal steroids. Recently, Segovia et al. (1982) found a dramatic decrease in the VNO volume, neuronal population and nuclear size of the receptor cells in adult rats that had been thyroidectomized on postnatal day 8 and treated with 100 µCi of [131]J on postnatal day 10. We should consider that gonadectomy may influence the action of thyroid hormones on the CNS. For example, when gonadectomy was performed on the third postnatal day, it reduced the accumulation of [131]triiodothyronine in the cerebellum, cerebral cortex and hypothalamus of both male and female rats, even though this reduction was greater in the female (Ford and Cramer 1977).

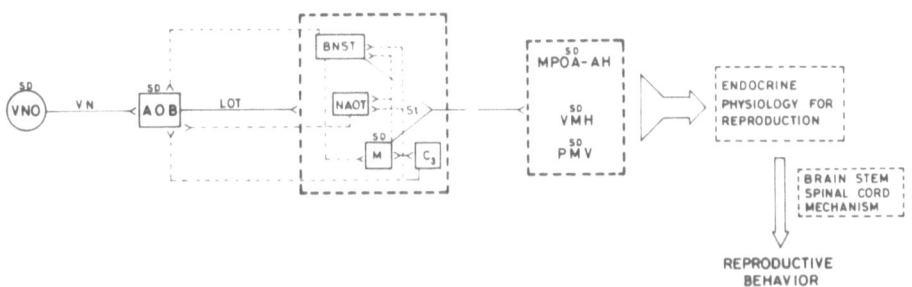

Fig. 1 Anatomical and functional hypothesis of the vomeronasal system in rodents. **VNO** vomeronasal organ; **VN** vomeronasal nerve; **SD** sexual dimorphism **AOB** accessory olfactory bulb; **LOT** lateral olfactory tract; **BNST** bed nucleus of the stria terminalis; **NAOT** nucleus of the accessory olfactory tract; **ST** stria terminalis; **M** medial nucleus of amygdala; **C3** posteromedial cortical nucleus of amygdala; **MPOA-AH** medial preoptic-area anterior hypothalamus; **VHM** ventromedial nucleus of the hypothalamus; **PMV** ventral premammillary nucleus

Despite these co-interactions, the existence of sexual dimorphism in VNO and AOB can be used as an experimental model for performing sophisticated investigations into the effect of sexual steroids on neurogenesis, neuronal plasticity and aging in the CNS. These structures are also reasonably accessible with electrophysiological and biochemical techniques. The special importance of such investigative approaches is further highlighted when one considers the specific role of this system in the regulation of reproductive behavior (Powers and Winans 1975; Fleming et al. 1979; Johns 1980; Teicher et al. 1984).

Summary

Several structures that receive vomeronasal input present sexual dimorphism and are also implicated in sexually dimorphic reproductive behavior. We hypothesize that the whole vomeronasal system may show sexual dimorphism in the rat. To verify this hypothesis a series of experiments was performed on the vomeronasal organ and the accessory olfactory bulb. Our results show that these structures present gonadal steroid controlled sexual dimorphism, at least during the early postnatal period.

Acknowledgments. This research was supported by grants from CAICYT (1925/1983) and FIS (83/0730).

References

Allison AC (1953) The morphology of the olfactory system in the vertebrates. Biol Rev 28: 195-244

Arnold AP, Gorski RA (1984) Gonadal steroids induction of structural sex differences in the central nervous system. Annu Rev Neurosci 7: 413-422

Dörner G (1976) Hormones and brain differentiation. Elsevier, Amsterdam

Dörner G, Staudt J (1968) Structural changes in the preoptic anterior hypothalamic area of the male rat, following neonatal castration and androgen treatment. Neuroendocrinology 3: 136-140

Dörner G, Staudt J (1969) Structural changes in the hypothalamic ventromedial nucleus of the male rat, following neonatal castration and androgen treatment. Neuroendocrinology 4: 278-281

Fleming A, Vaccarino F, Tombosso L, Chee PH (1979) Vomeronasal organ and olfactory systems modulation of maternal behavior in the rat. Science 203: 372-374

Ford DH, Cramer EB (1977) Developing nervous system in relation to thyroid hormones. In: Grave GD (ed) Thyroid hormones and brain development. Raven Press, New York, pp 1-18

Gorski RA, Gordon JH, Shryne JE, Southam AM (1978) Evidence for a morphological sex difference within the medial preoptic area of the rat brain. Brain Res 148: 333-346

Gorski RA, Harlan RE, Jacobson CD, Shryne JE, Southam AM (1980) Evidence for existence of a sexually dimorphic nucleus in the preoptic area in the rat. J Comp Neurol 193: 529-539

Goy RW, McEwen BS (1980) Sexual differentiation on the brain. MIT Press, Massachusetts

Graziadei PPC (1971) The olfactory mucosa of vertebrates. In: Beidler M (ed) Handbook of sensory physiology, vol IV. Springer, Berlin Heidelberg New York, pp 27-58

Johns MA (1980) The role of the vomeronasal system in mammalian reproductive physiology. In: Müller-Schwarce D, Silverstein RM (eds) Chemical signals. Plenum Press, New York, pp 341-364

Kevetter GA, Winans SS (1981) Connections of the corticomedial amygdala in the golden hamster. I. Efferents of the "vomeronasal amygdala". J Comp Neurol 197: 81-98

Larsson K (1979) Features of the neuroendocrine regulation of masculine sexual behavior. In: Beyer C (ed) Endocrine control of sexual behavior. Compr Endocrinol Ser. Raven Press, New York, pp 77-163

Marques DM (1979) Roles of the main olfactory and vomeronasal systems in the response of the female hamster to young. Behav Neurol Biol 26: 311-329

Montero MT, Guillamón A, Azuara MC, Ambrosio E, Segovia S, Orensanz LM (1983) Sex differences in the protein content of membrane fractions from several regions of the rat CNS. IRCS Med Sci 11: 317

Morishita H, Nagamachi N, Kawamoto M, Tomioka M, Higuchi Y, Hashimoto T, Tanaka T, Kuroiwa S, Nakago K, Mitani H, Miyauchi Y, Azasa T, Adachi H (1978) The effect of prepuberal castration on the development of the nuclear sizes of the neurons in the hypothalamic nuclei of female rats. Brain Res 146: 388-391

Orensanz LM, Guillamón A, Ambrosio E, Segovia S, Azuara MC (1982) Sex differences in alpha-adrenergic receptors in the rat brain. Neurosci Lett 30: 275-278

Pfaff DW (1980) Estrogens and brain function. Springer, Berlin Heidelberg New York

Powers JB, Winans SS (1975) Vomeronasal organ: Critical role in mediating sexual behavior of the male hamster. Science 187: 961-963

Ramón y Cajal S (1901-1902) Textura del lóbulo olfactivo accesorio. Trabajos Lab Inves Biol 1: 143-150

Scalia F, Winans SS (1976) New perspectives on the morphology of the olfactory system: Olfactory and vomeronasal pathways in mammals. In: Doty RL (ed) Mammalian olfaction, reproductive processes, and behavior. Academic Press, New York, pp 7-28

Segovia S, Guillamón A (1982) Effects of sex steroid on the development of the vomeronasal organ in the rat. Dev Brain Res 5: 209-212

Segovia S, Del Cerro MCR, Guillamón A (1982) Effects of neonatal thyroidectomy on the development of the vomeronasal organ in the rat. Dev Brain Res 5: 206-208

Segovia S, Orensanz LM, Valencia A, Guillamón A (1984a) Effects of sex steroids on the development of the accessory olfactory bulb in the rat: a volumetric study. Dev Brain Res 16: 312-314

Segovia S, Paniagua R, Nistal M, Guillamón A (1984b) Effects of post-puberal gonadectomy on the neurosensorial epithelium of the vomeronasal organ in the rat. Dev Brain Res 14: 289-291

Smith CG (1935) The change in volume of the olfactory and accessory olfactory bulbs of the albino rat during postnatal life. J Comp Neurol 61: 477-508

Staudt J, Dörner G (1976) Structural changes in the medial and central amygdala of the male rat, following neonatal castration and androgen treatment. Endocrinology 67: 296-300

Stumpf WE, Sar M (1982) The olfactory system as a target organ for steroid hormones. In: Breipohl W (ed) Olfaction and endocrine regulation. IRL Press, London, pp 11-21

Teicher MH, Shaywitz BA, Lumia AR (1984) Olfactory and vomeronasal system mediation of maternal recognition in the developing rat. Devel Brain Res 12: 97-110

Valencia A, Segovia S, Guillamón A (1985) Effects of sex steroids on the development of the accessory olfactory bulb-mitral cells in the rat. Devel Brain Res (accepted, July)

Yanay J (1979) Strain and sex differences in the rat brain. Acta Anat 103: 150-158

C. MORPHOLOGICAL MATURATION OF THE OLFACTORY PERIPHERY

Development of Olfactory Epithelium in the Rat

Al Farbman and BPhM Menco

Department of Neurobiology and Physiology, Northwestern University, Evanston, Illinois 60201, USA

1 Introduction

Several papers have been written on studies of the development of olfactory epithelium in vertebrates, including frogs (Klein and Graziadei 1983), chicks (e.g., Disse 1897; Van Campenhout 1937; Breipohl and Fernandez 1977; Breipohl and Ohyama 1981; Mendoza et al. 1982), mice (e.g., Smart 1971; Breipohl et al. 1973; Cuschieri and Bannister 1975a,b; Kerjaschki and Hörandner 1976; Kerjaschki 1977; Noda and Harada 1981), rats (Menco and Farbman 1985a,b; Farbman 1986), hamsters (Waterman and Meller 1973; Taniguchi et al. 1982) and humans (O'Rahilly 1967; Bossy 1980; Pyatkina 1982; Nakashima et al. 1984). Most of these reports are primarily descriptive rather than analytical. In this chapter, we shall try to correlate descriptive morphological data on the ontogeny of the rat olfactory epithelium with some biochemical, immunological and physiological studies on early development and maturation

2 Gross Anatomy of Nasal Cavity Development

The first sign of olfactory development in the rat embryo is seen at approximately the 11th embryonic day (E11) when the olfactory placodes become apparent bilaterally as oval-shaped ectodermal thickenings on the antero-lateral aspect of the head. (E1 is the day when the dam is sperm-positive). By E12, each placodal region becomes a shallow depression, the so-called nasal pit, outlined by a horseshoe-shaped ridge of tissue. The epithelium of the placode, as well as that of the pit, is a pseudostratified columnar epithelium having three to four layers of nuclei. The frequent presence of mitotic figures in the epithelium (Fig. 1) indicates that it is a rapidly growing tissue.

Up to E15 there is an increase in the surface area of the olfactory epithelium as the nasal pit deepens to form the nasal cavity. The lateral walls of the early nasal cavity begin to fold into the ridges that will ultimately develop into the turbinates (Fig. 4). Throughout this early development the nasal cavity is continuous with the oral cavity, and the nasal septum is in virtual contact with the dorsal surface of the tongue.

At about E17, the lateral palatine processes of the rat embryo fuse in the midline, resulting in the separation of the nasal cavity from the oral cavity (Vidic et al. 1972). The ventral aspect of the nasal septum fuses with the newly formed palate and this effectively separates the left from the right nasal cavity. In the lateral wall of each nasal cavity, the turbinates become more elaborate, thus increasing further the total surface area of the olfactory epithelium. At this point, the general form of the nasal cavity is similar to that of the adult.

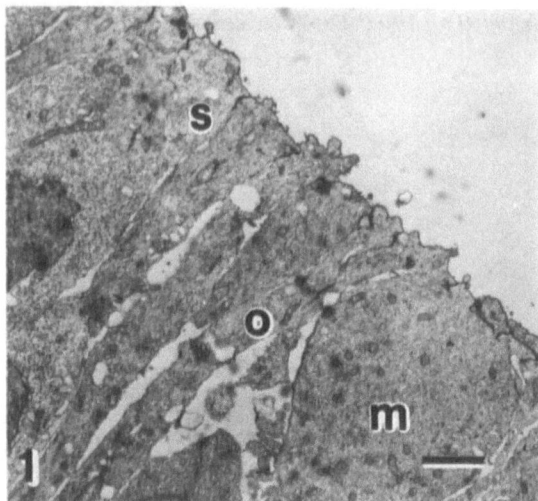

Fig. 1 Electron micrograph of the surface of olfactory epithelium from an E11 rat embryo. The rounded cell **m** near the surface is in mitosis.
s presumptive supporting cell, **o** presumptive olfactory receptor cell. Marker = 2µm

3 Cellular Structure of Developing Olfactory Epithelium

At the placodal stage of development, it is possible to distinguish between presumptive receptor and supporting cells on the basis of their morphology as seen with the electron microscope (Farbman 1977). The distal process of a young receptor cell terminates in a knob which usually bears a single primary cilium. The primary cilium is not considered a true olfactory cilium because it is much shorter, and is most likely a temporary structure, as is seen in many cell types during differentiation (Sorokin 1968; Tucker and Parsdee 1982; Wheatley 1982; Menco and Farbman 1985a). Primary cilia are also found in supporting cells at the placodal stage (Fig. 2; Menco and Farbman 1985 a,b).

In general, it may be said that development of both receptor and supporting cells in the posterior portion of the olfactory region is chronologically somewhat ahead of that in the more anterior portions (Menco and Farbman 1985a).

3.1 Dendrite Development

The early stages of dendrite development on the nasal pit are characterized by slender distal processes often occurring in clusters (Fig. 3; Menco and Farbman 1985a). The number of clusters diminished with age as the receptor cells become isolated from one another. The density of olfactory knobs remains fairly constant during embryonic development, at about 4 to 5 x 10^6 per cm^2 (Menco and Farbman 1985b). After birth, the areal density increases gradually to 10 x 10^6 per cm^2 at 18 month, and then declines (Hinds and McNelly 1981).

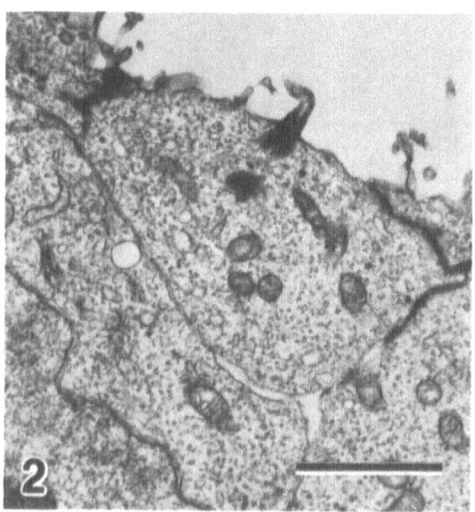

Fig. 2 Electron micrograph of a primary cilium projecting from a supporting cell taken from an E11 embryo. Marker = 2µm

<u>Ciliogenesis</u>.The distal processes of olfactory receptor neurons in E12 to E14 embryos contain many ribosomes, mitochondria, and a few microtubules. Formation of secondary, i.e., proper olfactory, cilia is preceded by centriole accumulation in the knob; many more of these structures can be discerned at E15 and E16 than at E14. The migration of centrioles to the apical terminal of the cell is accompanied by a reduction in the amount of ribosomes in the supranuclear cytoplasm and an increase in the number of microtubules (Menco and Farbman 1985a).

From E16 onwards, the distal process of the young olfactory receptor cell begins to resemble an olfactory dendrite as the terminal knob becomes multiciliated (Fig. 6). Morphometric studies on the genesis of cilia in developing rat olfactory epithelia have shown the average number of cilia per dendritic knob increases linearly at the rate of approximately one per day, from about 1 at E16 to about 7 at E22 (Menco and Farbman 1985a,b). This value is approximately 50% of the adult complement (Menco 1980, 1983). At E22, more than 90% of the dendrites are multiciliated (Menco and Farbman 1985b).

Unlike the primary cilia, which are usually only about 1 µm long, olfactory cilia continue to grow and are estimated to achieve length ultimately of 50 µm or more in adult rats (Seifert 1970; Menco 1983). The cilium tapers beyond the first 1 - 2 µm, i.e., and the cilium continues to grow at a narrower diameter than at the base and even more distal the number of axonemal microtubules is reduced. Tapering becomes a prominent feature of olfactory cilia at E18 in the rat (Menco and Farbman 1985a).

It has been suggested that growth of cilia may, in some way, depend upon an influence from the bulb because the cilia first appear after the growing olfactory axons reach the bulb (Cuschieri and Bannister 1975b). However, olfactory epithelium grown in organ or tissue culture in the absence of the bulb do express olfactory cilia (Farbman 1977; Chuah et al. 1985; Gonzales et al. 1985). When olfactory bulb tissue is added to a culture of olfactory epithelium the number of multiciliated dendritic knobs is increased by a factor of approximately two (Chuah et al. 1985). Thus it appears that cilia may be expressed without the bulb but the presence of the bulb enhances ciliogenesis.

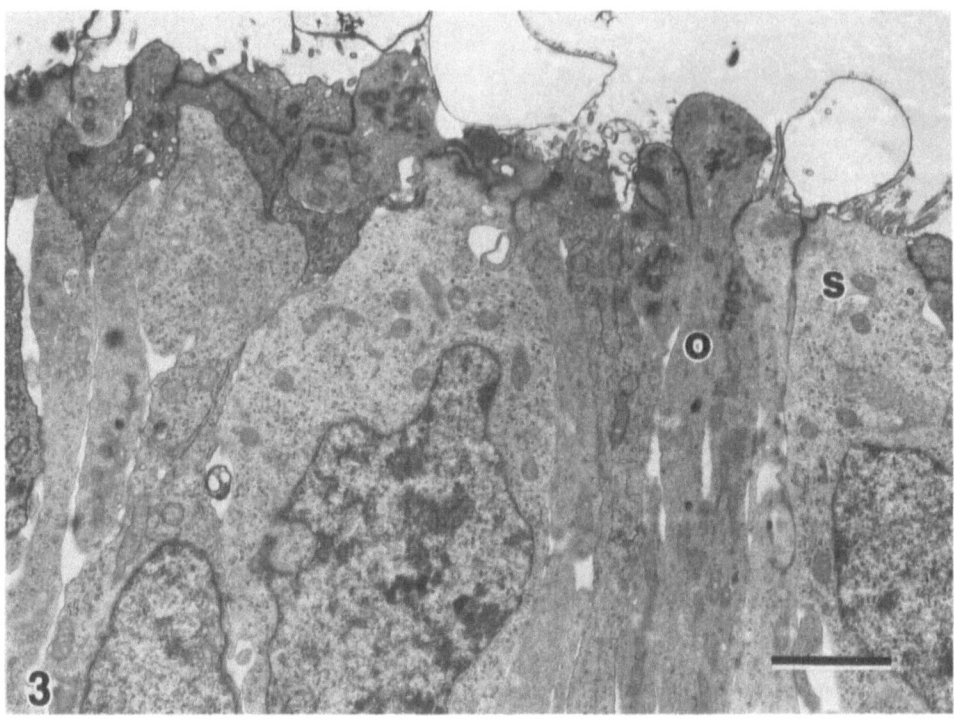

Fig. 3 Electron micrograph of olfactory epithelium from an E14 rat embryo. Olfactory dendrites **o** have centrioles and mitochondria and are clearly distinguishable from supporting cells **s**. Marker = 2μm

3.2 Development of Axons

The olfactory axons begin to develop as narrow processes growing out of the basal pole of the perikarya. At E13 and E14 the axons break through the basal lamina as they continue their growth toward the central nervous system. Axons are accompanied by cells that migrate out of the epithelial compartment into the underlying lamina propria (Fig. 5; Farbman 1986; Farbman and Squinto 1985). The phenomenon of epithelial cell migration accompanying growth of olfactory axons has been described in other vertebrates, including chick (e.g., Disse 1897; Mendoza et al. 1982; Romanoff 1960; Van Campenhout 1937), golden hamsters (Taniguchi et al. 1982) and humans (Bossy 1980). The ultimate fate of these epithelial cells is not known, but three possibilities have been suggested (Farbman and Squinto 1985; Farbman 1986):

a) they may become olfactory nerve Schwann cells. Although this must be considered a possibility, Farbman and Squinto (1985) have shown that the migrating epithelial cells do not express a positive immunocytological reaction for glial fibrillary acidic protein (GFAP), a protein found in olfactory nerve Schwann cells (Barber and Lindsay 1982). On the other hand, experiments on chick-quail chimeras suggest that olfactory Schwann cells have the same origins as the epithelium. These data support the notion that these migrating epithelial cells could be Schwann cells (Couly and Le Douarin 1985).

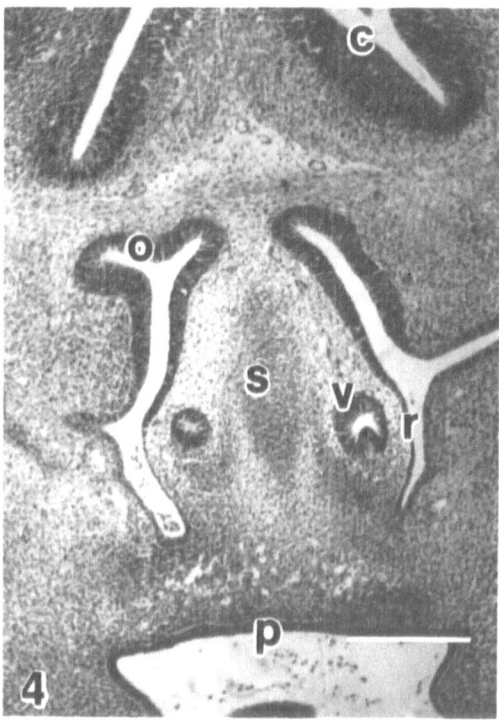

Fig. 4 A photomicrograph of a coronal section through the head of an E15 embryo. The irregular shape of the nasal cavity is evident as the turbinates have begun to form. **c** cerebral vesicle, **o** olfactory epithelium, **p** primary palate, **r** respiratory region of nasal cavity, **s** nasal septum, **v** vomeronasal organ. Marker = 1 mm

b) At least some of the migrating cells may become ganglion cells of the nervus terminalis (Graziadei and Monti Graziadei 1978; Mendoza et al. 1982).

c) at least some of the migrating cells may function as guide cells that direct the axons toward their target (Farbman and Squinto 1985).

The first axons reach the presumptive olfactory bulb at E15, although most axons arrive later. Even when they reach the bulb, no morphologically defined synaptic junctions are found until E18 (Farbman 1986). Similary, in the mouse a period of approximately 2 days elapses between the time when the olfactory axon terminal reaches its target organ and the time when synapses can be demonstrated (Hinds 1972).

3.3 Supporting Cells

The supporting cells also exhibit morphological changes during development. Small microvilli appear on the surface of supporting cells at E14. At E18, the time when many olfactory cilia are becoming tapered, there is a growth spurt in the length of supporting cell microvilli. Between E18 and E 19 their average length increases from about 0.4 µm to about 1.3 µm. Virtually no growth in average microvillus length occurs after E19 (Menco and Farbman 1985b). It should be noted, however, that some supporting cells bear long microvilli while others are covered with short ones, even in adult rats (Menco 1984).

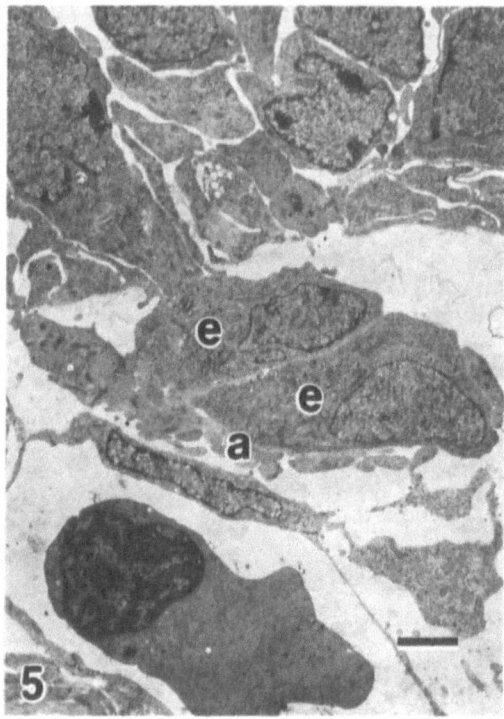

Fig. 5 At the base of the olfactory epithelium from an E 14 fetus some migrating epithelial cells **e** and axons **a** are leaving the main part of the receptor sheet. Marker = 2μm

3.4 Biochemical Differentiation of Receptor Neurons

At about the same time as ciliogenesis begins on E16, the receptor cell expresses a unique macromolecule, olfactory marker protein (OMP) (Allen and Akeson 1985b). OMP is soluble, acidic protein synthesized exclusively by olfactory receptor cells and found throughout their cytoplasm from dendrite to axon terminal (Margolis 1972, 1980, 1982; Farbman and Margolis 1980; Monti Graziadei et al. 1980). Although the function of this protein has not been identified, it is known that the protein is expressed by the more mature cells in olfactory epithelium (Farbman and Margolis 1980; Monti Graziadei et al. 1980). In a study on postnatal mice, it was shown that OMP is expressed in differentiating receptor cells 7 - 8 days after cell division (Miragall and Monti Graziadei 1982).

Other neurochemical studies in developing rat olfactory tissue have shown that a dipeptide, carnosine (beta-alanyl-histidine), first appears at E16 (Margolis et al. 1985). Carnosine is a major constituent of olfactory receptor cells where it has been postulated to play a role as a neurotransmitter or neuromodulator (Margolis 1980).

Allen and Akeson (1985a) have identified a glycoprotein immunoreactive with a monoclonal antibody, referred to as 2B-8, on some olfactory receptor cells beginning at E14. The significance of this observation is that it suggests there may be subclasses of olfactory receptor cells, some that contain antigens

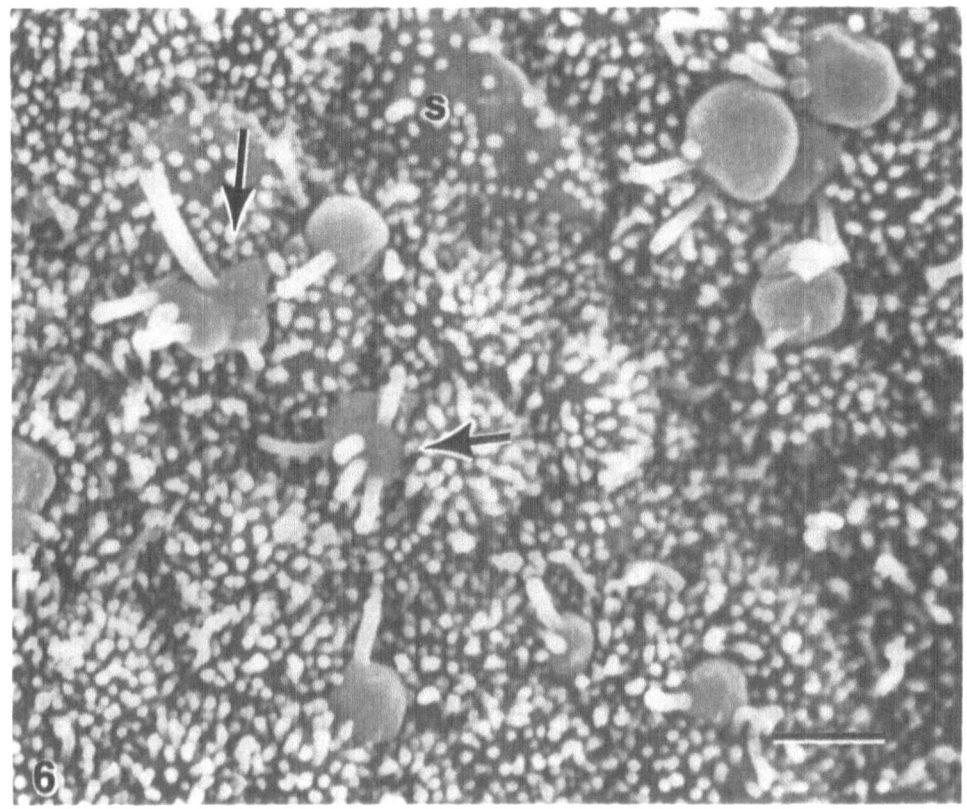

Fig. 6 This scanning electron micrograph, taken from an E16 animal, shows some olfactory dendrites with multiple cilia (arrows). **s** supporting cell with primary cilium. At this age, olfactory cells are still found in clusters, e.g., in upper right corner of micrograph. Marker = 2μm

immunoreactive with 2B-8 and some that do not. The possibility that there are subgroups of olfactory receptor cells is supported by data from other immunocytochemical studies. Mollicone et al.(1985) showed that two blood-group antigens are demonstrable on receptor cell surfaces. The H antigen first appears on E15, and the B antigen appears on E16 in some but not all the cells expressing the H antigen. Other studies have provided evidence for the presence of subclasses of olfactory cells on the basis of different immunoreactivities to monoclonal antibodies (Mori et al. 1985; Schwob and Gottlieb 1985).

The intermediate filament protein vimentin is generally found in young neurons throughout the developing nervous system, but is replaced, as development proceeds, by neurofilament protein. Olfactory neurons differ from other neurons in their retention of this protein (Schwob et al. 1986). The retention of what has been called the juvenile state (Schwob et al. 1986) may be related to the fact that neurogenesis continues in olfactory epithelium may be related to the fact that neurogenesis continues in olfactory epithelium throughout life (cf. Graziadei and Monti Graziadei 1978).

4 Physiological Studies

It is interesting to examine the morphological and biochemical observations in light of some physiological data from a study by Gesteland et al. (1982). In this study on fetal rats, electro-olfactograms could be recorded on or after E14, and, by E19, the magnitude of the response is essentially similar to that in adult animals. Single unit responses were first recorded extracellularly at E16. From this day until E18 – E19 virtually all of the cells tended to respond indiscriminatily to many odorants, whereas later, more cells became responsive to a smaller number of odorants, i.e., they became selective. An analysis of these data suggested that cells responding nonselectively may be those that had no cilia or only primary cilia, whereas the selectively responsive cells were probably multiciliated (Menco and Farbman 1985b).

In considering development of olfactory epithelium it is important to note that neurogenesis and differentiation of receptor cells continues in the vertebrate olfactory epithelium throughout the life of the animal (cf. Graziadei and Monti Graziadei 1983). It is generally assumed (cf. Graziadei and Monti Graziadei 1983), through not proven beyond doubt (cf. Hinds et al. 1984; Farbman 1986), that the receptors become fully differentiated and functional and that there is a true turnover of cells. If this were so, at any given age, embryo or adult, receptor cells in all stages of differentiation, from postmitotic to fully mature, would be found on the olfactory epithelium. Recent work suggests that receptor cells may live for quite a long time in an environment protected from trauma (Hinds et al. 1984); in other words, turnover or cell replacement may not be physiological but a kind of wound healing phenomenon. In either event, the fact that receptor cells of all ages are present in the olfactory epithelium is an important consideration in the interpretation of electrophysiological data because the responsivity of young cells may differ from that of older cells (Gesteland et al. 1982).

Summary

Descriptive morphological data on the development of the rat olfactory epithelium are considered in the light of some biochemical, immunological and physiological studies on early development and maturation. Olfactory cilia are formed in young receptor cells at the same embryonic age when 1) olfactory marker protein is first demonstrable, 2) carnosine is first synthesized by receptor cells, and 3) the receptor cells become selectively responsive to odorant stimuli. Expression of cilia is enhanced by the contact of olfactory axons with the bulb. Axon growth is accompanied by the migration of cells from the epithelium. Synapses are found between olfactory cells and bulbar second order neurons about 2 days after axons reach the bulb. Immunological data suggest there are subsets of olfactory receptor cells.

Acknowledgment. The work reported in this paper was supported by grants NS 06181, NS 18490, and NS 21555 from The National Institutes of Health.

References

Allen WK, Akeson R (1985a) Identification of an olfactory receptor neuron subclass: Cellular and molecular analysis during development. Dev Biol 109: 393-401

Allen WK, Akeson R (1985b) Identification of a cell surface glycoprotein family of olfactory receptor neurons with a monoclonal antibody. J Neuroscience 5: 284-296

Barber PC, Lidsay RM (1982) Schwann cells of the olfactory nerves contain glial fibrillary acidic protein and resemble astrocytes. Neuroscience 7: 2687-2695

Bossy Y (1980) Development of olfactory and related structures in staged human embryos. Anat Embryol 161: 225-236

Breipohl W, Fernandez M (1977) Scanning electron microscopical investigations of olfactory epithelium in the chick embryo. Cell Tissue Res 183: 105-114

Breipohl W, Ohyama M (1981) Comparative and developmental SEM studies on olfactory epithelia in vertebrates. (Biomedical aspects and speculations). Biomed Res (Suppl) 2: 437-448

Breipohl W, Mestres P, Meller K (1973) Licht- und elektronmikroskopische Befunde zur Differenzierung des Riechepithels des weissen Maus. Verh Anat Ges Jena 67: 443-449

Campenhout E van (1937) Le dévellopement du système nerveux cranien chez le poulet. Arch Biol (Liege) 48: 611-666

Chuah MI, Farbman AI, Menco BPhM (1985) Influence of olfactory bulb on dendritic knob density of rat olfactory receptor neurons in vitro. Brain Res 338: 259-266

Couly GF, Le Douarin NM (1985) Mapping of the early neural primordium in quail-chick chimeras. I. Developmental relationships between placodes, facial ectoderm and prosencephalon. Devel Biol 110: 422-439

Cuschieri A, Bannister LH (1975a) The development of the olfactory mucosa in the mouse: light microscopy. J Anat 119: 277-286

Cuschieri A, Bannister LH (1975b) The development of the olfactory mucosa in the mouse: electron microscopy. J Anat 119:471-498

Disse J (1897) Die erste Entwicklung der Riechnerven. Anat Hefte 9: 257-300

Farbman AI (1977) Differentiation of olfactory receptor cells in organ culture. Anat Rec 189: 187-200

54

Farbman AI (1986) Prenatal development of mammalian olfactory receptor cells. Chem Senses (in press).

Farbman AI, Margolis FL (1980) Olfactory marker protein during ontogeny: Immunohistochemical localization. Dev Biol 74: 205-215

Farbman AI, Squinto LM (1985) Early development of olfactory receptor cell axons. Dev Brain Res 19: 205-213

Gesteland RC, Yancey RA, Farbman AI (1982) Development of olfactory receptor neuron selectivity in the rat fetus. Neuroscience 7: 3127-3136

Gonzales F, Farbman AI, Gesteland RC (1985) Cell and explant culture of olfactory chemoreceptor cells. J Neurosci Methods 14: 77-90

Graziadei PPC, Monti Graziadei GA (1978) Continuous nerve cell renewal in the olfactory system. In: Jacobson M (ed) Development of sensory systems. Springer, Berlin, Heidelberg New York pp 55-83

Graziadei PPC, Monti Graziadei GA (1983) Regeneration in the olfactory system of vertebrates. Am J Otolaryngol 4: 228-233

Hinds JW (1972) Early neuron differentiation in the mouse olfactory bulb. II. Electron microscopy. J Comp Neurol 146: 253-276

Hinds JW, McNelly NA (1981) Aging in the rat olfactory system: correlation of changes in the olfactory epithelium and olfactory bulb. J Comp Neurol 203: 441-453

Hinds JW, Hinds PL, McNelly NA (1984) An autoradiographic study of the mouse olfactory epithelium: evidence for long lived receptors. Anat Rec 210: 375-383

Kerjaschki D (1977) Some freeze-etching data on the olfactory epithelium. In: Le Magnen J, Mac Leod P (eds) Olfaction and taste vol VI. Information Retrieval, London, Washington DC, pp 75-85

Kerjaschki D, Hörandner H (1976) The development of mouse olfactory vesicles and their cell contacts: A freeze-etching study. J Ultrastruct Res 54: 420-444

Klein SL, Graziadei PPC (1983) The differentiation of the olfactory placode in Xenopus laevis: A light and electron microscope study. J Comp Neurol 217: 17-30

Margolis FL (1972) A brain protein unique to the olfactory bulb. Proc Natl Acad Sci USA 69: 1221-1224

Margolis FL (1980) An olfactory neuropeptide. In: Barker JL, Smith T (eds) Role of peptides in neuronal function. Dekker, New York, pp 545-572

Margolis FL (1982) Olfactory marker protein (OMP). Scand J Immunol (Suppl 9) 15: 181-199

Margolis FL, Grillo M, Kawano T, Farbman AI (1985) Carnosine synthesis in olfactory tissue during ontogeny: Effect of exogenous b-alanine. J Neurochem 44: 1459-1464

Menco BPhM (1980) Qualitative and quantitative freeze-fracture studies on olfactory and nasal respiratory structures of frog, ox, rat, and dog. Cell Tissue Res 207: 183-209

Menco BPhM (1983) The ultrastructure of olfactory and nasal respiratory epithelium surfaces. In: Reznik G, Stinson SF (eds) Nasal tumors in animals and man, vol I. Anatomy, physiology and epidemiology. CRS Press, Boca Raton, pp 45-102

Menco BPhM (1984) Ciliated and microvillous structures of rat olfactory and nasal respiratory epithelia. A study using ultra-rapid cryo-fixation followed by freeze substitution or freeze-etching. Cell Tissue Res 235: 225-241

Menco BPhM, Farbman AI (1985a) Genesis of cilia and microvilli of rat nasal epithelia during prenatal development. I. Olfactory epithelium, qualitative studies. J Cell Sci 78: 283-310

Menco BPhM, Farbman AI (1985b) Genesis of cilia and microvilli of rat nasal epithelia during prenatal development. II. Olfactory epithelium, a morphometric analysis. J Cell Sci 78: 311- 336

Mendoza AS, Breipohl W, Miragall F (1982) Cell migration from the chick olfactory placode: a light and electron microscopic study. J Embryol Exp Morphol 69: 47-59

Miragall G, Monti Graziadei GA (1982) Experimental studies on the olfactory marker protein. II Appearance of the olfactory marker protein during differentiation of the olfactory sensory neurons of the mouse: An immunohistochemical and autoradiographic study. Brain Res 239: 245-250

Mollicone R, Trojan J, Oriol R (1985) Appearance of H and B antigens in primary sensory cells of the rat olfactory apparatus and inner ear. Dev Brain Res 17: 275-279

Monti Graziadei GA, Stanley RS, Graziadei PPC (1980) The olfactory marker protein in the olfactory system of the mouse during development. Neuroscience 5: 1239-1252

Mori K, Fujita SC, Imamura K, Obata K (1985) Immunohistochemical study of subclasses of olfactory nerve fibers and their projections to the olfactory bulb in the rabbit. J Comp Neurol 242: 211-229

Nakashima T, Kimmelman CP, Snow JB Jr (1984) Structure of human fetal and adult olfactory neuroepithelium. Arch Otolaryngol 110: 641-646

Noda M, Harada Y (1981) Development of olfactory epithelium in the mouse: Scanning electron microscopy. Biomed Res (Suppl) 2: 449-454

O'Rahilly R (1967) The early development of the nasal pit in staged human embryos. Anat Rec 157: 380

Pyatkina GA (1982) Development of the olfactory epithelium in man. Z Mikrosk Anat Forsch, Leibzig 96: 361-372

Romanoff AL (1960) The avian embryo, structural and functional development. MacMillan, New York, p 317

Schwob JE, Gottlieb DI (1985) The primary olfactory projection has two chemically distinct zones. Neurosci Abstr 11: 970

Schwob JE, Farber NB, Gottlieb DI (1986) Neurons of the olfactory epithelium in adult rats contain vimentin. J Neurosci 6: 208-217

Seifert K (1970) Die Ultrastruktur des Riechepithels beim Makrosmatiker. Eine elektronenmikroskopische Untersuchung. In: Bargmann W, Doerr W (eds) Normale und pathologische Anatomie. Monographien in zwangloser Folge. Thieme, Stuttgart, pp 1-99

Smart IHM (1971) Location and orientation of mitotic figures in the developing mouse olfactory epithelium. J Anat 109: 243-251

Sorokin SP (1968) Reconstructions of centriole formation and ciliogenesis in mammalian lungs. J Cell Sci 3: 207-230

Taniguchi K, Taniguchi K, Mochizoki K (1982) Comparative developmental studies on the fine structure of the vomeronasal sensory and the olfactory epithelia in the golden hamster. Jpn J Vet Sci 44: 881-890

Tucker RW, Pardee AB (1982) Primary cilia and their role in the regulation of DNA replication and mitosis. In: Nicolini C (ed) Cell growth. Plenum Press, New York, pp 365-376

Vidic B, Greditzer HG, Litchy WJ (1972) The structure and prenatal morphogenesis of the nasal septum in the rat. J Morphol 137: 131-148

Waterman RE, Meller SM (1973) Nasal pit formation in the hamster: A transmission and scanning electron microscopic study. Dev Biol 34: 255-266

Wheatley DN (1982) The centriole: a central enigma of cell biology. Elsevier Biomedical Press, Amsterdam New York Oxford

Morphological Maturation of the Olfactory Epithelia of Australian Marsupials

JE Kratzing

Department of Veterinary Anatomy, University of Queensland, St.Lucia 4067, Australia

1 Introduction

In the early phase of independent mammalian life olfaction is reported to play an important part in nipple search, nesting behaviour and identification and bonding between mother and young (Hudson and Distel 1983; Schmidt et al. 1983; Grau 1976; Doty this Vol). Most of the studies of the functional development of the olfactory system have been carried out in rodents (Alberts 1976); the ontogeny of the pheripheral system has been studied in mice (Breipohl et al. 1973; Cuschieri and Bannister 1975) and the hamster (Waterman and Meller 1973) and that of the olfactory bulb in the mouse (Hinds and Hinds 1976; Mair and Gesteland 1982; Benson et al. 1984). The general picture emerging from these studies is that the peripheral system which develops from the olfactory placode, reaches a developmental stage at birth capable of transmitting chemosensory information to the central system. However, central development of the olfactory circuits may be only partially complete at birth (Mair and Gesteland 1982; Greer et al. 1982) and may require input from the peripheral system in the early neonatal period for full development (Schmidt et al. 1983; Constanzo 1982). Postnatal importance of olfactory sensory experience may be different for marsupials for whom a short period in utero is characteristic and the young at birth differ in many respects from their eutherian counterparts, even in those mammals with a comparable gestation period. Moreover, their immediate postnatal conditions are very different. The marsupial neonate needs only to succeed in a brief period of independent existence before it settles to a further development in the protected environment of the pouch.

Immediately after birth the young marsupial must make its way unaided to the pouch, locate a teat and secure it in its mouth. To this end it must achieve in utero sufficient forelimb development, a fuctional respiratory system, and presumably, sufficient sensory equipment to reach its objective. Since the young are born with eyes and ears closed, tactile and olfactory senses seem more likely to provide sensory guidance. Olfactory competence is suggested because the nostrils are well developed and wide open (Sharman 1973) and embryological studies by a number of authors have noted the presence of olfactory epithelium, nerve bundles and olfactory bulb at birth (see Table 1).

Table 1   Some references to olfactory studies in marsupial and monotreme embryology

Species	Developmental stage	Gestation period (days)	Observations	References
Dasyurus viverrinus	Birth		Olfactory sense cells present in the olfactory portion of nasal epithelium which is by far more extensive than the respiratory portion. Olfactory portion is composed of olfactory and supporting cells; basal cells are not recognizable. There are no Bowman's glands. Jacobsen's organ forms a thick-walled groove, closing behind to a short tube; total length 0.06 mm	Hill and Hill 1955
Schinobates volans (Kerr)	18 somites (3.4 mm)	25–27	Anlagen of olfactory, optic and otic organs were differentiated	Bancroft 1973
	Near term (5.0 mm)		External nares open	
Tachyglossus sp.	Hatching	Uncertain; up to 27 days uterine, about 10 days incubation in the pouch	Olfactory bulbs, nerves and epithelium well developed and apparently functional	Griffith 1978
Atechinus stuarti (Macleay)	26	27.2	Open nostrils	Selwood 1980
Macropus rufogriseus	25[a]	26[a]	Olfactory apparatus well developed	Walker and Rose 1981
Trichosurus vulpecula	Birth	17.5	Olfactory epithelium a well developed layer of pseudostratified epithelium; a small number af filium olfactorium and Bowman's glands	Hughes and Hall 1984
Didelphys[b] *virginiana*[b]	10 11 12	13	Olfactory placode well developed and noticeably depressed Olfactory nerve represented by a few fibres from the olfactory pit to the forebrain Jacobsen's organ represented as a diverticulum into the septum	McCrady 1938

[a]Days after removal of pouch young, i.e., after diapause

[b]Figures for the American opossum are included for comparison

Once the neonate has located the pouch and secured a teat it will undergo further development while firmly attached by structural modifications of the teat and oral cavity which ensure a constant food supply in a protected environment. Russell (in press) considers that this period in the pouch eliminates the need for early development of recognition between mother and young. This is in direct contrast to conditions for most eutherian neonates who must repeatedly find a teat, learn to recognize and return to the nest, and identify mother and litter mates. Marsipual young may use the early period in the pouch to achieve full olfactory activity, or they may be able to postpone it to a later period of pouch development when they spend some time away from the mother but return to the pouch for food and shelter. Thus the ontogeny of olfaction in marsupials need clarifying at three stages:

1. In utero development and capability at birth.

2. Development during the portion of pouch life attached to a teat.

3. Development in the transition period from pouch life to independent living.

First investigations focussed to clarify these aspects are presented here for the peripheral olfactory system in Isoodon macrourus, the northern bandicoot, from the late fetal stage through the first 35 days of pouch life.

2 Material and Methods

Abbreviations used
OEP   main olfactory epithelium
SOO   septal olfactory organ of Rodolfo-Masera
VNO   vomeronasal organ

Specimens were obtained from two individuals of Isoodon macrourus at each of the following ages: 10.5 days fetal development, 1,2,5,11,16 and 35 days postnatal. The gestation period for Isoodon macrourus is 12.5 days (Lyne 1974). The young are attached to a teat in the pouch for about 50 days, and leave it at about 60 days (Gordon 1974). Tissues from one of each pair were prepared for TEM. The head of the other specimen was examined by light microscopy.

3 Results

3.1 10.5 Days Gestation

The olfactory pit is lined by pseudostratified epithelium 5-7 nuclei in depth (Fig 1a). Mitoses are more frequent towards the lumen, a distibution that reminds on the situation in rodents (Smart 1971; Breipohl et al. this Vol). Perinuclear cytoplasm contains dense ribosomes clusters, few mitochondria and short lengths of rough endoplasmic reticulum (Fig. 1b). Some distal processes have a single cilium or centriole (Fig. 1c), and neurotubules are apparent in some axonic processes (Fig 1d). Between the olfactory pit and telencephalon, there are aggregations of cells whose cytoplasmic organization resembles that of olfactory

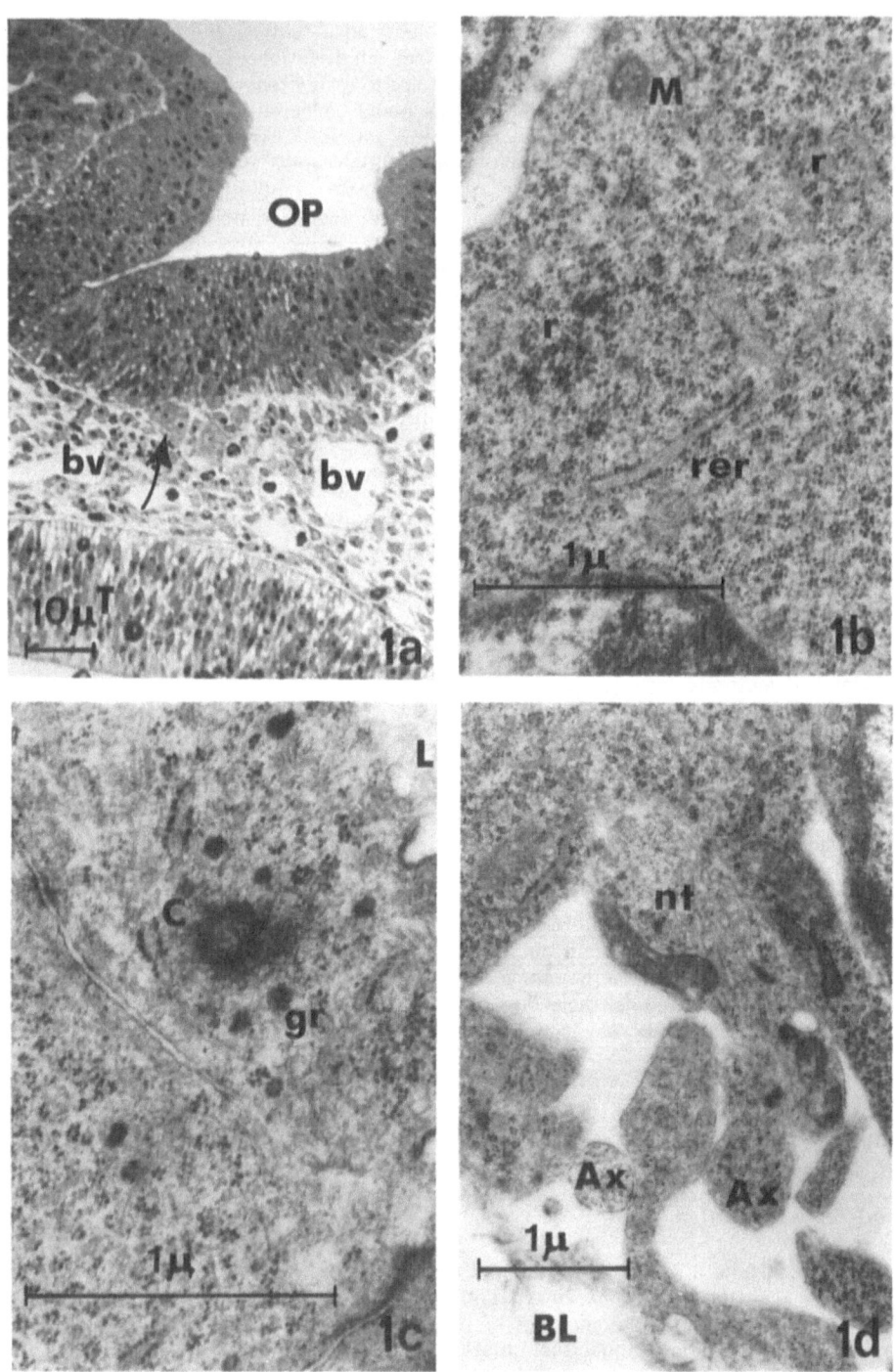

cells. Cellular density per area in the telencephalon is less than in the olfactory pit, but the cytoplasmic organization of cells is similar (Fig 1a). Differentiation between the prospective OEP, SOO and VNO could not be made at this developmental stage.

Fig. 1 Olfactory neuroepithelium - 10.5 days gestation. a) Thick (1 µm) section of olfactory pit **OP** and telencephalon **T**. Neuroepithelium lining the pit is denser than that of the telencephalon. In loose mesenchyme below the pit, note abundant blood vessels **bv** and discrete mass of cells (arrow) associated with the neuroepithelium. b) Cytoplasm of cell at mid-epithelial level shows ribosomal clusters **r**, some rough endoplasmic reticulum **rer** and sparse mitochondria **m**. c) Distal process of olfactory cell. A single centriole **C** and scattered dark granules **gr** lie close to the lumen **L**. d) Base of neuroepithelium. Most axonic processes **AX** contain ribosome clusters and neurotubules **nt** in one area. **BL** basal lamina.

◄───

3.2 Day 1

Main olfactory epithelium (OEP). The newborn bandicoot has a simple nasal cavity lacking turbinates. Neuroepithelium lines most of the dorsal wall and septum. Some receptor cells are clearly distinguishable, but not other cell types. Receptor cell perikarya show little differentiation but most dendritic processes and in a rounded rod containing centrioles (Fig 2a); few rods have cilia. Microtubules are differentiated in dendritic and axonic processes. Loose connective tissue under the epithelium contains bundles of nerve fibers and small blood vessels; the latter are never close enough to indent the basal lamina. There are no secretory cells.

Septal Olfactory Organ of Rodolfo-Masera (SOO). The septum does not reach the floor in the caudal two thirds of the nasal cavity and its free ventral edge is wide but there is no evidence of a septal organ.

Vomeronasal Organ (VNO). The vomeronasal organ opens the incisive duct as it enters the nasal cavity and the organ is in communication with oral and nasal cavities. Receptor and supporting cells are less readily distinguished than in the OEP and no basal cells can be identified. Vomeronasal receptor cells have more extensive rough endoplasmic reticulum than main olfactory sensory cells (Fig. 2c) and some dendrites have a cluster of centrioles and neurotubules. Cells with a single cilium or a pair of surface centrioles are probably supporting cells. Vomeronasal nerve fibres are seen on the nasal septum in association with ganglionic cells.

3.3 Day 2

OEP lines the dorsal septum and dorsolateral caudal two thirds of the nasal cavity. The ratio of receptors to supporting cells has increased but their perikarya show little change from that of the newborn. Most olfactory rods have clusters of centrioles but many lack cilia. Vesicles are often evident in the cytoplasm towards the surface of supporting cells. There are no basal cells or secretory cells.

SOO. No SOO receptors could be identified.

VNO cells resemble those seen at birth except for better definition of neurotubules in the axonic processes.

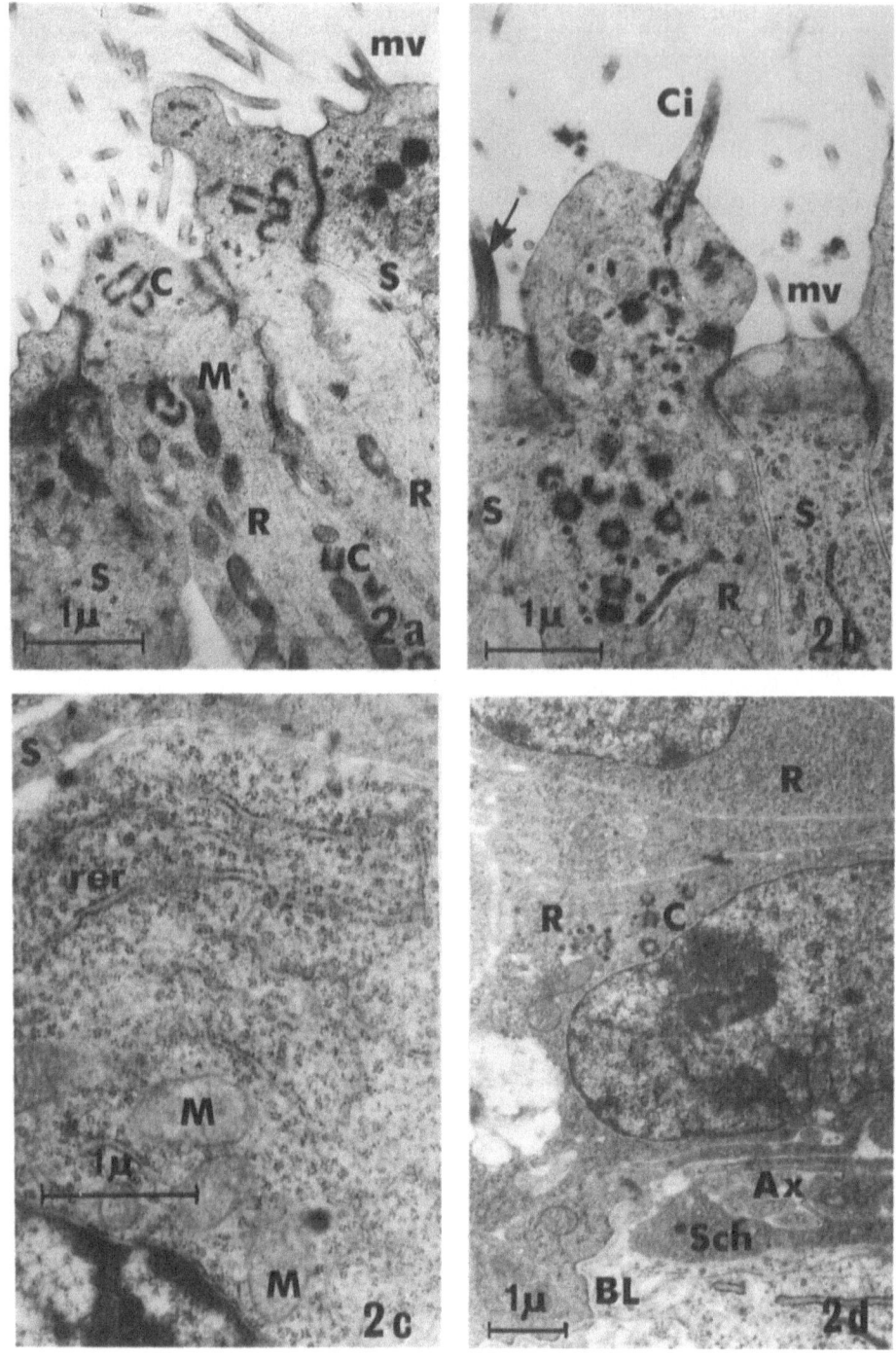

Fig. 2 OEP a) Distal processes of receptor **R** and supporting S cells. Receptor cells have clusters of centrioles **C** at several levels but lack cilia, mitochondria **M** are frequent and neurotubules well developed. Supporting cells have few of microvilli **mv** that may branch b) OEP surface. The receptor cell dendrite R has a number of centrioles and centriole precursor bodies and dark granules and ends in a well-defined rod. Cilium **Ci.** Supporting cells **S** with cilium (arrow) and microvilli **mv.** c) Receptor cell perikaryon. Cytoplasm resembles that in OEP receptor cells but rough endoplasmic reticulum **rer** and mitochondria **M** are more abundant here. **S** supporting cell. d) Base of neuroepithelium. A VNO receptor cell perikaryon **R** shows a cluster of centrioles **C.** Axonic processes **AX** form bundles wrapped by Schwann cells **Sch** below the basal lamina **BL.**

◄─────────────────────

3.4 Day 5

OEP. There is an increase in the proportion of receptor cells to supporting cells, ands most rods have cilia (Fig. 2b). A few basal cells can be distinguished by the presence of fibrillar bundles. Below the epithelium there are a few secretory cells.

SOO. The septal organ appears on the free ventrolateral edge of the septum as a crowded patch of cells with numerous surface mitoses. Receptor and supporting cells can be distinguished but cytoplasmic differentiation is less advanced than that of the OEP at birth.

VNO perikarya remain little changed (Fig. 2d) but more dendritic processes have centrioles and granular clusters and in some cases a single cilium (Fig. 3b). A few supporting cells also have an apical centriole or cilium.

3.5 Day 11

OEP. By 11 days fine structure of receptor, supporting and dark basal cells resemble that in the adult (Fig. 3b). Olactory glands have a clear lumen and granules appear in secretory cells. Secretion is also apparent in glands associated with the respiratory epithelium.

SOO. The neuroepithelium resembles that in the OEP at 5 days. There are no olfactory gland cells.

VNO. The proportion of receptors to supporting cells has further increased. Both cell types have more microvilli than in previous stages, many receptors have a single cilium, and some supporting cells a cilium or pair of centrioles. There are no vomeronasal secretory cells.

3.6 Days 16 and 35

OEP. From day 16 onwards the density per area of receptors and total extent of the neuroepithelium increase as the turbinates grow in complexity. At 35 days the OEP resembles that of the adult in all structural aspects.

SOO secretory cells are evident by day 16. By day 35 the structure of neuroepithelium and glands resembles that of the adult.

64

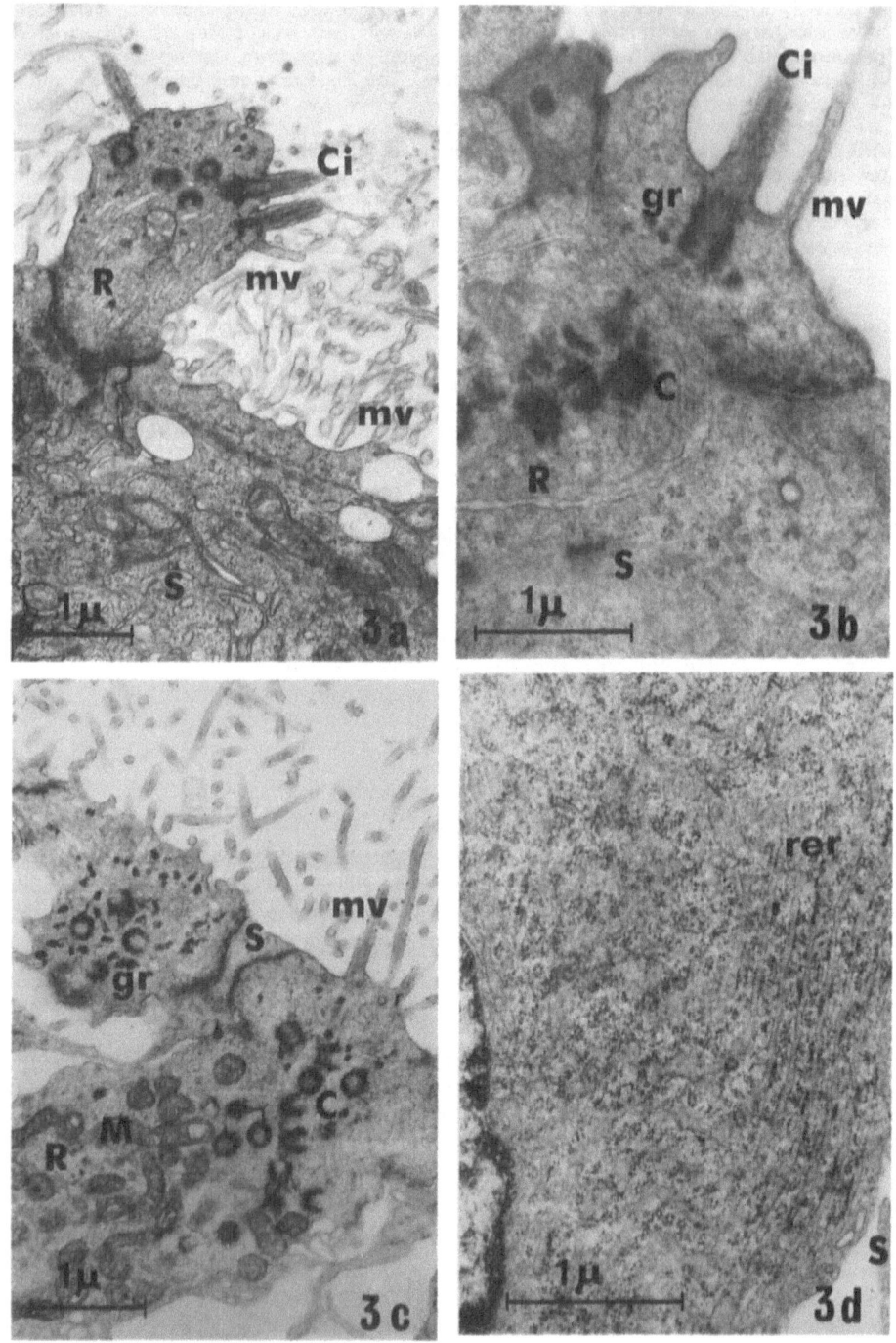

Fig. 3 a) OEP surface. An olfactory rod **R** shows centrioles and dark granules, and carries cilia **Ci** and microvilli **mv**. Supporting cells **S** have smooth and rough endoplasmic reticulum and numerous microvilli. b) VNO surface. A distal process **R** shows a number of centrioles **C**, dark granules **gr**, microvilli **mv** and a cilium **Ci**. **S** supporting cell. c) VNO receptor cell perikaryon day 35. Cytoplasm contains parallel arrays of rough endoplasmic reticulum **rer** and some smooth endoplasmic reticulum close to the nucleus. **S** supporting cell process. d) VNO Neuroepithelial surface 35 days. Receptor dendrites **R** show numerous mitochondria **M**, centrioles **C**, dark granules **gr** and microvilli **mv** which are slightly greater in diameter than those of supporting cells **S**.

VNO. At day 35, receptors with a single cilium remain more frequent but in other respects the neuroepithelium resembles that of the adult (Figs. 3c,d).

3.7 Main Olfactory Bulb

Maturation in olfactory bulb fine structure is not included in this study, but light microscopy shows that at birth the bulb has a dense layer of 8/10 neuroblast nuclei radially arranged around the ventricle. This layer merges with a more loosely organized zone of fibres and cells contacted rostroventrally by axons from the receptor cells. Five days after birth the incoming fibres surround the ventral half of the bulb and a narrow outer plexiform layer lies between them and the immature mitral cell layer. At day 11 the following regions are evident:

1. periventricular neuroblast layer 8/10 nuclei in depth

2. loosely arranged fibre/cell layer (prospective granule cell layer)

3. mitral cell layer 2/3 nuclei in depth

4. outer plexiform layer

5. zone of incoming olfactory nerve fibres

At 16 days the mitral cell layer is still more than one cell thick and the glomerular layer has developed with large and small nuclei at its inner margin. By 35 days a granule cell layer with concentrically arranged nuclei surrounds the ventricular lining; its most peripheral ring of nuclei is in contact with the mitral cells which now form a single cell layer.

4 Discussion

Very few developmental studies of marsupial olfaction have been carried out. The surge of interest in monotremes and marsupials in the late 19th and early 20th centuries provided basic embryological data which frequently remain the only published material. More recent work provides little specific information on olfactory systems (Table 1) and is limited to light microscopy, which restricts evidence of functional capacity. Fine structural studies of the peripheral system have been concerned with the adult form (Kratzing 1978, 1982, 1984; Biggins 1979) and reveal an extensive olfactory and vomeronasal neuroepithelium and in at least one species a septal olfactory organ. Nelson and Harman (1983) reported that visual, auditory and olfactory systems were poorly developed at birth in Dasyurus hallucatus and that almost all development of these systems and the anterior part of the brain occurred during pouch life. After 44 days in the pouch the olfactory bulb is 3.4% of total brain volume, rising to 5.3% at 68 days (about the end of the period of attachment to the teat) and reaching the adult proportion of 7.4% at 89 days (Nelson 1985).

This study indicates that the anatomical development of the peripheral olfactory system in Isoodon may be slower than in eutherian species. Direct comparisons of fetal stages are difficult since the physiological demands on the neonates differ between eutherians and non-eutherians. At 10.5 days gestation, 2 days before birth, the differentiation of neurotubules and centrioles in Isoodon resembles that in the mouse at 10 days (Cuschieri and Bannister 1975), though in Isoodon there is no differentiation into light and dark cells. Olfactory rods are not differentiated though they are evident at 9.5 days in the hamster (Waterman and Meller 1973).

If the degree of development at birth is used as a landmark for comparison, then the peripheral system in Isoodon differs from eutherians in that it does not undergo a rapid change to full differentiation just before birth. In Isoodon the OEP receptor cells have few cilia and lack the extensive Golgi apparatus and endoplasmic reticulum of the mature cell. Olfactory nerves have small numbers of axons of variable diameter loosely wrapped by Schwann cells. Dark basal cells first become apparent 5 days postnatally, by which time the cytoplasm of supporting cells has developed some of the rough and smooth endoplasmic reticulum seen in the mature cell. However, their surface microvilli are sparse for the first 10 postnatal days, which contrasts with full development at 17 fetal days in the mouse (Cuschieri and Bannister 1975) and the first 2 days after birth in the rat (Kratzing 1971).

VNO receptor cell dendrites are less developed at birth than those of the OEP receptor cells. Leon (1983) considers that the prenatal development of the rat VNO and its projection sites precedes that of the OEP. The first appearance of SOO receptor cells some five days after birth is surprising since Rodolfo-Masera (1943) found the organ in the fetal guinea pig and mouse which retain the organ in the adult, and also in the fetal rabbit which does not have an adult SOO. The MO is probably not a vestigial organ, since considerable postnatal morphological development has been described for it e.g. in mice (Breipohl and Naguro personal communication). The organ is not widely distributed in mammals and has not been reported in marsupials other than Isoodon and in a fetal Didelphis (Rodolfo-Masera 1943). Functions suggested for it include olfactory monitoring in quiet respiration (Adams and McFarland 1971; Kratzing 1978) and perception of pheromones (Wysocki et al. 1980). It is evident from its late time of appearance in Isoodon that it can not play a part in locating pouch and teat.

There are no olfactory or vomeronasal glands present for the first five days of pouch life; by contrast, secretion droplets appear in Bowman's glands in the mouse in the last quarter of gestation (Breipohl 1973) and in rats at 17 days fetal (Kratzing 1971). When secretion droplets appear in the olfactory glands at 16 days they are also apparent in other glands of the nasal cavity, none of which are active up till this time. The pouch provides a humid environment which minimizes the need for temperature and humidity control of inspired air. It is difficult to assess whether the receptors' ability to respond to chemical signals is reduced by the absence of special olfactory and vomeronasal secretions since their role has never been clarified unanimously (Moulton and Beidler 1967). The glands are present at birth in the brush-tailed possum <u>Trichosurus vulpecula</u> (Hughes and Hall 1984), present but sparse in a pouch-young koala at about 16 weeks and in an eastern grey kangaroo at about 8 weeks (unpublished observations).

At the light microscope level observations of the olfactory bulb confirm Nelsons' (1985 personal communication) view that differentiation is far from complete at birth. However, though the timetable of development differs, and may reflect reduced demands on the olfactory sensory systems once secure in the pouch, the sequence of differentiation appears similar to that reported in other mammals. That mitral cells are the first to be clearly recognizable and granule cells last agrees with the findings of Hinds (1968 a,b) and Altman (1969) in the mouse, and Mair and Gesteland (1982) in the rat, and the gradual spread of olfactory fibres around the bulb and subsequent differentiation of the glomerular layer agree with data of Monti-Graziadei et al. (1980) in the mouse. It is not clear whether the early crowded appearance of the mitral cell layer is due to an initial neuronal cell overproduction, or to rapid provision of the final population of cells to be spread over a wider layer as the bulb matures.

5 Conclusions

<u>Isoodon</u> completes anatomical differentiation of the OEP, SOO and VNO and of the olfactory bulb in the first 35 days of pouch life, well before it is free of the teat. The early stages in fetal development in <u>Isoodon</u> are similar to those in eutherians but there is no acceleration prior to birth in the former. Rather, a continuous process of differentiation for the peripheral system and bulb continues into the first half of pouch life. Thus the general morphological immaturity of the olfactory system at birth, particularly the bulb, indicates that olfaction is not likely to guide this neonate marsipual to the pouch and teat, and that the development of other senses at this stage, perhaps tactile and thermal sensitivity, should be explored. Further studies on the functional development of the olfactory system during the first 35 days and in the transitional period when the young marsipual is able to leave and explore surroundings but return to the pouch for food and shelter. They could help to elucidate environmental and inherent influences on the postnatal morphological and functional maturation of the olfactory system (see discussions in Breipohl et al., Coopersmith and Leon, Panhuber, Schmidt, all this Vol) and to compare with eutherian olfactory maturation. Investigations in the early postnatal stages in marsupials could well have model-like character for the ontogenetic development of this sensory system in vertebrates, and to further elucidate the influence of early sensory experience on the organization and function of the olfactory system in adults.

Summary

To elucidate of the importance of environmental and inherent factors towards the postnatal maturation of the olfactory system a light and electron microscopical study was performed in marsupials. The investigations presented here indicate differences in general maturity of the olfactory system between eutherians and marsupials, as exemplified for the bandicoot <u>Isoodon macrourus</u>. The results suggest that the olfactory system may not be sufficiently differentiated at birth to aid the newborn in locating the pouch and teat. The sequence of events in anatomical development is similar to that described in eutherians, but there is no rapid move to full differentiation of the peripheral neuroepithelia before birth. Instead, there is a steady sequence of development continuing into pouch life but completed well in advance of the time the young can leave the pouch even temporarily. The question of sensory guidance to pouch and teat for the marsupial neonate remains to be solved, and requires investigation of other sensory systems, in particular tactile and thermal sensitivity.

References

Altman J (1969) Autoradiographic and histological studies of postnatal neurogenesis. IV Cell proliferation and migration in the anterior forebrain with special reference to persisting neurogenesis in the olfactory bulb. J Comp Neurol 137: 433-458

Adams DR, McFarland LZ (1971) Septal olfactory organ in Peromyscus. Comp Biochem Physiol 40a: 971-974

Alberts JR (1976) Olfactory contributions to behavioural development in rodents. In: Doty RL (ed) Mammalian olfaction, reproductive processes and behaviour. Academic Press, London New York, p 67

Bancroft BJ (1973) Embryology of Schinobates volans (Kerr) Marsupialia: Petauridae. Aust J Zool 21: 33-52

Benson TE, Ryugo DK, Hinds JW (1984) Effects of sensory deprivation on developing mouse olfactory system: a light and electron microscope morphogenetic analysis. J Neurosci 4: 638-653

Biggins JG (1979) Olfactory communication in the brush-tailed possum Trichosurus vulpecula Kerr, 1972 (Marsupialia : Phalangeridae). Thesis, Monash Univ, Melbourne

Breipohl W (1973) Licht- und elektronmikroskopische Befunde zur Struktur der Bowmanschen Drüsen in Riechepithel der weissen Maus. Z Zellforsch 131: 329-346

Breipohl W, Mestres P, Meller K (1973) Licht- und elektronenmikroskopische Befunde zur Differenzierung des Riechepithels der weissen Maus. Verh Anat Ges 67: 443-449

Constanzo RM (1984) Comparison of neurogenesis and cell replacement in the hamster olfactory system with and without a target (olfactory bulb). Brain Res 307: 295-301

Cuschieri A, Bannister LH (1975) The development of the olfactory mucosa in the mouse: electron microscopy. J Anat 119: 471-498

Gordon G (1974) Movements and activity of the short-nosed bandicoot Isoodon macrourus Gould (Marsupialia). Mammalia 38: 405-431

Grau GA (1976) Olfaction and reproduction in ungulates. In: Doty RL (ed) Mammalian olfaction, reproductive processes and behaviour. Academic Press, London New York.

Greer CA, Stewart WB, Teicher MH, Shepherd GM (1982) Functional development of the olfactory bulb and a unique glomerular complex in the neonatal rat. J Neurosci 2: 1744-1759

Griffiths M (1978) The biology of Monotremes. Academic Press, New York

Hill JP, Hill WCO (1955) The growth-stages of the pouch-young of the Native Cat (Dasyurus viverrinus) together with observations on the anatomy of the new-born young. Trans Zool Soc London 28: 349-452

Hinds JW (1968a) Autoradiographic study of histogenesis in the mouse olfactory bulb. I Time of origin of neurons and neuroglia. J Comp Neurol 134: 287-304

Hinds JW (1968b) Autoradiographic study of histogenesis in the mouse olfactory bulb. II Cell proliferation and migration. J Comp Neurol 134: 305-322

Hinds JW, Hinds PL (1976) Synapse formation in the mouse olfactory bulb. 1. Quantitative studies. J Comp Neurol 169: 15-40

Hudson R, Distel H (1983) Nipple location by new-born rabbits: Behavioural evidence for pheromonal guidance. J Comp Physiol A 155: 13-17

Hughes RL, Hall LS (1984) Embryonic development in the common brushtail possum Trichosurus vulpecula. In: Smith AP, Hume ID (eds) Possums and Gliders. Aust Mammal Soc, Sydney

Kratzing JE (1971) Olfactory and vomeronasal organs in rats, sheep, and a lizard. Thesis, Univ Queensland, Brisbane

Kratzing JE (1978) The olfactory apparatus of the bandicoot (Isoodon macrourus): fine structure and presence of a septal olfactory organ. J Anat 125: 601-613

Kratzing JE (1982) The anatomy of the rostral nasal cavity and vomeronasal organ in Tarsipes rostratus (Marsupialia Tarsipedidae). Aust Mammal 5: 211-219

Kratzing JE (1984) The anatomy and histology of the nasal cavity of the koala (Phascolarctos cinereus). J Anat 138: 55-65

Leon M (1983) Chemical communication in mother-young interactions. In: Vandenbergh JG (ed) Pheromones and reproduction in mammals. Academic Press, London New York

Lyne AG (1974) Gestation period and birth in the marsupial Isoodon macrourus. Aust J Zool 22: 303-309

McCrady E (1938) The embryology of the opossum. Am Anat Mem 16: 1-233

Mair RC, Gesteland RC (1982) Response properties of mitral cells in the olfactory bulb of the neonatal rat. Neuroscience 7: 3117-3125

Monti-Graziadei GA, Stanley RS, Graziadei PPC (1980) The olfactory marker protein in the olfactory system of the mouse during development. Neuroscience 5: 1239-1252

Moulton DG, Beidler LM (1967) Structure and function in the peripheral olfactory system. Physiol Rev 47: 1-52

Nelson JE, Hartman A (1983) Developmental anatomy of Dasyurus hallucatus. Symp Ecol Wet-Dry Trop . Aust Mammal Soc, Darwin

Nelson (1985) personal communication.

Rodolfo-Masera T (1943) Sul esistenza di un particolare organo olfattivo nel setto nasale della cavia e di altri roditori. Arch Ital Anat Embryol 48: 157-213

Russel EM (in press) Scent glands and olfaction in the social behaviour of marsupials. In: Macdonald D, Brown RD (eds) Social odours in mammals. Oxford Univ Press, Oxford

Schmidt U, Eckert M, Schäfer HG (1983) Untersuchungen zur ontogenetischen Entwicklung des Geruchssinnes bei der Hausmaus (Mus musculus) Z Säugetierk 48: 355-362

Selwood L (1980) A timetable of embryonic development of the Dasyurid marsupial Antechinus stuartii (Macleay). Aust J Zool 28: 649-668

Sharman GB (1973) Adaptations of marsupial pouch young for extrauterine existence. In: Austin CR (ed) The mammalian fetus in vitro. Chapman and Hall, London, pp 67-90

Smart I (1971) Location and orientation of mitotic figures in the developing mouse olfactory epithelium. J Anat 109: 243-251

Walker MT, Rose R (1981) Prenatal development after diapause in the marsupial Macropus rufogriseus. Aust J Zool 29: 167-168

Waterman RE, Meller SM (1973) Nasal pit formation in the hamster: a transmission and scanning electron microscope study. Dev Biol 34: 255-266

Wysocki CT, Wellington TL, Beauchamp GK (1980) Access of urinary nonvolatiles to the mammalian vomeronasal organ. Science 207: 781-783

Ontogeny of the Secretory Elements in the Vertebrate Olfactory Mucosa

ML Getchell, B Zielinski and TV Getchell

Department of Anatomy and Cell Biology, Wayne State University School of Medicine, Detroit, Michigan 48201 USA

1 Introduction

The olfactory mucosa of vertebrates contains secretory elements that produce a layer of mucus that covers the surface of the olfactory epithelium (Graziadei 1971; Yamamoto 1982). The secretions presumably play a role in olfactory function by (a) providing the ionic and macromolecular environment in which odorants interact with their receptor sites on the olfactory receptor neurons (Bannister 1974; TV Getchell and ML Getchell 1977), (b) affecting access of odorants to the epithelial surface (Mozell 1971; Bostock 1974) and (c) facilitating odorant removal (Hornung and Mozell 1981; TV Getchell et al. 1985).

The olfactory mucosa of the tiger salamander, Ambystoma tigrinum, has been utilized extensively for the study of peripheral olfactory function. The large olfactory receptor neurons in this species have made the animal's olfactory mucosa an advantageous preparation for intracellular recording techniques (TV Getchell 1977; Masukawa et al. 1983; Trotier and MacLeod 1983; Masukawa et al. 1985 a, b). Several studies have described the organization and cellular morphology of this tissue (TV Getchell 1977; Graziadei and Monti Graziadei 1976; Kauer 1981; Simmons et al. 1981; Breipohl et al. 1982; Rafols and TV Getchell 1983). Recent investigations from our laboratory have focused on the histology, histochemistry and neuropharmacology of the secretory components (ML Getchell et al. 1984; TV Getchell et al. 1985; ML Getchell and TV Getchell 1984 a,b).

During maturation, the tigersalamander passes from an exclusively aquatic larval to a combined aquatic-terrestrial adult phase and from a nonsexually active to a sexually active phase. The organization of the larval olfactory mucosa on a series of ridges and grooves resembles that observed in fish (e.g., Kleerekoper 1969; Breipohl et al. 1973a; Zeiske et al. 1976; Yamamoto 1982). In addition, the larval olfactory mucosa contains olfactory glands that are the primordia of the extensive network of subepithelial Bowman's glands found in the adult (ML Getchell et al. 1984).

The purpose of this chapter is to review the literature on the ontogeny of the secretory elements of the olfactory system and to present our observations on the histological and histochemical maturation of the secretory elements in the tiger salamander olfactory mucosa.

2 Ontogeny of Secretory Elements: Review of the Literature

2.1 Goblet Cells

Goblet cells are found in fish (e.g., Breipohl et al. 1973a; Filyushina and Bakhmin 1975; Bakhtin 1976; Zeiske et al. 1976; Bertmar 1973; Theisen et al. 1980) and amphibian (ML Getchell et al. 1984) olfactory mucosae and in the nasal epithelium of higher vertebrates (e.g., Tos 1982). Histochemical analyses of goblet cells in olfactory, nasal and respiratory tissues reveal the presence of acidic and/or sulfated mucopolysaccharides (Spicer et al. 1971; Ojha and Kapoor 1972; Hafez 1977; Thaete et al. 1981). Goblet cells appear in the 11th or 12th week of embryonic development, before the maturation of the mucosal glands, in the human nose (Bang 1964) and trachea (Bucher and Reid 1961). By the 16th fetal week, the goblet cells of the human nasal epithelium stain positively for neutral mucopolysaccharides and are reported to secrete mucus for several months prior to birth (Bang 1964).

2.2 Sustentacular Cells

Sustentacular cells are found in the olfactory epithelium of all vertebrates (Graziadei 1971). They contain neutral and acidic mucopolysaccharides in fish, amphibians, reptiles, and birds and only neutral mucopolysaccharides in mammals (for review, see ML Getchell et al.1984, Table 1). Sustentacular cells in Xenopus embryos originate from the nonneural ectoderm covering the olfactory placode (Klein and Graziadei 1983). In mice, sustentacular cells appear to differentiate from placodal stem cells whose nuclei are located superficially (Cuschieri and Bannister 1975b); the origin of the stem cells was not investigated since the study commenced after cellular differentiation had begun. Breipohl et al. (1973b) also reported that sustentacular cells arise from mitoses of cells located distally in the olfactory epithelium. Sustentacular cells are recognizable in rainbow trout (Zielinski 1985) and Xenopus embryos (Klein and Graziadei 1983) at the same stage that axons from the receptor cells can be observed growing out of the mucosa. In chick embryos, sustentacular cells can be identified on about the 7th day of incubation when surfaces of most receptor cells show differentiation into olfactory knobs but remain unciliated (Breipohl and Fernandez 1977). In two different strains of mice, sustentacular cells appear on about the 15th day (NMRI strain, Breipohl et al. 1973b) to the 17th day (S.A.S. strain, Cuschieri and Bannister 1975b) of gestation, at about the same time that Bowman's glands begin to grow into the subepithelial region.

In NMRI mice, sustentacular cells in 15 1/2-day-old fetuses contain secretory vesicles in a frequency similar to that seen in adult mice (Breipohl et al. 1973b). In S.A.S. mice, sustentacular cells contain numerous secretory vesicles near their apical surfaces after the 17th day of development; shortly after birth, there is an increase in the amount of endoplasmic reticulum in the supranuclear region (Cuschieri and Bannister 1975b). In human embryos, sustentacular cells contain secretory material resembling that found in adults by the 9th week of development (Pyatkina 1982). In Xenopus, sustentacular cells are reported to secrete during late embryonic stages (Klein and Graziadei 1983). The time of the first appearance of mucus over the olfactory epithelium in higher vertebrates has not been precisely determined. Although it has been reported to be secreted first after birth (Breipohl 1972), the possibility that it is present earlier but is extremely difficult to preserve has been considered (Seifert 1971). Therefore, the time of onset of sustentacular cell secretion remains to be clarified.

2.3 Bowman's Glands

Bowman's glands are found in the lamina propria of the olfactory mucosa of all terrestrial vertebrates (Graziadei 1971). In amphibians and some reptiles, Bowman's glands contain only neutral mucopolysaccharides; in other reptiles and in birds and mammals, they also contain acidic mucopolysaccharides (ML Getchell et al. 1984). The glands originate from the basal layer of the embryonic epithelium as described by Tos and Poulsen (1975) in human fetuses and by Cuschieri and Bannister (1975 a,b) in mice. During the 15th week of human development (Tos and Poulsen 1975) and on about the 17th to 19th day in mice (Breipohl 1972; Cuschieri and Bannister 1975a), they begin to grow into the underlying mesenchyme, where they are found in proximity to olfactory nerve bundles. Subsequent to the appearance of the glands, ducts are observed in the embryonic mouse epithelium (Cuschieri and Bannister 1975b). By 2 days after birth there is a definite increase in the number of Bowman's glands that continues for 2 to 3 weeks after birth (Breipohl 1972).

At the time of their first appearance in the lamina propria of NMRI mice, the cells of Bowman's glands appear to lack secretory material (Breipohl 1972). In S.A.S. mice, "secretion vacuoles" were present at the earliest stage of recognition of the Bowman's glands (Cuschieri and Bannister 1975b). The time of onset of secretion is uncertain due to the lack of information about the precise time of appearance of the mucous covering (see above). Breipohl (1972) postulates that contact of the epithelial surface with air is the stimulus that initiates secretion. However, Bang (1964) reports that submucosal glands in the human nose secrete mucus several months prior to birth.

3 Secretory Elements in Several Different Developmental Stages of the Salamander Olfactory Mucosa.

3.1 Larval Mucosa

The olfactory mucosa of the larval tiger salamander (Fig. 1) is organized into buds of developing neuroepithelium whose surface is depressed to form olfactory grooves. The buds contain sustentacular cells, olfactory receptor neurons and basal cells. The nuclei of these cells are stratified in the lower two-thirds of the bud with large, elongated sustentacular cell nuclei located most superficially, the checkerboard-patterned olfactory receptor cell nuclei in the middle and the pale, closely packed basal cell nuclei in the deepest layer. At the surface, the buds are separated by ridges containing large ciliated cells, "mucoid" cells and goblet cells. Goblet cells tend to occur in small groups in close proximity to ciliated cells, as also noted by Breipohl et al. (1973a) in the goldfish. Histochemical techniques reveal that the goblet cells contain acidic and sulfated mucopolysaccharides; the mucoid cells contain these and neutral mucopolysaccharides (Getchell and Getchell, unpublished observations). These latter two types of cells presumably secrete into the mucus that covers the epithelial layer as in fish (e.g., Herberhold 1968;

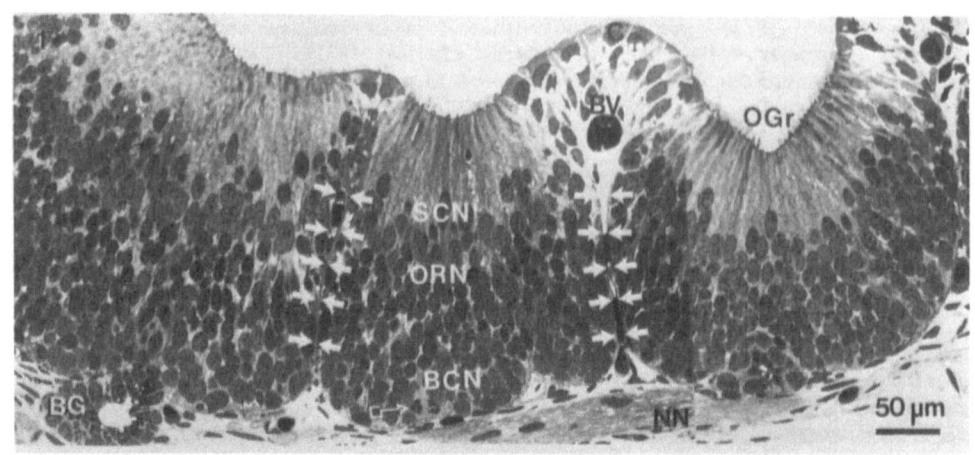

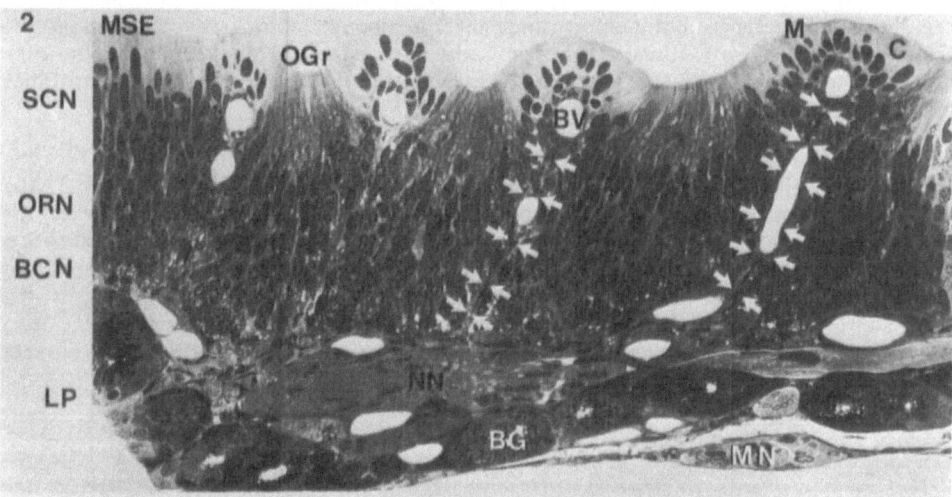

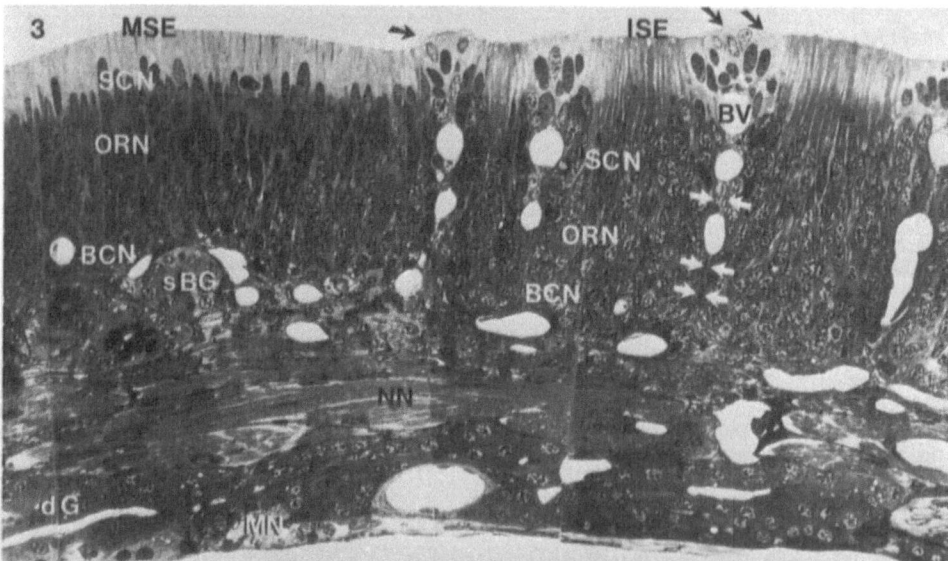

Fig. 1 Larval olfactory mucosa. Sensory epithelium is located in bud-like structures whose surfaces are depressed to form olfactory grooves **OGr**. Mitotic figure (open arrow) is seen occasionally in the region occupied by basal cell nuclei **BCN**. Buds are separated by connective tissue septa (white arrows). Ciliated cells **C**, blood vessels **BV** and loose connective tissue form ridges between olfactory grooves. Bowman's glands **BG** and nonmyelinated nerve bundles **NN** lie in the lamina propria. **ORN** olfactory receptor cell nuclei; **SCN** sustentacular cell nuclei. Calibration bar for Figs. 1-3 appears in Fig. 1.

Fig. 2 Adult olfactory mucosa. Immature sensory mucosa to the right closely resembles larval olfactory mucosa; it consists of grooves **OGr** of sensory epithelium separated by ridges composed of ciliated cells **C**, mucoid cells **M**, blood vessels **BV** and connective tissue septa (white arrows) which demarcate buds. Lamina propria **LP** contains blood vessels, nonmyelinated **NN** and mixed **MN** nerves, and Bowman's glands **BG**. Mature sensory epithelium **MSE** is shown at far left. Treatment of tissue with 1 mM guaiacol prior to fixation caused protrusion of mucoid cells and vacuolation near epithelial basement membrane. **SCN** sustentacular cell nuclei; **ORN** olfactory receptor nuclei; **BCN** basal cell nuclei.

Fig. 3 Adult olfactory mucosa. The immature sensory epithelium **ISE** shows signs of maturation compared to that shown in Fig. 2. The surface is flat, and ridges have become less prominent. Nuclei of epithelial cells (**SCN** sustentacular cell nuclei; **ORN** olfactory receptor nuclei; **BCN** basal cell nuclei) still lie deeper within the epithelium in the immature areas than in the mature sensory epithelium **MSE** at left. Lamina propria under MSE contains two layers of olfactory glands, superficial Bowman's glands, **sBG** and deep glands **dG**, and bundles of nonmyelinated **NN** and mixed **MN** nerves. Treatment of tissue with 0.01 mM galbazine prior to fixation caused secretory granule depletion from the glands, vacuolation and protrusions at apical surfaces of sustentacular cells and doming of mucoid cells (arrows). **BV** blood vessels; connective tissue septum between buds (white arrows).

Breipohl et al. 1973a; Zeiske et al. 1976). Deeper in the epithelium, the buds are separated from one another by connective tissue septa that enclose blood vessels and often extravascular cells of the immune system as well as loose connective tissue. The connective tissue septa appear to be continuous with the basement membrane separating the buds from the underlying lamina propria. The lamina propria contains nerve bundles, a single layer of Bowman's glands, blood vessels and connective tissue. The glands consist of acini that only very rarely have structural components suggestive of necks or ducts penetrating toward the surface.

The surface of the larval neuroepithelium (Fig. 4) is composed of three elements: the supranuclear region of the sustentacular cells, the apical olfactory dendrite and, very rarely, the opening of a duct from a subepithelial Bowman's gland. The supranuclear regions of the sustentacular cells contain small granules that often are seen also in the blebs protruding from the apical surfaces of these cells. These blebs are considered to be signs of secretory activity from the sustentacular cells (Okano and Takagi 1974; TV Getchell et al. 1985). Ducts of Bowman's glands are lined by flattened, elongated duct cells whose apical cytoplasm is adjacent to the duct lumen and contains a border of very fine granules. The secretory granules in the apical poles of the acinar cells of Bowman's glands are larger than those seen in the sustentacular cells or ducts cells. Release of the secretory granules from the larval acinar cells can be induced by the ß-adrenergic agonist, isoproterenol; the granules in the sustentacular cells appear to be largely unaffected by this agent (Getchell ML et al. 1984). This difference persists in the adult (ML Getchell and TV Getchell 1984b). Vacuoles suggestive of

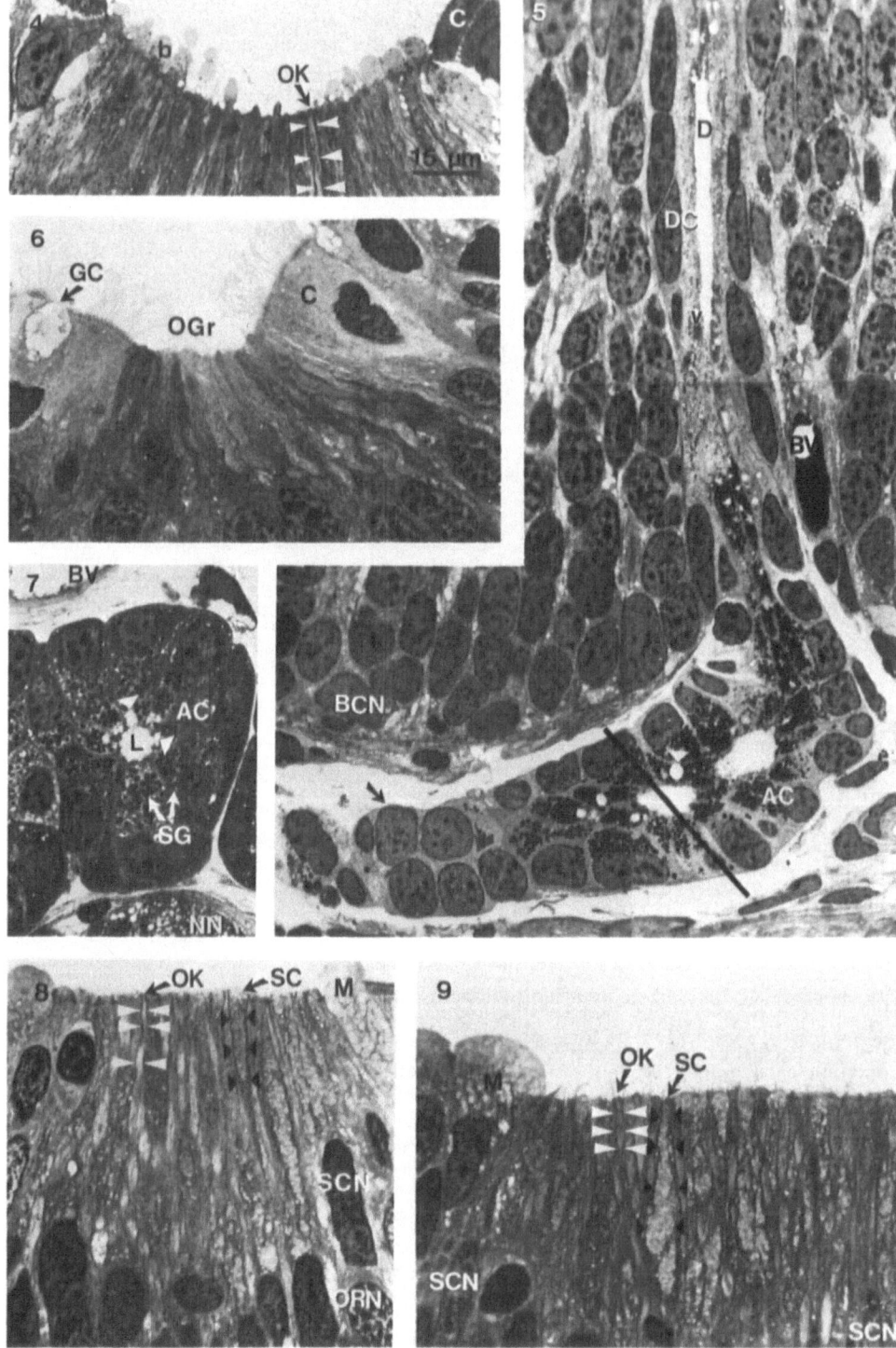

Fig. 4 Olfactory groove from larval salamander. Olfactory knobs **OK** of receptor dendrites (white arrowheads) and blebs **b** protruding from sustentacular cells (black arrowheads) project into olfactory groove. Note secretory granules in sustentacular cells and blebs. Large ciliated cells **C** of ridges lie at borders of the sensory epithelium. Calibration bar for Figs. 4-9 appears in Fig. 4

Fig. 5 Larval Bowman's gland. The acinus narrows as it enters the sensory epithelium to form the duct **D**. Some cells (arrow) within the gland appear undifferentiated, with large nuclei and thin rims of pale cytoplasm. AC acinar cell; **BCN** basal cell nucleus; **BV** blood vessel; **DC** duct cell; vacuole (white arrowhead). Black line indicates plane of section of Fig. 7

Fig. 6 Olfactory groove **OGr** from adult salamander. Immature is revealed by depth of the groove, presence of goblet cells **GC** in ridges and ciliated cells **C** located adjacent to sensory epithelium. Tissue was treated with 1 mM guaiacol prior to fixation

Fig. 7 Cross section of larval Bowman's gland acinus. Secretory granules **SG** and vacuoles (white arrowheads) occupy apical poles of acinar cells **AC**. **BV** blood vessel; **L** lumen; **NN** nonmyelinated nerve

Fig. 8 Olfactory groove from adult salamander. Indications of further maturation compared to Fig. 6 include broader supranuclear portions of sustentacular cells (**SC** black arrowheads) in proportion to olfactory dendrites (white arrowheads) and the presence of mucoid cells **M** rather than ciliated cells adjacent to sensory epithelium. **OK** olfactory knob; **ORN** olfactory receptor nucleus, SCN sustentacular cell nucleus

Fig. 9 Mature sensory epithelium adjacent to mucoid cell **M** of ridge from adult salamander. Supranuclear region of sustentacular cells (**SC** black arrowheads) appears broader in proportion to olfactory receptor dendrite (white arrowheads) than in immature sensory epithelium of Fig. 8. **OK** olfactory knob; **SCN** sustentacular cell nuclei

water and electrolyte secretion are often observed in the apical cytoplasm and blebs of the sustentacular cells, in the duct cells (Fig. 5) and in the acinar cells of Bowman's glands (Figs. 5 and 7).

3.2 Adult Immature Olfactory Mucosa

In the adult salamander, the olfactory cavity consists of a simple sac lined primarily by a flat sheet of olfactory mucosa. Using scanning electron microscopy, Breipohl et al. (1982) described a region on the lateral aspect of the sac floor containing three to seven prominent bands of tissue that consisted of ciliated, microvillar and microplicar cells and that separated grooves containing sensory neuroepithelium. The structure of this region has also been described by Graziadei and Monti Graziadei (1976).

Examination of this region, which we have called the adult immature mucosa, revealed that it resembles the larval mucosa as described by ML Getchell et al. (1984) (cf. Figs. 1 and 2). Both are organized into grooves and ridges. The surface of the bands or ridges contain similar cell types; the connective tissue septa beneath them contain blood vessels. The placement of the blood vessels at the level of the epithelial nuclei in the regions of immature sensory epithelium rather than below the epithelium as in the mature mucosa (see below) allows the recognition of the late developmental stages of these immature areas in older adults in which the ridges may no longer be identifiable and the epithelial surface generally is flat (Fig. 3, right). In the sensory areas, the nuclei of the sustentacular and olfactory receptor cells lie deep within the epithelium. In the lamina propria, there is generally a single layer of glands lying below the olfactory nerve bundles. The occurrence of ducts linking the superficial glands to the surface is rare in immature sensory epithelium, but occasionally regions of apparent proliferation of ducts are seen in the latest stages of development of the immature areas. Ducts from the superficial layer of glands appear to develop adjacent to the connective tissue septa and the associated blood vessels.

There are several notable differences between the larval mucosa and the adult immature mucosa. In the larval form, the sustentacular cells contain small secretory granules as described above (Fig. 4); in the adult immature sensory epithelium, the sustentacular cells contain vesicular material (Figs. 6, 8). Histochemically, the sustentacular cells at both stages stain positively for the presence of neutral, acidic and sulfated mucopolysaccharides (ML Getchell et al. 1984; Getchell and Getchell, unpublished observations). In contrast, in Bowman's glands in the larval mucosa, an occasional faint positive reaction for acidic (but not sulfated) mucopolysaccharides is seen in addition to a strong positive reaction for the presence of neutral mucopolysaccharides (Getchell and Getchell, unpublished observations) while in the adult immature mucosa, the glands show the presence of only neutral mucopolysaccharides (ML Getchell et al. 1984).

Another difference between the larval and adult immature mucosae can be seen in the cell types of the ridges between the buds. In the larval form, there are fewer goblet and mucoid cells relative to ciliated cells in the ridges; the ciliated cells usually lie adjacent to the cells of the sensory epithelium (Fig. 4). In the immature adult, mucoid cells are generally adjacent to the cells of the neuroepithelium (Figs. 8, 9) as noted also by Breipohl et al. (1973a) in the goldfish. The secretory nature of these cells is revealed when the epithelium is treated with odorants such as guaiacol or galbazine that cause the formation of "domes" at the surface of the mucoid cells (Figs. 2, 3, 9).

3.3 Adult Mature Mucosa

Figures 2 and 3 show sections of olfactory mucosa taken from areas where mature and immature olfactory mucosa lie adjacent to one another. The mature sensory epithelium (Figs. 2, 3, left) is characterized by a relatively flat surface. However, as can be seen in Fig. 2 (right), in the smallest and presumably youngest adults, the immature sensory mucosa is organized into grooves and ridges that closely resemble those seen in the larval mucosa. In the mature epithelium, the nuclei of the sustentacular cells, the olfactory receptor neurons and the basal cells occupy a greater proportion of the depth of the epithelial layer than in the immature regions. Blood vessels in the mature mucosa are located below the epithelial basement membrane in the lamina propria.

Two distinct layers of olfactory glands (Fig. 3, left) can be identified in the lamina propria (Getchell ML et al. 1984): a superficial layer whose acini send ducts to the surface (Bowman's glands) and a deep layer, lying below the olfactory nerve bundles, whose acini do not appear to be connected to the overlying surface directly via ducts. The acini of the single layer of glands in the immature mucosa (Figs. 2, 3, right) closely resemble the acini of the deep layer of glands in the lamina propria of the mature mucosa.

4 Maturation of the Secretory Components

Several changes in the secretory components can be documented as the aquatic larval salamander transforms into the aquatic/terrestrial adult. Various stages in the maturation of the goblet cells can be seen, with goblet cells containing scanty secretory product seen in the larval and most immature adult mucosae (Fig. 6) and well-developed, flask-shaped goblet cells seen in older animals. Athough it is possible that these differences represent phases of the secretory cycle of the goblet cells (see Breipohl et al. 1973a), they appear to occur predictably in immature and mature mucosae respectively. In the immature region of the adult mucosa, goblet and mucoid cells of the ridges increase in number, and mucoid cells are more often seen adjacent to the sensory epithelium when compared to the arrangement in the larval mucosa. Maturation of the sustentacular cells appears to involve a change in the packaging of the secretory product from granules to vesicles but their mucopolysaccharide content remains the same.

As the adult matures, several additional changes can be seen in its olfactory mucosa. The epithelial surface flattens, and the nuclei of the sustentacular cells appear to lie closer to the epithelial surface. Two layers of Bowman's glands are present under the mature sensory epithelium rather than the one layer seen under the immature region. The two layers may arise in a manner similar to that described for the seromucous glands in the nasal epithelium (Tos 1982): the deep glands under the mature sensory epithelium are the glands that developed when the overlying epithelium was still immature. During the maturation of the lamina propria, blood vessels and nerve fibers form a layer beneath these first glands; glands that subsequently arise from the basal layers of the epithelium remain above this obstruction to downward growth and form a superficial layer, within the lamina propria. Another possibility is that the deep layer of glands is continuous with the superficial layer, each gland tube being a continuous spiral structure. On rare occasions, an acinus near the middle of the lamina propria will contain cells characteristic of both superficial and deep glands (ML Getchell et al. 1984). Further studies are required to clarify this point. Ducts from the superficial layer of glands to the epithelial surface increase in number but the deep glands do not appear to be connected to the overlying surface. Two possible explanations can be proposed: if the deep glands do represent the earliest glands formed under the larval and immature epithelium, their duct orifices may remain in the anterior region of the nasal cavity where the glands first developed and may possess elongated ducts far anterior to the region of mature sensory mucosa; if, on the other hand, the deep glands are continuous with the superficial glands, then the ducts of the superficial glands would carry the deep gland secretions to the surface.

In summary, the secretory elements in the salamander olfactory mucosa undergo maturational changes that reflect the ontogenic transition from an aqueous environment to contact with with both water and air. The most dramatic example is the proliferation of Bowman's glands, which may be required to provide a stable environment for olfactory function in terrestrial vertebrates. A review of the literature reveals that in mammalian species, the secretory elements differentiate during fetal development. The differentiation of the sensory epithelium precedes the appearance of Bowman's glands, an ontogenetic phenomenon in the olfactory mucosa that may reflect the phylogenetic transition between aquatic and terrestrial vertebrates.

Acknowledgments. This research was supported by Grant # NIH-NS-16340 to TVG an by Grant # NSF-BNS-07949 to MLG.

References

Bakhtin YK (1976) Morphology of the olfactory organ of some fish species and a possible functional interpretation. J Ichthyol 16: 786-804

Bang BG (1964) The mucous glands of the developing human nose. Acta Anat 59: 297-314

Bannister LK (1974) Possible functions of mucus at gustatory and olfactory surfaces. In: Poynder TM (ed) Transduction mechanisms in chemoreception. IRL Press, London, pp 39-46

Bertmar G (1973) Ultrastructure of the olfactory mucosa in the homing Baltic Sea trout Salmo trutta trutta. Mar Biol 19: 74-88

Bostoch H (1974) Diffusion and the frog EOG. In: Poynder TM (ed) Transduction mechanisms in chemoreception. IRL Press, London, pp 27-38

Breipohl W (1972) Licht- und elektronenmikroskopische Befunde zur Struktur der Bowmanschen Drüsen in Riechepithel der weißen Maus. Z Zellforsch 131: 329-346

Breipohl W, Fernandez M (1977) Scanning electron microscopic investigations of olfactory epithelium in the chick embryo. Cell Tissue Res 183: 105-114

Breipohl W, Bijvank GJ, Zippel HP (1973a) Die Oberflächenstruktur der olfaktorischen Drüsen des Goldfisches (Carassius auratus). Eine rastermikroskopische Studie. Z Zellforsch 140: 567-582

Breipohl W, Mestres P, Meller K (1973b) Licht- und elektronenmikroskopische Befunde zur Differenzierung des Riechepithels der weißen Maus. Verh Anat Ges 67: 443-449

Breipohl W, Moulton D, Ummels M, Matulionis DH (1982) Spatial pattern of sensory cell terminals in the olfactory sac of the tiger salamander. I. A scanning electron microscope study. J. Anat 134: 757-769

Bucher U, Reid L (1961) Development of the mucus-secreting elements in human lung. Thorax 16: 219-225

Cuschieri A, Bannister LH (1975a) The development of the olfactory mucosa in the mouse: light microscopy. J Anat 119: 277-286

Cuschieri A, Bannister LH (1975b) The development of the olfactory mucosa in the mouse: electron microscopy. J Anat 119: 471-498

Filyushina YY, Bakhmin YK (1975) Electron-microscopic investigation of the olfactory epithelium of the Baikal omul Coregonus autumnalis migratorius. J Ichthyol 15: 284-290

Getchell ML, Getchell TV (1984a) ß-Adrenergic regulation of the secretory granule content of acinar cells in olfactory glands of the slamander. J Comp Physiol A 155: 435-443

Getchell ML, Getchell TV (1984b) Cholinergic control of glandular secretion in the olfactory mucosa of the salamander. In: Abstr Proc Annu Meet Assoc Chemoreception Sci, Sarasota, FL

Getchell ML, Rafols JA, Getchell TV (1984) Histological and histochemical studies of the secretory components of the salamander olfactory mucosa: Effects of isoproterenol and olfactory nerve section. Anat Rec 208: 553-565

Getchell TV (1977) Analysis of intracellular recordings from salamander olfactory epithelium. Brain Res 123: 275-286

Getchell TV, Getchell ML (1977) Histochemical localization and identification of secretory products in salamander olfactory epithelium. In: LeMagnen J, MacLeod P (eds) Olfaction and taste, vol VI. IRL Press, London, pp 105-112

Getchell TV, Zielinski B, Getchell ML (1985) Pyrazine-mediated neural and secretory activity in the olfactory mucosa of the salamander (abstr). Chem Sens 10: 398

Graziadei PPC (1971) The olfactory mucosa of vertebrates. In: Beidler LM (ed) Handbook of sensory physiology, vol IV. Chemical senses 1. Olfaction. Springer, Berlin Heidelberg New York, pp 27-58

Graziadei PPC, Monti Graziadei GA (1976) Olfactory epithelium of Necturus maculosus and Ambystoma tigrinum. J Neurocytol 5: 11-32

Hafez ESE (1977) Functional anatomy of mucus-secreting cells. In: Elstein M, Parke DV (eds) Mucus in health and disease. Plenum Press, New York, pp 19-34

Herberhold C (1968) Vergleichende histochemische Untersuchungen am peripheren Riechorgan von Säugetieren und Fischen. Arch Klin Exp Ohr Nas Kehlk Heilkd 190: 166-182

Hornung DE, Mozell MM (1981) Accessibility of odorant molecules to the receptors. In: Cagan RH, Kare MR (eds) Biochemistry of taste and olfaction. Academic Press, London New York, pp 33-45

Kauer JS (1981) Olfactory receptor cell staining using horseradish peroxidase. Anat Rec 200: 331-336

Kleerekoper H (1969) Olfaction in Fishes. Indiana Univ Press, Bloomington

Klein SL, Graziadei PPC (1983) The differentiation of the olfactory placode in Xenopus laevis: a light and electron microscope study. J Comp Neurol 217: 17-30

Masukawa LM, Kauer JS, Shepherd GM (1983) Intracellular recordings from two cell types in an in vitro preparation of the salamander olfactory epithelium. Neurosci Lett 35: 59-64

Masukawa LM, Hedlund B, Shepherd GM (1985a) Electrophysiological properties of identified cells in the in vitro olfactory epithelium of the tiger salamander. J Neurosci 5: 128-135

Masukawa LM, Hedlund B, Shepherd GM (1985b) Changes in the electrical properties of olfactory epithelial cells in the tiger salamander after olfactory nerve transection. J Neurosci 5: 136-141

Mozell MM (1971) Spatial and temporal patterning. In: Beidler LM (ed) Handbook of sensory physiology, vol IV. Chemical senses 1. Olfaction. Sringer, Berlin Heidelberg New York, pp 205-215

Ojha PP, Kapoor AS (1972) Histochemistry of the olfactory epithelium of the fish, Canna punctatus Bloch. Acta Anat 83: 540-555

Okano M, Takagi SF (1974) Secretion and electrogenesis of the supporting cell in the olfactory epithelium. J Physiol (London) 242: 353-370

Pyatkina GA (1982) Development of the olfactory epithelium in man. Z Mikrosk Anat Forsch Leibzig 96: 361-372

Rafols JA, Getchell TV (1983) Morphological relations between the receptor neurons, sustentacular cells and Schwann cells in the olfactory mucosa of the salamander. Anat Rec 206: 87-101

Seifert K (1971) Licht- und elektronmikroskopische Untersuchungen der Bowman-Drüsen in der Riechschleimhaut makrosmatischer Säuger. Arch Klin Exp Ohr Nas Kehlk Heilkd 200: 252-274

Simmons PA, Rafols JA, Getchell TV (1981) Ultrastructural changes in olfactory receptor neurons following olfactory nerve section. J Comp Neurol 197: 237-257

Spicer SS, Chakrin LW, Wardell JR Jr, Kendrick W (1971) Histochemistry of mucosubstances in the canine and human respiratory tract. Lab Invest 25: 483-490

Thaete LG, Spicer SS, Spock A (1981) Histology, ultrastructure, and carbohydrate cytochemistry of surface and glandular epithelium of human nasal mucosa. Am J Anat 162: 243-263

Theisen B, Breucker H, Zeiske E, Melinkat R (1980) Structure and development of the olfactory organ in the garfish Belone belone (L.) (Teleostei, Atheriniformes) Acta Zool (Stockholm) 61: 161-170

Tos M (1982) Goblet cells and glands in the nose and paranasal sinuses. In: Proctor DF, Andersen I (eds) The nose. Upper airway physiology and the atmospheric environment. Elsevier Biomedical Press, Amsterdam New York Oxford, p 99

Tos M, Poulsen J (1975) Mucous glands in the developing human nose. Arch Otolaryngol 101: 367-372

Trotier D, MacLeod P (1983) Intracellular recordings from salamander olfactory cells. Brain Res 268: 225-237

Yamamoto M (1982) Comparative morphology of the peripheral olfactory organ in teleosts. In: Hara TJ (ed) Chemoreception in fishes. Elsevier, Amsterdam, pp 39-59

Zeiske E, Melinkat R, Breucker H, Kux J (1976) Ultrastructural studies of the epithelia of the olfactory organ of cyprinodonts (Teleostei, Cyprinodontoidea). Cell Tissue Res 172: 245-267

Zielinski B (1985) Morphological and physiological studies of the development of the peripheral olfactory organ in rainbow trout, Salmo gairdneri. Thesis, Univ. Manitoba, Winnipeg

Developmental Studies on Enzyme Histochemistry of the Three Olfactory Epithelia in the Golden Hamster.

Kazuyuki Taniguchi, Kazumi Taniguchi* and Shin-ichi Mikami

Department of Veterinary Anatomy, Faculty of Agriculture, Iwate University, 3-18-8 Ueda, Morioka, Iwate 020, Japan

*Department of Anatomy, Iwate Medical University, 19-1 Uchimaru, Morioka, Iwate 020, Japan

1 Introduction

Olfactory receptor cells can occur at three sites in mammals, i.e., olfactory epithelium proper (OEP), vomeronasal organ (VNO) and septal organ of Masera (MO) (Adams 1972; Ciges et al. 1977; Graziadei 1977; Kratzing 1978; Rodolfo-Masera 1943). Although the significance of these three olfactory systems is not completely understood, morphological and experimental data do suggest functional specializations (Aronson and Cooper 1974; Estes 1972; Murphy and Schneider 1970; Powers and Winans 1975; Winans and Scalia 1970). Differences in the enzymatic maturation of the three sensory epithelia may reflect ontogenetic shifts in both the relative importance of these tissues and/or enzymes.

The olfactory receptor cells of all three epithelia are embryologically derived from the stem cells of the olfactory placode (Cushieri and Bannister 1975; Cushieri and Bannister 1975; Taniguchi et al. 1982), but undergo some modifications according to the sites where they occur (Ciges et al. 1977; Graziadei 1977; Kratzing 1978). These modifications may also be taken as an indication of functional specification of the different olfactory sensory epithelia. Differences in the sensitivity to odoriferous compounds have been reported (Miragall et al. 1984; Wysocki et al. 1980). There are, however, very few reports in the literature on differences in enzyme histochemistry between the different epithelia, especially during development. The histochemistry of the MO has even been completely neglected in the fetus and adult.

The purpose of the present study was therefore to check differences in the appearance and localization of several enzymes in the three olfactory sensory epithelia of the golden hamster during development and to discuss whether such differences reflect both metabolic and functional differences (Barady and Bourne 1953; Cushieri 1974; Shantha and Nakajima 1970; Taniguchi et al. 1981).

2 Materials and Methods

Abbreviations used:
AChE	acetylcholinesterase
ACP	acid phosphatase
ALD	aldolase
ALP	alkaline phosphatase
ATPase	adenosine triphosphatase
CCO	cytochrome oxidase

ChE	cholinesterase
ß-GL	ß-glucuronidase
GB	(olfactory) glands of Bowman
G6PD	glucose-6-phosphate dehydrogenase
LDH	lactate dehydrogenase
MAO	monoamine oxydase
MO	septal organ of Masera
NRE	nasal respiratory epithelium
NsE	non-specific esterase
OEP	olfactory epithelium proper
SDH	succinate dehydrogenase
TPP	thiamine pyrophosphatase
VNO	vomeronasal organ
VNO-NE	vomeronasal neuroepithelium
VNO-RFE	vomeronasal receptor-free epithelium

Golden hamsters maintained in our laboratory as a closed colony were used in the present study. Fetuses were obtained on days 12, 13, 14, and 15 of gestation from two mothers at each stage. Days of gestation were counted from the end of witnessed copulation, which was regarded as 0 h. After decapitation the animal's head and nose were cut transversally. The histochemical procedures for the enzymes examined in the present study are detailed in Table 1. Control sections were incubated in substrate-free media for NsE, LDH, and SDH. Some inhibitors were used as shown in Table 1 for ACP, ALP and ATPase.

3 Results and Discussion

The first chronological appearance of the enzymes (as determined by enzymatic activities) during intrauterine life of the golden hamster is summarized in Table 2. This table also has results from previously published studies using other species. In Table 3 the enzyme histochemical properties of the mature OEP and/or VNO are listed for some species as taken from the literature.

3.1 Acid Phosphatase

Although ACP is subdivided into lysosomal and extra-lysosomal ACP (Miyayama et al. 1975), the present study is restricted to lysosomal ACP, well known as a marker enzyme of the lysosome. This enzyme has previously been reported to be present (Cushieri 1974; Shantha and Nakajima 1970; Taniguchi et al. 1981) in the supranuclear cytoplasm of receptor cells in the OEP and VNO-NE, where lysosomes are known to be abundant (Taniguchi and Mikami 1985; Taniguchi and Mochizuki 1983; Vaccarezza et al. 1981). Cushieri (1974) alluded the participation of ACP in the lysosomal destruction of aged or damaged organelles and subsequent cell repair.

In early reports on the histochemistry of the OEP during development, the activity of ACP was believed to occur at very early stages in the olfactory placode (Lejour 1967; Lejour-Jeanty 1965; Milaire 1974), prior to its differentiation in the three olfactory sensory epithelia. Although this activity was assumed to have derived from lysosomes, the enzymes may have been localized either in the

Table 1 Histochemical methods used in the present study

Enzyme	Fixative	Fixation	Section	Substrate	Incubation	Inhibitor	Reference
ACP	2% Glutar-aldehyde	4°C, 30 min	Frozen, 7 μm	Sodium-ß-glycero-phosphate	1 h at 37°C	10^{-4} M sodium fluoride	Gomori 1952; Pearse 1972 (lysosomal ACP)
ALP	80% Ethanol	-20°C, overnight	Paraffin, 5μm	Sodium-ß-glycero-phosphate	1 h at 37°C	0.05 M L-phenyl-alanine	Gomori 1952; Watanabe and Fishman 1964
ATPase	Calcium-formol	4°C, overnight	Frozen, 7 μm	Adenosine-tri-phosphate	1 h at room temperature	2.5×10^{-3} M p-mercuri-benzoate	Padykula and Hermann 1955; Pearse 1972 (calcium-activated ATPase)
NsE	80% Ethanol	-20°C, overnight	Paraffin, 5 μm	α-Naphtyl acetate	15 min at room temperature	Not used	Gomori 1952; Pearson and Defendi 1957
LDH	Acetone	-20°C, 1 h	Frozen, 7 μm	Sodium-DL-lactose	30 min at 37°C	Not used	Hess et al. 1958
SDH	Acetone	-20°C, 1 h	Frozen, 7 μm	Sodium-succinate	1 h at 37°C	Not used	Nachlass et al. 1957

Table 2 First appearance of enzymatic activities during ontogeny

Species	Epithelia	ACP	ALP	ATPase	G6PD	NsE	AChE	LDH	SDH	Reference
Golden hamster (Mesocricetus auratus)	OEP	–	–	–	x	–	x	13 d	14 d	Present study
	MO	–	–	–	x	–	x	14 d	14 d	
	VNO-NE	–	–	–	x	–	x	14 d	14 d	
Rat (Rattus norvegicus)	OEP	x	11–12 d	11.5 d	14 d	x	x	10–12 d	16.5 d	Lejour 1967
	VNO-NE	x	11–12 d	x	14 d	x	x	10–12 d	16.5 d	Shapiro 1970
Rat (Epimys norvegicus)	OEP	x	x	x	x	x	13–14 d	x	x	Filogamo and Robecci 1969
Mouse (Mus musculus)	OEP	9.5 d	11–12 d	x	14 d	x	x	10–12 d	16.5 d	Milaire 1974
	VNO-NE	x	11–12 d	x	14 d	x	x	10–12 d	16.5 d	Shapiro 1970
Guinea pig (Cavia cobaja)	OEP	x	x	x	x	x	13–14 d	x	x	Filogamo and Robecci 1969
Rabbit (Lepus caniculus)	OEP	x	x	x	x	x	13–14 d	x	x	Filogamo and Robecci 1969

– negative before birth;   x not investigated

Table 3 Enzymatic activities investigated in the olfactory and vomeronasal epithelia of adult animals

Species	Epithelia	Enzymes														Reference
		ACP	ALP	ATPase	TPP	APL	ß-GL	ALD	CCO	MAO	NsE	AChE	G6PD	LDH	SDH	
Golden hamster (Mesocricetus auratus)	OEP	+++	+	-	x	x	x	x	x	-[a]	-	-[a]	x	+++[a]	+++[a]	Taniguchi et al. 1981
	VNO-NE	+++	+/-	+	x	x	x	x	x	-[a]	++	-[a]	x	++[a]	+++[a]	
	VNO-RFE	+/-	+++	++	x	x	x	x	x	-[a]	+/-	-[a]	x	+++[a]	++[a]	
Mouse (SAS ICI mouse)	OEP	+++	+++	+++	x	x	+	x	++	x	+++	x	x	x	+++	Cushieri 1974
	VNO-NE	+++	+++	+++	x	x	++	x	++	x	+++	x	x	x	+++	
	VNO-RFE	+/-	-	+++	x	x	++	x	+++	x	++	x	x	x	+++	
Rhesus monkey (Macaca mulatta)	OEP	+++	++	++	+++	+	x	++	x	+/-	+/-	+/-	+++	+++	+++	Shantha and Nakajima 1970

- negative; +/- slight; + mild; ++ moderate; +++ intense; x not investigated

[a] Tanigushi, unpublished

stem cells of the olfactory placode or in migrating phagocytic cells. Since stem cells contain only few membranous organelles except mitochondria (Taniguchi et al. 1982), it is more likely that ACP activity originates from migrating phagocytic cells.

In the presenct study, the activity of lysosomal ACP was not observed in OEP, MO or VNO-NE during intrauterine life after 12 days of gestation, although it was intense in the OEP and VNO-NE in our preliminary study in the adult golden hamster (Taniguchi et al. 1981) and in the olfactory placode (Lejour 1967; Lejour-Jeanty 1965; Milaire 1974). Lysosomal ACP in the olfactory system is duly interpreted in connection with autophagic function of lysosomes. The results appear to reflect the absence of lysosomal autophagic function in these epithelia before birth in the golden hamster.

3.2 Alkaline Phosphatase

No activity of ALP is found in the receptor cells of the OEP, although some activity is present in the basal cells (Shantha and Nakajima 1970; Taniguchi et al. 1981). In the VNO, however, results vary among investigators. Cushieri (1974) reported ALP activity only in the uppermost layer of the VNO-NE, and not in the VNO-RFE of the mouse. Shapiro (1970), on the other hand, observed its basal localization in the VNO-NE of fetal rats and mice. In our preliminary investigations in the adult golden hamster (Taniguchi et al. 1981), we found intense ALP activity only at the brush border of the VNO-RFE which has both cilia and microvilli. Although such differences appear to be species related, further investigation is warranted.

In the present study, the activity of ALP was first observed at the brush border of the NRE at 15 days of gestation, i.e., about 1 day before birth, but it was absent in the OEP, MO, VNO-NE, and VNO-RFE. In the adult golden hamster, however, activity is intense at the brush border of the VNO-RFE (Taniguchi et al. 1981) and NRE (unpublished observations). The lack of ALP activity in the VNO-RFE at 15 days of gestation suggests that the VNO is not fully developed in relation to the function of ALP before birth.

Although the physiological significance of ALP is still unknown, ALP is often observed on the plasma membrane in various tissues. In our present and previous studies, its activity is intensely positive at the brush border of VNO-RFE and NRE, i.e., on the cilia, but negative on the olfactory cilia of the OEP. These results highlight the functional differences between cilia of OEP and VNO-RFE, and the functional similarity of the latter with the NRE, as previously suggested by Breipohl et al. (1979).

3.3 Adenosine Triphosphatase

Although there are several kinds of ATPase with different responses to inhibitors and activators, the present study is restricted to calcium-activated ATPase, which is localized in myosin, plasma membrane, or mitochondria (Pearse 1972).

Previous reports on the activity of the calcium-activated ATPase are not consistent. According to Shantha and Nakajima (1970), the activity of calcium-activated ATPase is slight in the olfactory vesicles, absent in the rest of the receptor or supporting cells of the OEP, and intense in the basal cells of the rhesus monkey. In mice (Cushieri 1974), ATPase activity is intense in the superficial zone of the OEP where mitochondria are aggregated, moderate in the basal cells, and intense in the supporting cell bodies and distal parts of receptor cells in the VNO-NE, while it is absent from the basal layer.

In our preliminary studies on the adult golden hamster (Taniguchi et al. 1981) we have used calcium-formol as fixative to suppress the activity of mitochondrial ATPase and to demonstrate extra-mitochondrial ATPase. The activity of calcium-activated ATPase was completely absent in the OEP. The superficial localization of extra-mitochondrial ATPase in the VNO-NE may reflect different functions between this part of the VNO-NE and that of the OEP.

No activity of this enzyme was found in the OEP, MO, and VNO-NE during intrauterine life of the golden hamster. Since this enzyme is present in the VNO-NE of the adult golden hamster (Taniguchi et al. 1981) it is likely that ATPase activity is not functionally important in the VNO-NE before birth.

3.4 Nonspecific Esterase

Esterase is subdivided into specific and non-specific esterase according to substrate specificities. While NsE is localized in various organelles such as endoplasmic reticulum, Golgi apparatus, mitochondria and lysosomes (Deimling and Böcking 1976), its physiological significance is still unclear. Among specific esterase, ChE and AChE are important in neurotransmission (Deimling and Böcking 1976; Eckenstein and Sofroniew 1983).

The activity of NsE is reported to be moderate to intense in the superficial zone of the VNO-NE (Cushieri 1974; Taniguchi et al. 1981). In the OEP, however, intense activity is found in this zone in the mouse (Cushieri 1974), while it is only slight or absent in all layers in the rhesus monkey and golden hamster (Shantha and Nakajima 1970; Taniguchi et al. 1981). Intense activity is also present in the GB in the mouse and golden hamster (Cushieri 1974; Taniguchi et al. 1981), but not in the rhesus monkey (Shantha and Nakajima 1970). These inconsistencies may reflect differences in species, fixatives, histochemical procedures and perhaps the nature of substrate.

Although the significance of the activity of NsE in the VNO-NE and GB of the golden hamster is still unknown, NsE in the superficial zone of the VNO-NE may relate to the transmission of olfactory stimuli. In the present study the activity of NsE was examined using α-naphthyl acetate as a substrate. No activity was found in the OEP, MO, and VNO-NE during development of the golden hamster. Because the activity of NsE is present in the VNO-NE and GB in the adult golden hamster (Taniguchi et al. 1981), its absence during intrauterine life of the golden hamster again indicates that the functions (as yet unknown) related to NsE do not manifest before birth in the VNO-NE and GB.

Since α-naphthyl acetate is a highly non-specific substrate resolved by almost all kinds of esterase including NsE, ChE and AChE, a negative result indicates that not only NsE but also ChE and AChE are absent in the tissue examined. We have also shown that activity is very slight to absent in the OEP and VNO-RFE of the adult golden hamster (Taniguchi et al. 1981). On the other hand, Filogamo and Robecchi (1969) have reported for the rabbit, rat, and guinea pig that AChE activity is present in the olfactory pit at very early stages of development, but disappears before birth. They suggested that AChE was involved in the neuronal

90

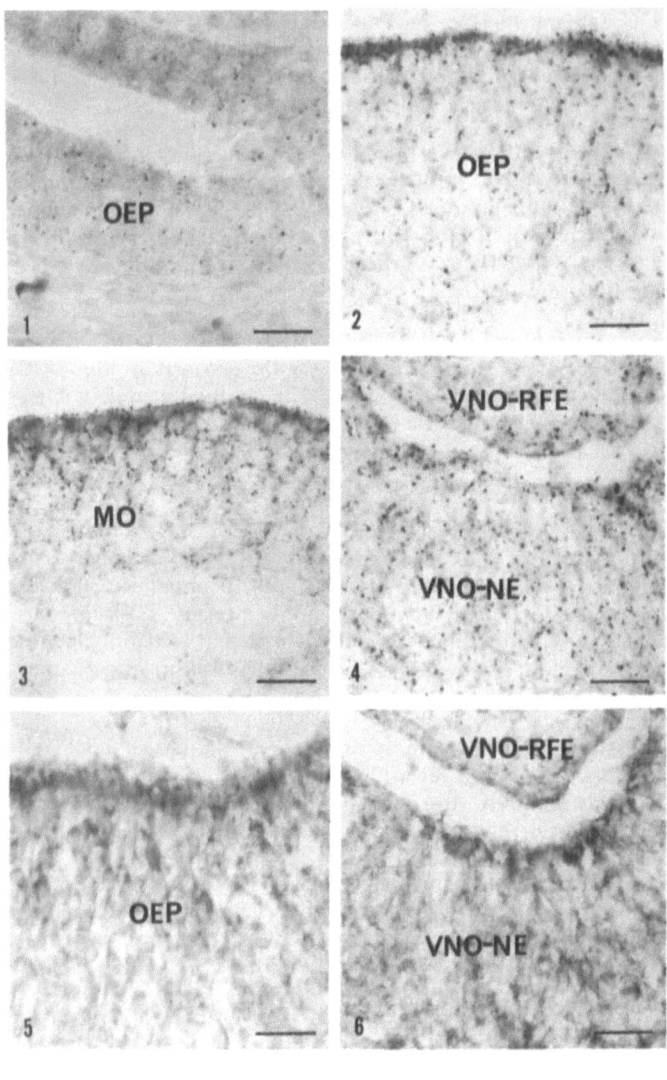

Fig. 1 OEP at 13 days of gestation. The activity of LDH is sporadically observed in the OEP. Scale bar = 10 μm

Fig. 2 OEP at 14 days of gestation. The activity of LDH is rather intense and tends to show the apical localization in the OEP. Scale bar = 10 μm

Fig. 3 MO at 15 days of gestation. The activity of LDH is rather intense and tends to show the apical localization in the MO. Scale bar = 10 μm

Fig. 4 VNO at 15 days of gestation. The activity of LDH is still diffuse in the VNO-NE. Scale bar = 10 μm

Fig. 5 OEP at 15 days of gestation. The activity of SDH is still not so intense, but tends to show the apical localization in the OEP. Scale bar = 10 μm

Fig. 6 VNO at 15 days of gestation. The activity of SDH is rather diffuse in the VNO-NE. Scale bar = 10 μm

differentiation of avian olfactory receptors. Our results do not confirm that for rodents, but suggest that esterase function is important only after birth in the golden hamster.

3.5 Lactate Dehydrogenase

LDH takes part in the metabolism of carbohydrates and is localized in the ribosomes. The activity of this enzyme is reported in the rhesus monkey to be intense in the upper part of the OEP and diffusely positive in the rest of the OEP (Shantha and Nakajima 1970). We have shown that it is also intense in the upper part of the OEP while only moderate in the upper part of the VNO-NE in the adult golden hamster (unpublished observations). As ribosomes are abundant in the upper part of the OEP and VNO-NE (Bhatnagar et al. 1982; Ciges et al. 1977; Graziadei 1977; Taniguchi and Mikami 1985; Taniguchi and Mochizuki 1983; Vaccarezza et al. 1981), the distribution of LHD activity is consistent with that of the ribosomes. Intense activity of LHD in the OEP may suggest more active metabolism or cellular turnover in this epithelium than that of the VNO-NE.

In the present study LHD activity first occurred sporadically in the OEP at 13 days of gestation (Fig. 1). At this stage, it was not observed in either MO or VNO-NE. At 14 days of gestation, it also became aggregated in the apical part of the OEP (Fig. 2) as in the adult, but it was rather weak and diffuse in the MO and VNO-NE. At 15 days of gestation, the activity of LDH became intense and also aggregated in the upper part of the MO (Fig. 3) while remaining diffuse in the VNO-NE (Fig. 4). These results may suggest that the metabolic activity of the OEP is achieved earlier than the MO and VNO-NE.

3.6 Succinate Dehydrogenase

SDH is a well-known mitochondrial enzyme marker. Mitochondria are abundant in the apical part of the receptor cells of the OEP, MO and VNO-NE (Bhatnagar et al. 1982; Ciges et al. 1977; Graziadei 1977; Miragall et al. 1984; Taniguchi and Mikami 1985; Vaccarezza et al. 1981). Consistent with these results are observations of intense SDH activity in the superficial zone of the OEP and VNO-NE of the mouse and rhesus monkey (Cushieri 1974; Shantha and Nakajima 1970). Our results in the adult golden hamster also support these findings (unpublished observations).

During intrauterine life of the golden hamster, the activity of SDH first occurred diffusely at 14 days of gestation in the OEP, MO and VNO-NE. This activity became moderate and tended to show the apical localization similar to the adult OEP at 15 days of gestation (Fig. 5), but it was still diffuse in the MO and VNO-NE (Fig. 6). These results confirm morphological studies in the golden hamster demonstrating mitochondrial aggregation in the apical part of the OEP and their diffuse distribution in the VNO-NE at 15 days of gestation (Taniguchi et al. 1982). However, the activity of SDH did not become intense before birth and its apical localization was not so distinct as in the adult. It appears that the metabolic rate of mitochondria in the three olfactory sensory epithelia is lower before birth.

4 Conclusion

The importance of specific enzyme systems in olfaction is still unclear, but the physiological significance of enzymes can partly be inferred from their cyto-chemical localizations: ACP is localized in the lysosomes and takes part in the autophagic function; ATPase is localized in the mitochondria, associated with myosin or plasma membrane and is related to energy metabolism; LDH is localized in the ribosomes and participates in the metabolism of carbohydrates; SDH is localized in the mitochondria and is related to energy metabolism. Although ALP and NsE are believed to be localized in the plasma membrane and some cytoplasmic organelles, their physiological significance is unknown.

As is demonstrated, the histochemical distribution of these enzymes in the three olfactory sensory epithelia varies considerably during development and adolescence. ACP, LDH, and SDH all show similar distributions among these epithelia in accordance with that of the organelles in which they are localized. However, ALP, ATPase and NsE exhibit different distributions among these epithelia. These results suggest that differences exist in either the metabolic of functional activities, at least at the distribution sites of these enzymes, between the epithelia.

The time sequence in the appearance of specific enzyme systems in the olfactory receptor cells during development may reflect the relative functional importance of these enzymes in olfaction especially with regard to intrauterine olfaction. As LDH and SDH appear in the three olfactory sensory epithelia during intrauterine life, these enzymes may be essential for cellular metabolism in these tissues.. This is also consistent with the presence of abundant ribosomes in olfactory stem cells (Taniguchi et al. 1982). Since the other enzymes are not present during intrauterine life, they may take part in more specialized or sophisticated functions which occur only after the olfactory sensory epithelia are fully differentiated.

Summary

The specific role of the various olfactory epithelia in general olfaction and especially during ontogeny was investigated histochemically in the golden hamster. The activities of six enzymes: acid phospatase (ACP), alkaline phosphatase (ALP), adenosine triphosphatase (ATPase), non-specific esterase (NsE), lactate dehydrogenase (LDH) and succinate dehydrogenase (SDH), were studied in the olfactory epithelium proper (OEP), septal organ of Masera (MO) and vomeronasal organ (VNO). Histochemically defined differences among the three epithelia are believed to reflect both differences in their stage of development and the relative functional importance of these enzymes.

References

Adams DR (1972) Olfactory and non-olfactory epithelia in the nasal cavity of the mouse Peromyscus. Am J Anat 133: 37–49

Aronson LR, Cooper ML (1974) Olfactory derivation and mating behavior in sexually experienced male cats. Behav Biol 11: 459–480

Barady AF, Bourne GH (1953) Gustatory and olfactory epithelia. Int Rev Cytol 2: 289–330

Bhatnagar KP, Matulionis D, Breipohl W (1982) Fine structure of the vomeronasal neuroepithelium of bats. A comparative study. Acta Anat 112: 158–177

Breipohl W, Bhatnagar KP, Mendoza A (1979) Fine structure of the receptor-free epithelium in the vomeronasal organ of the rat. Cell Tissue Res 200: 383–395

Ciges M, Labella T, Gayoso M, Sanchez G (1977) Ultrastructure of the organ of Jacobson and comparative study with olfactory mucosa. Acta Otolaryngol 83: 47–58

Cushieri A, (1974) Enzyme histochemistry of the olfactory mucosa and vomeronasal organ in the mouse. J Anat 118: 477-490

Cushieri A, Bannister LH (1975a) The development of the olfactory mucosa in the mouse: light microscopy. J Anat 119: 277-286

Cushieri A, Bannister LH (1975b) The development of the olfactory mucosa in the mouse: electron microscopy. J Anat 119: 471-498

Deimling OV, Böcking A (1976) Esterases in histochemistry and ultrahisto-chemistry. Histochem J 8: 215-252

Eckstein F, Sofroniew MW (1983) Identification of central cholinergic neurons containing both choline acetyltransferase and acetylcholinesterase and of central neurons containing only acetylcholinesterase. J Neurosci 3: 315-341

Estes PD (1972) The role of the vomeronasal organ in mammalian reproduction. Mammalia 36: 315-341

Filogamo G, Robecchi MG (1969) Neuroblasts in the olfactory pits in mammals. Acta Anat Suppl 56: 182-187

Gomori G (1952) Microscopic histochemistry. University of Chicago Press, Chicago

Graziadei PPC (1977) Functional anatomy of the mammalian chemoreceptor system. In: Müller-Schwarze D, Mozell MM (eds) Chemical signals in vertebrates. Plenum Press, London New York, pp 435-454

Hess R, Scarpelli DG, Pearse AGE (1958) Cytochemical localization of oxidative enzymes. II. Pyridine nucleotide-linked dehydrogenase. J Biophys Biochem Cytol 4: 753-760

Kratzing JE (1978) The olfactory apparatus of the bandicoot (Isoodon macrourus): fine structure and presence of a septal olfactory organ. J Anat 125: 601-613

Lejour M (1967) Activité de quatre enzymes dephosphorylants au cours de la morphogenèse du palais primaire chez le rat (phosphatase alcaline, phosphatase acide, ATP- et AMP-ases).

Lejour-Jeanty M (1965) Etude morphologique et cytochemique du dévelopment du palais primaire chez le rat (complétée par des observations chez le lapin et l'homme). Arch Biol (Liège) 76: 97-168

Milaire J (1974) Histochemical aspects of organogenesis in vertebrates. III. The sense organs. In: Handbuch der Histochemie,vol VIII, Suppl 3. Fischer, Stuttgart, pp 61-74

Miragall F, Breipohl W, Naguro T, Voss-Wermbter G (1984) Freeze-fracture study of the plasma membranes of the septal olfactory organ of Masera. J Neurocytol 13: 111-125

Miyayama H, Solomon R, Sasaki M, Lin C-W, Fishman WH (1975) Demonstration of lysosomal and extralysosomal sites for acid phosphatase in mouse kidney tubule cells with p-nitrophenylphosphate lead-salt technique. J Histochem Cytochem 23: 439-451

Monteiro-Riviere NA, Popp JA (1984) Ultrastructural characterization of the nasal respiratory epithelium in the rat. Am J Anat 169: 31-44

Murphy MR, Schneider GE (1970) Olfactory bulb removal eliminates mating behavior in the male golden hamster. Science 167: 302-304

Nachlas MM, Tsou K-C, Souza EJ de, Cheng C-S, Seligman AM (1957) Cyto-chemical demonstration of succinic dehydrogenase by the use of a new p-nitro-phenyl substituted ditetrazole. J Histochem 5: 420-436

Padykula HA, Hermann E (1955) Factors affecting the activity of adenosine triphosphatase and other phosphatases as measured by histochemical tech-niques. J Histochem Cytochem 3: 161-169

Pearse AGE (1972) Histochemistry, technical and applied. 3rd ed. Churchill, London

Pearson B, Defendi V (1957) A comparison between the histochemical demonstra-tion of non-specific esterase activity by 5-bromo-indoxyl acetate, α-naphtyl acetate and naphtol AS acetate. J Histochem Cytochem 5: 72-83

Rodolfo-Masera T (1943) Sul' esistenza di un particolare organo olfattivo nel setto nasale della cavia e di altri roditori. Arch Ital Anat Embriol 48: 157-213

Powers, JB, Winans SS (1975) Vomeronasal organ: critical role in mediating sexual behavior of the male hamster. Science 187: 961-963

Shantha TR, Nakajima Y (1970) Histological and histochemical studies on the rhesus monkey (Macaca mulatta) olfactory mucosa. Z Zellforsch 103: 292-319

Shapiro BL (1970) Enzyme histochemistry of embryonic nasal mucosa. Anat Rec 166: 87-98

Taniguchi K, Mikami S (1985) Fine structure of the epithelia of the vomeronasal organ of horse and cattle. A comparative study. Cell Tissue Res 240: 41-48

Taniguchi K, Mochizuki K (1983) Comparative morphological studies on the vomeronasal organ in rats, mice, and rabbits. Jpn J Vet Sci 45: 67-76

Taniguchi K, Mamba K, Sagi M, Taniguchi K, Mochizuki K (1981) Enzyme histo-chemical studies on the vomeronasal sensory and the olfactory epithelium in the golden hamster. St Marianna Med J 9: 241-246

Taniguchi K, Taniguchi K, Mochizuki K (1982) Comparative developmental studies on the fine structure of the vomeronasal sensory and the olfactory epithelia in the golden hamster. Jpn J Vet Sci 44: 881-890

Vaccarezza OL, Sepich LN, Tramezzani JH (1981) The vomeronasal organ of the rat. J Anat 132: 167-185

Watanabe K, Fishman W (1964) Application of stereospecific inhibitor L-phenyl-alanine to the enzymology of intestinal alkaline phophatase. J Histochem Cytochem 12: 252-260

Winans SS, Scalia F (1970) Amygdaloid nucleus: new afferent input from the vomeronasal organ. Science 170: 330-332

Wysocki CT, Wellington TL, Beauchamp GK (1980) Access of urinary nonvolatiles to the mammalian vomeronasal organ. Science 207: 781-783

Immunocytochemical Studies on the Maturation of the Rodent Olfactory Mucosa.

J Morgan

Roche Institute of Molecular Biology, Roche Research Center, Nutley, New Jersey 07110, USA

1 Introduction

The rodent olfactory neuroepithelium is remarkable amongst neural tissues inasmuch as it has the capacity for regeneration (Graziadei 1971, 1973; Graziadei and Metcalf 1971; Graziadei and Monti-Graziadei 1978; Moulton 1974; Thornhill 1970). Thus, following loss of adult receptor neurons by physical or chemical damage, new neurons are generated by the proliferation and differentiation of precursor neuroblasts, termed basal cells, located adjacent to the basement membrane (Graziadei et al. 1979; Harding et al. 1977; Monti-Graziadei 1983; Samanen and Forbes 1984). While constant proliferation of basal cells appears to occur in normal adult animals, there is no direct evidence to show that they ever develop functional synapses in the olfactory bulb. Indeed, it is presently a matter of some controversy as to whether receptor neurons have a finite, predetermined, lifespan (Graziadei and Monti-Graziadei 1979; Moulton 1974; Moulton and Fink 1972), or whether they are long-lived and only replaced upon elimination by exogenous factors such as viruses or bacteria (Hinds et al. 1984; Breipohl et al. 1985; Breipohl et al. this Vol).

In addition to these aspects of the cellular dynamics of the olfactory epithelium, there are several other salient features of this structure that relate to its ontogeny that are worthy of mention. Firstly, the embryonic olfactory placode appears to be able to organize the forebrain (Clairambault 1971; Harrison 1958; Stout and Graziadei 1980). Secondly, axons from the olfactory receptor neurons find their way from the epithelium to the olfactory bulb in a spatially and functionally organized manner (Jastreboff et al. 1984; Lancet et al. 1982). Thirdly, arriving olfactory receptor neuron axons have the ability to organize glomeruli in olfactory bulb, and in ectopic brain regions (Graziadei et al. 1978, 1979). The aim of this review is to summarize what has been learned concerning these various aspects of cell dynamics in the olfactory epithelium using immunocytochemical approaches.

2 Materials and Methods

In the studies emanating from this laboratory, the following strategy has been employed to raise monoclonal antibodies (MAbs) to the rat olfactory epithelium. Female BALB/c mice were immunized with 100 mg of homogenate from adult rat olfactory epithelium in Freunds' complete adjuvant. The animals were boosted four weeks later with 50 mg homogenate in Freunds' incomplete adjuvant. Four days later spleens were taken and fused according to standard protocols to yield antibody-secreting hybridomas (Hempstead and Morgan 1983a, 1985; Morgan 1984). Supernatants were screened on cryostat sections of fixed olfactory epithelium of rat using immunofluorescence as described previously (Hempstead and Morgan 1983 a,b, 1985). In recent studies, MAbs have been produced to membrane fractions from adult and neonated olfactory epithelium of rat as well as homogenate of embryonic day 14 epithelium (Farbman, Hempstead, and Morgan unpublished). Some results derived from these latter studies will be mentioned here as well.

3 Polyclonal Antibodies to the Olfactory Epithelium

3.1 Antibodies to the Olfactory Marker Protein (OMP)

OMP is a 18.5 kD acidic cytoplasmic protein unique to the olfactory pathway (Margolis 1972). Antibodies developed to purified OMP (Keller and Margolis 1975) have shown the protein to be restricted to the mature olfactory receptor neurons of rats (Farbman and Margolis 1980; Hartman and Margolis 1975), mice (Monti-Graziadei et al. 1977), rabbits (Hempstead and Morgan unpublished), hamsters (Samanen and Forbes 1984), gerbils (Nakashima et al. 1984) and humans (Nakashima et al. 1985). Specifically the antiserum stains the perikaryon, apical dendrite and axon of the mature olfactory neuron (Farbman and Margolis 1980; Hartman and Margolis 1975; Monti-Graziadei et al. 1977; Samanen and Forbes 1984). In contrast, precursor basal cells and less mature neurons are OMP-negative (Farbman and Margolis 1980; Hartman and Margolis 1975; Monti-Graziadei et al. 1977; Samanen and Forbes 1984). Indeed, studies in mice show that a period of seven days is required following terminal division of the basal cell before OMP is expressed (Miragall and Monti-Graziadei 1982). A lag time of 8 days has been determined in the hamster (Samanen and Forbes 1984).

In addition to staining primary olfactory receptor neurons, the OMP antiserum also reacts with the neuronal population of the vomeronasal organ (Farbman and Margolis 1980) and their axonal projections to the accessory olfactory bulb (Hempstead and Morgan 1985).

During the normal development of the olfactory epithelium OMP immunoreactivity appears between embryonic days 15 (Allen and Akeson 1985a) and 18 (Farbman and Margolis 1980) in the rat and at embryonic day 14 in the mouse (Farbman and Margolis 1980; Monti-Graziadei et al. 1980). In organotypic cultures of embryonic rat olfactory epithelium, OMP was only expressed in explants removed from the animal after embryonic day 15 (Chuah and Farbman 1983; Farbman and Margolis 1980). The study of Chuah and Farbman (1983) also established that approximately twice as many olfactory neurons were OMP-positive in organotypic cultures where the olfactory bulb was co-explanted with the epithelium. This has led them to postulate that the olfactory bulb might play a modulatory role in olfactory neuron maturation.

OMP antibodies have also been used to elucidate the dynamics of olfactory neuron turnover in both decentralized epithelium and epithelial transplants. Removal of the olfactory bulb (Graziadei et al. 1979; Monti-Graziadei 1983) or axotomy (Samanen and Forbes 1984) leads to the dead of all mature OMP-positive neurons. Subsequently there ensues the proliferation and differentiation of basal cells which repopulate the epithelium with OMP-positive neurons (Graziadei et al. 1979; Monti-Graziadei 1983; Samanen and Forbes 1984). In the case of axotomy, the regenerated epithelium is indistinguishable from normal (Samanen and Forbes 1984). However, where bulbectomy has been performed, and no target now exists for new olfactory neurons, the reconstituted neuroepithelium has fewer OMP-positive cells (Costanzo 1984; Monti-Graziadei 1983). These results have been interpreted as meaning that the olfactory bulb does not control OMP expression per se, but rather that the organ modifies the overall dynamics of neuroblast proliferation and differentiation (Monti-Graziadei 1983).

In the studies cited above, the intracranial cavity was plugged with gelfoam following bulbectomy, in the absence of this procedure olfactory axons enter the frontal cortex and organize ectopic glomeruli that are OMP-positive (Graziadei et al. 1979). It has not been established whether these inappropriate synapses can substitute for the olfactory bulb in maintaining normal olfactory neuron differentiation and OMP expression. The case for OMP being expressed in the absence of the olfactory bulb has been further supported by the observation that olfactory receptor neurons retain the ability to synthesize OMP when transplanted into various ectopic sites including the anterior chamber of the eye (Barber et al. 1982; Heckroth et al. 1983), the parietal cortex (Morrison and Graziadei 1983) and cerebral ventricle (Morrison and Graziadei 1983). These explant experiments highlight the value of this antiserum in unambiguously marking olfactory neurons.

3.2 Antibodies to Carnosinase

The dipeptide carnosine (ß-alanyl-L-histidine) is present in high concentration in the olfactory mucosa (Margolis 1980), where it may function as a neurotransmitter (Rochel and Margolis 1982). Carnosinase, an enzyme capable of hydrolyzing carnosine has been isolated from mouse kidney and used to generate polyclonal antibodies (Margolis et al. 1983). These antibodies have revealed that in the mouse olfactory epithelium, carnosinase is abundant within Bowman's glands with a lower level of expression being observed in sustentacular cells (Margolis et al. 1983). The neurons and axons of the vomeronasal organ (VNO) were also reactive with this antiserum (Margolis et al. 1983). In the rat the same antibody only gave staining of Bowman's glands (Hempstead and Morgan 1983b). The VNO has not been examined in the rat so far. The chapter by Margolis (this Vol) discusses the ontogenetic aspects of carnosine metabolism.

3.3 Antibodies to Glial Fibrillar Acidic Protein (GFAP)

The cells ensheathing the olfactory axonal bundles, often referred to as perineuronal sheath cells, react with antibodies to the astrocyte marker GFAP (Barber 1982). This fact distinguishes the sheath cells from the Schwann cells of the peripheral nervous system that do not express this protein. The presence of this marker may also account for the presence of GFAP-positive cells in cultures of rat olfactory epithelium (Hempstead and Morgan 1983c).

4 Monoclonal Antibodies to the Rat Olfactory Epithelium

4.1 The 2B8 Monoclonal Antibody

Allen and Akeson (1985b) have reported the production of a MAb to PC12 phaeo-chromocytoma cells that cross-reacts with membranes of olfactory receptor neurons and neurons of the VNO. In the adult rat the 2B8 antibody binding co-localizes with OMP immunoreactivity, although only 25-30% of all OMP-positive cells express the 2B8 antigen (Allen and Akeson 1985a). These data are interpreted as indicating the presence of a subset of olfactory neurons that are OMP+ and 2B8+. This contention is further supported by the finding of glomeruli in the olfactory bulb that are either OMP+ and 2B8+ or OMP+ and 2B8- (Allen and Akeson 1985a). During normal development the appearance of 2B8 (embryonic day 13) precedes the expression of OMP (embryonic day 15) thereafter the percentage of OMP+ and 2B8+ cells increases until it reaches adult levels around birth (Allen and Akeson 1985a).

As a part of these studies the molecular characteristics of the 2B8 antigen were described. Immunoblots revealed immunoreactive material in several non-neuronal tissues as well as in olfactory epithelium, olfactory bulb and brain (Allen and Akeson 1985b). The situation in olfactory epithelium and bulb suggested that there was an embryonic pattern of 2B8 proteins up to fetal day 19, whereafter a shift occurred towards an adult spectrum of proteins.

4.2 Monoclonal Antibodies Revealing an Organization of Olfactory Neuron Projections to the Olfactory Bulb

Recently monoclonal antibodies have been described that show a spatial organization of axons entering glomeruli of the olfactory bulb (Fujita et al. 1985). It is known that each glomerulus receives axons from several regions of the olfactory epithelium (Jastreboff et al. 1984). Further, glomeruli may represent functionally discrete units of the olfactory system (Lancet et al. 1982). Thus, such MAbs may be detecting antigens important for gross organization and guidance, of olfactory neurons or molecules involved in more discrete olfactory functions. However, taken together with the results derived from the 2B8 MAb analysis, as well as those derived below using the NEU-9 MAb, these studies indicate the possibility of subclasses of olfactory neurons. Such a conclusion is also supported by the finding of a subset of carbonic anhydrase-positive neurons in the olfactory epithelium (Brown et al. 1984).

4.3 Sustentacular Cell Antibodies (SUS)

A number of clones have been derived that stain the somata and basal processes of sustentacular cells (Hempstead and Morgan 1983a, 1985). At least one of these clones, SUS-1, also reacted with some of the cells around the acini of Bowman's glands as well as occasional groups of cells that appeared to penetrate the basal lamina (Hempstead and Morgan 1983a). We were unable to establish whether this indicated some functional homology between the two cell populations, or even a lineage relationship as suggested in one previous study (Mulvany and Heist 1971). We have recently observed a similar staining distribution with antibodies to the cytochrome P-450 enzymes (Farbman, Morgan, and Thomas unpublished).

The SUS-1 antigen appears in sustentacular cells between fetal days 17 and 19, and is marked by fetal day 20 (Farbman, Hempstead and Morgan unpublished). The antigen is absent in the supporting cells of the adult VNO (Hempstead and Morgan unpublished). In unilaterally bulbectomized rats where there is a marked reduction in the number of OMP-positive neurons, SUS-1 reactivity is not detectably altered (Hempstead and Morgan 1983a), indicating that neither the sustentacular cell population nor the content of the SUS-1 antigen is affected by this procedure.

4.4 Neuronal Antibodies (NEU)

The most revealing group of MAbs reacted with receptor neurons and/or axonal bundles (Hempstead and Morgan 1985). For brevity, only three antibodies from the library, NEU-4, NEU-5 and NEU-9, will be discussed here. All three MAbs stain olfactory receptor neuron perikarya to some extent but give intense staining of axonal bundles (Hempstead and Morgan 1985) and the bundle layer of the olfactory bulb (Hempstead and Morgan unpublished). NEU-4 and NEU-9 MAbs are similar to OMP antibodies inasmuch as they give only weak staining of axons of the vomeronasal nerve (Hempstead and Morgan 1985), in contrast the NEU-5 MAb stains both the primary olfactory axons and those of the VNO equally well (Hempstead and Morgan 1985). However, of the three antibodies only NEU-9 gives significant staining of the perikarya of the VNO neurons (Hempstead and Morgan unpublished).

The three MAbs are markedly different when they are compared on olfactory neuroepithelium from unilaterally bulbectomized rats. When regeneration was complete, staining on the lesioned epithelium with the NEU-5 MAb was indistinguishable from the control mucosa (Hempstead and Morgan in press). In contrast staining with NEU-4 and NEU-9 MAbs was greatly attenuated in the ipsilateral mucosa. The same study also revealed that the NEU-9 MAb detected a subset of neurons in the ipsilateral mucosa that were absent or only infrequently observed in the contralateral epithelium.

A further aspect of olfactory epithelium structure following bulbectomy is the presence of apparently misplaced axonal bundles in the neuroepithelium (Hempstead and Morgan in press). These structures are highly reactive with antibodies to OMP and the three MAbs discussed here, but do not stain with non-neuronal MAbs. The bundles always seem to occupy the receptor neuron layer and only rarely invade the sustentacular cell layer, suggesting some mechanism of selective adhesion. Indeed, the NEU-5 antigen has been identified by us as a 210 kD protein with properties similar to the rat neural cell adhesion molecule (Kwiatkowski, Hempstead and Morgan unpublished), which could clearly be of relevance for autoadhesion mechanisms.

Of the three MAbs described here, NEU-9 is the earliest to be expressed in the fetal rat (Farbman, Hempstead, and Morgan unpublished). This might suggest that the new class of neurons seen following bulbectomy may represent a general shift towards a more embryonic state of the epithelium. Preliminary evidence suggests that the NEU-9 antigen appears around fetal day 13 in rats followed closely by NEU-5 on days 14-15, whereas NEU-4, like OMP, is only first clearly present at fetal day 16 (Farbman, Hempstead, and Morgan unpublished). Yet another MAb, 1A6, has been isolated that reacts with E13 and E14 olfactory neuroepithelium (Farbman, Hempstead, and Morgan unpublished). This antibody reacts not only with olfactory epithelium, but also with neural tube. Its staining is not coincident with either NEU-5 or NEU-9 and it disappears by birth in rat. This antigen may be important in the initial pathfinding of olfactory axons (Farbman, Hempstead, and Morgan unpublished).

4.5 Luminal Boundary Antibodies (LUM)

One group of clones secrete antibodies that give a distinctive band of staining at the level of the olfactory cilia and the sustentacular cell brush border (Hempstead and Morgan 1985). Some of these antibodies also react with respiratory cilia (Hempstead and Morgan 1985) and cilia of cultured respiratory cells (Hempstead and Morgan 1983c), perhaps indicating antigens common to many types of cilia. Clones that specifically react with the olfactory cilia may thus be of some value in probing the role of these structures in olfaction. During development the LUM-type antibodies first stain about fetal day 18 when ciliogenesis has been initiated (Farbman, Hempstead, and Morgan unpublished).

4.6 Antibodies to Bowman's Glands (GLA)

Several clones reacted with Bowman's glands and their contents (Hempstead and Morgan 1985). At least one of these MAbs, GLA-2, also stained presumptive secretory cells in the respiratory mucosa, suggesting an antigenically common secretory product (Hempstead and Morgan 1985).

5 Conclusions and Perspectives

The ultimate goal of our study is to clone individal basal cells in vitro. The MAbs described here will be useful in this endeavor both as markers and as purification probes. Basal cell lines would permit us to address several questions pertaining to olfaction and the regulation of olfactory neuron differentiation in more detail.
1. How is the normal balance of proliferation and differentiation of basal cells maintained?
2. What are the molecular sequelae of the induction of basal cells?
3. What is the molecular basis of olfactory diversity?

The data generated using immunocytochemical methods, provides compelling evidence for the existence of subsets of olfactory receptor neurons (Allen and Akeson 1985a,b; Fujita et al. 1985). This fact further confuses the issue of programmed olfactory neuron death (Graziadei and Monti-Graziadei 1979; Moulton 1974; Moulton and Fink 1972) versus long-lived receptor neurons (Hinds et al. 1984), since it is conceivable that there are both long-lived and short-lived neuronal subpopulations. Our own studies point to antigenic determinants that fail to be expressed in olfactory neurons lacking a target. Such antigens will act as markers to establish whether newly generated neurons generally form synapses or whether they perish without forming functional connections. The recently described MAbs revealing spatial organization of olfactory neurons (Fujita et al. 1985), will lead directly to the molecules involved in olfactory diversity and topographical organization. Such MAbs used together with others described here may then prove of immense importance in analyzing olfactory neuron development under normal conditions and in transplants.

Acknowledgments. The author wishes to thank a number of colleagues for their cooperation in preparing this text. First, Drs. G. Monti Graziadei, P. Graziadei and R.Akeson for the provision of preprints and reprints of their work. Drs. F. Margolis and A. Farbman for helpful discussion and Jim Hempstead who generated much of the data included in this review. Finally, Drs. W. Breipohl and R. Apfelbach for the opportunity of presenting our results in this form and for a most pleasant workshop in Tübingen.

References

Allen WK, Akeson R (1985a) Identification of an olfactory receptor neuron subclass: cellular and molecular analysis during development. Dev Biol 109: 393-401

Allen WK, Akeson R (1985b) Identification of a cell surface glycoprotein family of olfactory receptor neurons with a monoclonal antibody. J Neurosci 5: 284-296

Barber PC (1982) Regeneration of olfactory sensory axons into transplanted segments of peripheral nerve. Neuroscience 7: 2677-2685

Barber PC, Jensen S, Zimmer J (1982) Differentiation of neurons containing olfactory marker protein in adult rat olfactory epithelium transplanted to the anterior chamber of the eye. Neuroscience 7: 2687-2695

Breipohl W, Rehn B, Molyneux GS, Grandt D (1985) Plasticity of neuronal cell replacement in the main olfactory epithelium of mouse. XII Int Anat Congr, London, Abstract

Brown D, Garcia-Segura LM, Orci L (1984) Carbonic anhydrase is present in olfactory receptor cells. Histochemistry 80: 307-309

Chuah M, Farbman AI (1983) Olfactory bulb increases marker protein in olfactory receptor cells. J Neurosci 3: 2197-2205

Clairambault P (1971) Les effets de l'ablation bilaterale de la placode nasale sur la morphogenèse du télencéphale des Anoures. Acta Embryol Exp 2: 61-92

Costanzo RM (1984) Comparison of neurogenesis and cell replacement in the olfactory system with and without a target (olfactory bulb). Brain Res 307: 295-301

Farbman AI, Margolis FL (1980) Olfactory marker protein during ontogeny: immunohistochemical localization. Dev Biol 74: 205-215

Fujita SC, Mori K, Imamura K, Obata K (1985) Subclasses of olfactory receptor cells and their segregated central projections demonstrated by a monoclonal antibody. Brain Res 326: 192-196

Graziadei PPC (1971) The olfactory mucosa in vertebrates. In: Beidler K (ed) Handbook of sensory physiology, vol IV. Chemical senses, Sect 1. Olfaction. Springer, Berlin Heidelberg New York, pp 27-58

Graziadei PPC (1973) Cell dynamics in the olfactory mucosa. Tissue cell 5: 113-131

Graziadei PPC, Levine RR, Monti Graziadei GA (1978) Regeneration of olfactory axons and synapse formation in the forebrain after bulbectomy in neonatal mice. Proc Natl Acad Sci USA 75: 5230-5234

Graziadei PPC, Levine RR, Monti Graziadei GA (1979) Plasticity of connections of the olfactory sensory neuron: regeneration into the forebrain following bulbectomy in the neonatal mouse. Neuroscience 4: 713-727

Graziadei PPC, Metcalf JF (1971) Autoradiographic and ultrastructural observations on the frog's olfactory mucosa. Z Zellforsch Mikrosk Anat 116: 305-318

Graziadei PPC, Monti Graziadei GA (1978) The olfactory system. A model for the study of neurogenesis and axon regeneration in mammals. In: Cotman C (ed) Neuronal plasticity. Raven Press, New York, pp 131-153

Graziadei PPC, Monti Graziadei GA (1979) Neurogenesis and neuron regeneration in the olfactory system of mammals. I. Morphological aspects of differentiation and structural organization of the olfactory sensory neurons. J Neurocytol 8: 1-18

Harding J, Graziadei PPC, Monti Graziadei GA, Margolis FL (1977) Denervation in the primary olfactory pathway of mice. IV. Biochemical and morphological evidence for neuronal replacement following nerve section. Brain Res 132: 11-28

Harrison JL (1958) Some hypoplastic modifications of the telencephalon following unilateral excision of the nasal placode in Rana pipiens embryos. Diss Abstr 19: 605

Hartman BK, Margolis FL (1975) Immunofluorescence localization of the olfactory marker protein. Brain Res 96: 176-180

Heckroth JA, Monti Graziadei GA, Graziadei PPC (1983) Intraocular transplants of olfactory neuroepithelium in rat. Int J Devl Neurosci 1: 273-287

Hempstead JL, Morgan JI (1983a) Monoclonal antibodies to the rat olfactory sustentacular cell. Brain Res 288: 289-295

Hempstead JL, Morgan JI (1983b) Fluorescent lectins as cell specific markers for the rat olfactory epithelium. Chem Sens 8: 107-120

Hempstead JL, Morgan JI (1983) Culture and immunocytochemical characterization of the rat olfactory epithelium. Soc Neurosci Abstr 9: 464

Hempstead JL, Morgan JI (1985) A panel of monoclonal antibodies to the rat olfactory epithelium. J Neurosci 5: 438-449

Hinds JW, Hinds PL, McNelly NA (1984) An autoradiographic study of the mouse olfactory epithelium: evidence for long-lived receptors. Anat Rec 210: 375-383

Jastreboff PJ, Pedersen PE, Greer CA, Stewart WB, Kauer JS, Benson TE, Shepherd GM (1984) Specific olfactory receptor populations projecting to identified glomeruli in the rat olfactory bulb. Proc Natl Acad Sci USA 81: 5250-5254

Keller A, Margolis FL (1975) Immunological studies of the rat olfactory protein. J Neurochem 24: 1101-1106

Lancet D, Greer CA, Kauer JS, Shepherd GM (1982) Mapping of odor-related neuronal activity in the olfactory bulb by high resolution 2-deoxyglucose autoradiography. Proc Natl Acad Sci USA 79: 670-674

Margolis FL (1972) A brain protein unique to the olfactory bulb. Proc Natl Acad Sci USA 69: 1221-1224

Margolis FL (1980) In: Barker JL, Smith T (eds) Role of peptides in neuronal function. Decker, New York, pp 545-572

Margolis FL, Grillo M, Granot-Reisfeld N, Farbman AI (1983) Purification, characterization and immunocytochemical localization of mouse kidney carnosinase. Biochim Biophys Acta 744: 237-248

Miragall F, Monti Graziadei GA (1982) Experimental studies on the olfactory marker protein. II. Appearance of the olfactory marker protein during differentiation of the olfactory sensory neurons of mouse: an immunohistochemical and autoradiographic study. Brain Res 239: 245-250

Monti Graziadei GA (1983) Experimental studies on the olfactory marker protein. III. The olfactory marker protein in the olfactory neuroepithelium lacking connections with the forebrain. Brain Res 262: 303-308

Monti Graziadei GA, Margolis FL, Harding JW, Graziadei PPC (1977) Immunocytochemistry of the olfactory marker protein. J Histochem Cytochem 25: 1311-1316

Monti Graziadei GA, Stanley RS, Graziadei PPC (1980) The olfactory marker protein in the olfactory system of the mouse during development. Neuroscience 5: 1239-1252

Morgan JI (1984) Monoclonal antibody production. In: Spector S, Bach N (eds) Modern methods in pharmacology, vol II. Liss, New York, pp 29-67

Morrison EE, Graziadei PPC (1983) Transplants of olfactory mucosa in the rat brain. I. A light microscopic study of transplant organization. Brain Res 279: 241-245

Moulton DG (1974) Cell renewal in the olfactory epithelium of the mouse. Ann N Y Acad Sci 237: 52-61

Moulton DG, Fink RP (1972) Cell proliferation and migration in the olfactory epithelium. In: Schneider D (ed) Olfaction and Taste, vol IV. Wiss Verlagsges MHB, Stuttgart, pp 20-26

Mulvany BD Heist HE (1971) Regeneration of rabbit olfactory epithelium. Amer J Anat 131: 241-252

Nakashima T, Kimmelman CP, Snow JB (1984) Effects of olfactory nerve section and hemorrhage on the olfactory neuroepithelium. Am College Surg Forum, XXXV: 562-564

Nakashima T, Kimmelman CP, Snow JB (1985) Olfactory marker protein in the human olfactory pathway. Arch Otolaryngol 110: 641-646

Rochel S, Margolis FL (1982) Carnosine release from olfactory bulb synaptosomes is calcium-dependent and depolarization-stimulated. J Neurochem 38: 1505-1514

Samanen DW, Forbes WB (1984) Replication and differentiation of olfactory receptor neurons following axotomy in the adult hamster: a morphometric analysis of postnatal neurogenesis. J Comp Neurol 225: 201-211

Stout RP, Graziadei PPC (1980) Influence of the olfactory placode on the development of the brain in Xenopus laevis. Neuroscience 5: 2175-2186

Thornhill RA (1970) Cell division in the olfactory epithelium of the lamprey, Lampetra fluviatilis. Z Zellforsch Mikrosk Anat 109: 147-157

D. POSTNATAL MATURATION OF THE OLFACTORY BULB

Ontogeny of Carnosine, Olfactory Marker Protein and Neurotransmitter Enzymes in Olfactory Bulb and Olfactory Mucosa of the Rat

FL Margolis, T Kawano* and M Grillo

Roche Institute of Molecular Biology, Roche Research Center, Nutley, New Jersey 07110 USA

*Department of Neurosurgery, Nagasaki University, School of Medicine, 7-1 sakamoto-machi, Nagasaki 852, Japan

Introduction

Both the olfactory and vomeronasal pathways are neuronal systems in which the sensory neurons are thought to be capable of continuous replenishment and re-innervation of their central targets (Graziadei and Monti-Graziadei 1978; Simmons and Getchell 1981; Hinds et al. 1984; Wang and Halpern 1982). This implies the occurrence of very dynamic morphogenetic and metabolic changes within the epithelium where the cell bodies reside, as well as in the bulb where synaptogenesis takes place. Alterations of the synaptic input of the primary olfactory neurons to their target neurons in the olfactory bulb has demonstrated that the afferent olfactory neurons can modulate the biochemical phenotype expressed by their juxtaglomerular target neurons (Baker et al. 1983, 1984; Kream et al. 1984).

The similarity of these events to normal ontogenesis of the olfactory system is not yet clear, although the influence of afferent olfactory neurons on embryonic bulbar development has been reported (Hinds 1972). In order to evaluate these questions it is necessary to be able to describe both the events occurring in normal ontogeny and those associated with the response to normal and induced cellular turnover. This laboratory has been addressing the biochemical changes associated with both of these aspects, and in this paper we will present some of our recent studies on the normal biochemical ontogeny of the rodent olfactory system.

2 Olfactory Marker Protein

Over a decade ago we reported on the presence of a small cytosolic protein subsequently shown to be uniquely restricted to mature olfactory neurons (for review see Margolis 1980a) in many species. This protein is called olfactory marker protein (OMP). Immunoassays with antibody prepared against the rodent OMP indicate it is phylogenetically conserved over a wide range of species. In the rat, OMP can be detected at birth and its content in bulb and mucosa increases during the first few postnatal weeks after which it plateaus at adult levels (Fig. 1). These data are quite similar to those we reported several years ago for the ontogeny of OMP in the mouse (Margolis 1975).

When studied by immunocytochemistry it is evident that the OMP appears in the olfactory mucosa during the last trimester of gestation (Farbman and Margolis 1980; Monti-Graziadei et al. 1980) apparently earlier in the mouse than in the rat.

More recent studies (Allen and Akeson 1985) indicate that this age discrepancy is more apparent than real and is probably dependent upon technical details. Thus, in both rodents OMP is first demonstrable at E14 to E15 in the chemoreceptor neurons and subsequently in their terminals in the bulb.

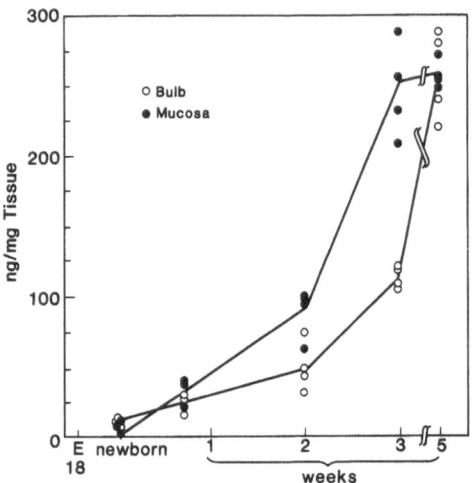

Fig. 1. Ontogeny of olfactory marker protein (OMP) in rat bulb and mucosa. OMP was determined by RIA according to Margolis (1980a).

When olfactory explants at E15 were assayed for the presence of OMP by radio immuno assay no reactivity was observed. However, when maintained in culture for several days OMP was demonstrable in these explants by immunocytochemistry and was quantifiable by RIA (Chuah and Farbman 1983). Co-culturing of these mucosal explants in the presence of presumptive olfactory bulb but not other tissues elevated the level of OMP by twofold and doublet the number of cells manifesting OMP immunocytochemically (Chuah and Farbman 1983).

These results suggest that the developmental program of OMP expression is partially target dependent. Quite possible the initiation of OMP synthesis is solely dependent upon the developmental state of the neuron. However, unless the target is contacted, stabilization of cellular development and continued OMP expression may be compromised. The influence of the developmental state of these neurons on OMP biosynthesis is also indicated by the fact that olfactory neurons activated in response to a lesion or olfactory neurons which are part of an expanding population in young mice manifest more rapid turnover of OMP than is seen in adult mice (Kream and Margolis 1984).

The mechanisms regulating these events are unknown but the applicability of molecular biological approaches to this tissue will now facilitate their exploration (Margolis et al. 1985b).

3 Ornithine Decarboxylase

In many tissues undergoing growth or regeneration, the level of ornithine decarb-oxylase (ODC) activity is high. This is true during tumorigenesis, ontogeny, or recovery from trauma (for references see Rochel and Margolis 1980 and Laitinen et al. 1985). Since normal mature cerebral tissue is reported to have very low levels of this enzyme activity, consistent with its lack of cellular replication, it was of interest to quantify the level of this enzyme in olfactory mucosa. Curiously, in both olfactory mucosa and bulb, the level of ODC in young rats was very high. At every age studied ODC activity exceeded that seen in cerebellum and hippo-campus, areas known to undergo postnatal neurogenesis (Rochel and Margolis 1980).

This high ODC activity was consistent with the known phenomenon of continuing neurogenesis and synaptic plasticity in the olfactory pathway. However, attempts to correlate ODC activity and neuronal turnover in the olfactory system were less than satisfactory (Rochel and Margolis 1980). A recent report (Laitinen et al. 1985), suggests that the level of ODC activity measured in neural tissue may not accurately reflect the amount of ODC enzyme protein due to the presence of a macromolecular ODC inhibitor. This suggests that ODC activity and protein levels should be re-evaluated in both developing and regenerating olfactory tissue.

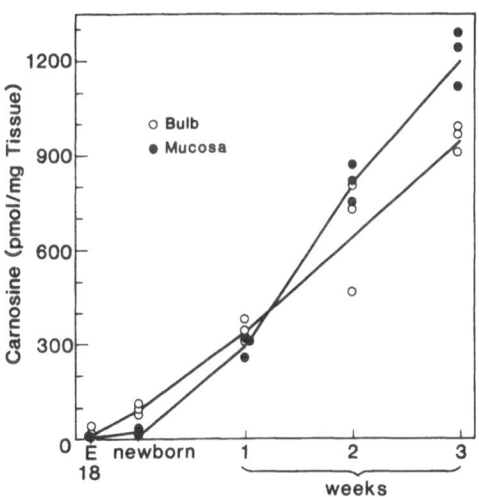

Fig. 2. Carnosine in rat olfact-ory bulb and mucosa at different ages. Carnosine was determined by HPLC separation and quantitation of the MDPF fluorophore as described by Margolis and Grillo (1984)

4 Carnosine

The nature of the transmitter utilized by the primary olfactory neurons at their synapses with target neurons in the olfactory bulb has been a subject of extens-ive study in this laboratory (for review see Margolis 1981). These studies led to the discovery of high levels of the dipeptide carnosine in the olfactory bulb and mucosa (Neidle and Kandera 1974; Margolis 1974). This compound has many of the biochemical properties associated with a transmitter (Margolis 1980b) but the neurophysiological data ara equivocal (Frosch and Dichter 1984; MacLeod and Straughan 1979; Gonzales-Estrada and Freeman 1980; Nicoll et al. 1980; Tonosaki and Shibuya 1979) suggesting that the identity of the primary olfactory transmitter is still in question and leaving in doubt the function of carnosine.

Since the olfactory receptor neurons contain both carnosine and OMP it was of interest to compare their appearance during ontogeny. Carnosine levels rise rapidly in the postnatal olfactory bulb and mucosa of the rat (Fig. 2). Similar observations have been reported for the postnatal olfactory bulb of the mouse (Neidle and Kandera 1974; Ferriero and Margolis 1975) and rat (Barbaro et al. 1978). The levels of carnosine in olfactory tissue are very low before birth (Fig. 2), as they are for OMP. The prenatal ontogeny of carnosine accumulation, carnosine synthetase activity and its dependence upon endogenous levels of precursor was therefore evaluated.

This was of interest, since we had shown earlier that in adults the level of olfactory carnosine responded to administration of exogenous ß-alanine but not histidine (Margolis 1981). This observation was consistent with the endogenous tissue levels of these two amino acids relative to their respective K_m values determined for the partially purified synthetase (Horinishi et al. 1978). Thus, in vivo the synthetase should be saturated with regard to histidine but not ß-alanine.

Table 1 Incorporation of $(1-^{14}C)$ß-alanine into (^{14}C)carnosine by embryonic olfactory mucosa in vitro

	Days in culture			
	0		3	
	(cpm (^{14}C)carnosine/set of explants)			
13-Day embryo	711:	900	10,209:	17,358
14-Day embryo	3,693:	3,687	20,877:	15,696
15-Day embryo	7,566:	6,838	48,012:	26,625
16-Day embryo	3,357:	4,233	23,361:	28,563
17-Day embryo	10,206:	11,481	29,919:	10,659

Olfactory mucosa was explanted at embryonic ages indicated and maintained in culture for 0 or 3 days. Cultures, in duplicate, each contained six explants. For 20 h prior to harvest, the culture medium was supplemented with $(1-^{14}C)$ß-alanine ($2 \mu Ci$ / ml) and (^{14}C)carnosine was isolated. Data are presented as total cpm (^{14}C)carnosine for each pool of explants. In all cases, > 90% of the radioactive carnosine was hydrolyzed back to to radioactive ß-alanine by purified carnosinase. (After Margolis et al. 1985a).

By chemical means we were able to demonstrate the presence of carnosine in olfactory tissue at least as early as embryonic day 16. When these explants were evaluated for their ability to incorporate radioactive ß-alanine we could demonstrate the formation of radioactive carnosine as early as 13-14 days of embryonic development (Table 1). This suggested that the carnosine synthetase activity may already be present prior to the appearance of OMP in this tissue. These observations suggested that even though the synthetase activity was present, carnosine synthesis was not taking place due to insufficient levels of the ß-alanine precursor. It was also possible that synthesis was occurring but the tissue was incapable of storing the carnosine product.

Both in vivo and in vitro experiments were performed to address these questions (Margolis et al. 1985a). Olfactory tissue was explanted from 14 and 15 day rat embryos and maintained in vitro in the presence and absence of 1 mM ß-alanine added to the standard culture medium. Much more carnosine was synthesized and stored in those explants grown in the presence of the added ß-alanine (Fig. 3) indicating that the enzyme activity was present but that availability of ß-alanine substrate was the limiting factor.

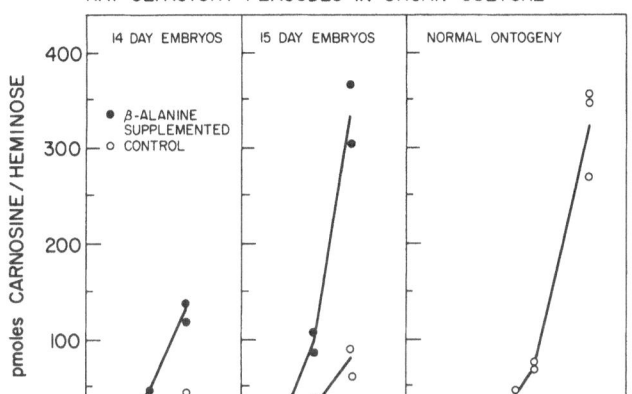

INFLUENCE OF β-ALANINE ON CARNOSINE CONTENT OF RAT OLFACTORY PLACODES IN ORGAN CULTURE

Fig. 3 Normal ontogeny of carnosine in olfactory tissue in vitro and the effect of the addition of 1 mM ß-alanine to culture medium on the carnosine content of rat olfactory mucosa in vitro. Each data point in culture represents a single determination of six pooled explants extirpated at the age indicated and held in culture for the desired period of time. For normal ontogeny, the number of animals per determination ranged from six at early embryonic stages to one for the newborns. (After Margolis et al. 1985a)

Elevated carnosine levels were also observed when pregnant female rats were treated with ß-alanine for 2 days and bulb and mucosa removed from fetuses at 18 and 20 days of gestation (Margolis et al. 1985a). Administration of exogenous ß-alanine also raised carnosine levels in adults, but was without significant effect on the level of carnosine synthetase activity. Thus, taken together, these data argue that carnosine synthetase activity is present during ontogeny prior to the appearance of OMP, suggesting its possible presence in the olfactory neuronal stem cell population. Further, they suggest that during development the availability of the ß-alanine substrate is a critical factor in determining the delay between the appearance of synthetase activity and chemically detectabe carnosine. Our recently generated monoclonal antibodies to this enzyme (Grillo and Margolis 1983, and unpublished data) should permit more detailed studies of these phenomena.

5 Transmitter Amines and Biosynthetic Enzymes

The distribution of various neuropeptides and transmitters, their binding sites, synthetic and degradative enzymes in the olfactory system has been recently reviewed (Halasz and Shepherd 1983; Margolis 1980b, 1981; Macrides and Davis 1983). In the olfactory bulb dopamine is virtually exclusively located in inter-neurons while norepinephrine and serotonin are contained in nerve fibers originat-ing in the locus coeruleus and raphe respectively.

To gain some insight into when these different pathways become biochemically functional, we began a study of their ontogeny and of some of the enzymes related to their metabolism.

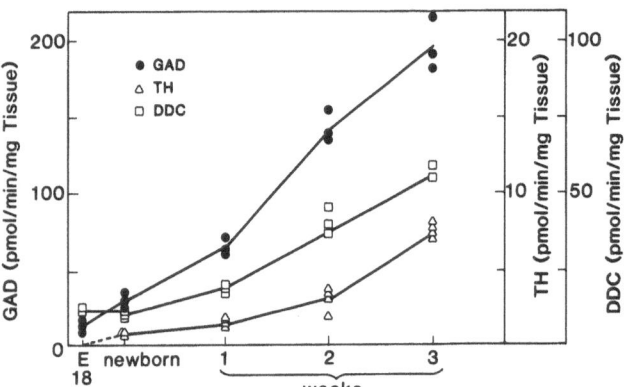

Fig. 4 Developmental changes of neurotrans-mitter enzymes in rat olfactory bulb. Enzyme activities were deter-mined as described in Baker et al. (1984). Glutamic acid decarb-oxylase GAD, tyrosine hydroxylase TH and DOPA-decarboxylase DDC. TH activity was undetectable in bulbs from 18-day embryos

The catecholamines and tyrosine hydroxylase, the rate-limiting enzyme involved in catecholamine synthesis, could not be detected in 18 day embryonic olfactory bulb although DOPA decarboxylase activity (aromatic amino acid decarboxylase activity) could be measured at this age (Fig. 4). TH was first detectable in the newborn and increased progressively thereafter. DOPA decarboxylase activity also rose nearly in parallel but the activity of the latter enzyme was several fold higher than seen for TH. This is consistent with the occurrence of DOPA decarboxylase in a variety of cell types and with TH being the rate-limiting enzyme in catechol-amine biosynthesis. In confirmation of our quantitative measurements is the report of Specht et al. (1981) that in the prenatal rat bulb TH cannot be detected immunocytochemically until embryonic day 21. Our inability to demonstrate the presence of any of the amines prior to birth is also consistent with these observa-tions on the biosynthetic enzymes. The catecholamines are first detectable in the newborn and rise progressively thereafter. In contrast, serotonin is not detect-able in the newborn but seems to plateau while the bulb content of catecholamines is still rising (Fig. 5). Dihydroxyphenylacetic acid, a major metabolite of DA, is seen to accumulate with a somewhat slower time frame than dopamine itself as might be expected. These observations are indicative of selective patterns of bio-chemical differentiation and olfactory tissue innervation by those cell types syn-thesizing the various amines.

The role of these biochemical events in regulating olfactory mediated behavior in developing rats has been studied by the laboratory of Cornwell-Jones (1977, 1979).

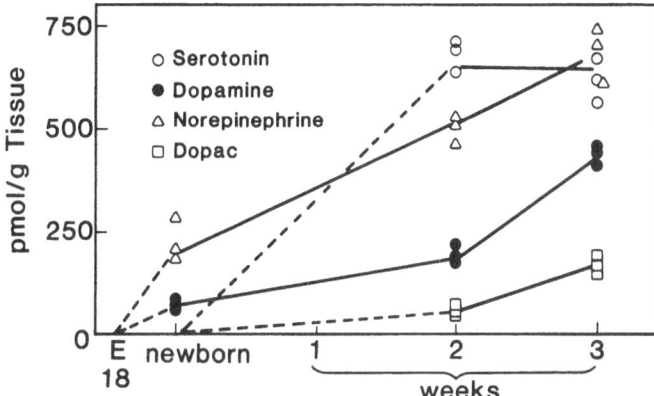

Fig. 5 Transmitter amines in rat olfactory bulbs as a function of age. None of these compounds could be detected in 18-day embryos and serotonin and dopac (dihydroxyphenylacetic acid) were still undetectable in the newborn bulb. Amines were determined as described in Kawano and Margolis (1982)

Additional studies characterizing the ontogeny in olfactory bulb of dopamine and serotonin both biochemically and anatomically are in progress in several laboratories. These include those of H.Baker and M. Shipley. Most recently, Brunjes et al. (1985) have reported on the postnatal development of dopamine and norepinephrine in the rat olfactory bulb. They have also observed a large increase in dopamine levels during early postnatal development. However, in contrast to our data, they find that after the first week postnatal dopamine levels exceed norepinephrine levels.

6 Glutamic Acid Decarboxylase

In contrast to dopamine which is almost exclusively contained in the juxta-glomerular cells GABA is primarily present in the granule cells of the bulb (Ribak et al. 1977; Quinn and Cagan 1981) a cellular compartment which is largely of postnatal origin (Altman 1969). Therefore, it was of interest to monitor the levels of glutamic acid decarboxylase (the enzyme responsible for GABA biosynthesis) during ontogeny for comparison with the catecholamine biosynthetic enzymes. It is evident in Fig. 4 that glutamic acid decarboxylase is detectable as early as embryonic day 18 and increases in a nearly linear fashion over the entire time period studied, during which time the granule cell population is expanding. Thus, the biochemical measures are in concert with the morphological events.

7 Conclusion

Specific biochemical indices of the olfactory bulb and epithelium can be used to characterize regulatory events during ontogeny. On the basis of this the principles of developmental regulation can be compared with those of reconstruction following NS lesions. Such a comparison ultimately could permit the identification of the responsible agents and lead to therapeutic benefits.

Summary

Studies on the ontogeny of the olfactory bulb and mucosa manifest a series of distinct developmental biochemical parameters profiles. This will enable us to evaluate the extent ontogeny and reconstitution following lesion are a reflection of each other.

During ontogeny dopamine and tyrosine hydroxylase activity appear about one week after the olfactory neurons first arrive in the bulb and exert their morphogenetic effects. Similarly, the reinnervation following olfactory nerve lesion is accompanied by a reconstitution of dopaminergic function in the bulb. Clearly the olfactory neurons are capable of inducing dopaminergic expression in susceptible target neurons. As this could be an indication of a more general capability of olfactory neurons one might consider the implications of these observations with regard to development of potential therapeutic benefits, e.g., for Parkinson's disease.

References

Allen WK, Akeson R (1985) Identification of a cell surface glycoprotein family of olfactory receptor neurons with a monoclonal antibody. J Neurosci 5: 284–296

Altman J (1969) Autoradiographic and histological studies of postnatal neurogenesis. IV. Cell proliferation and migration in the anterior forebrain, with special reference to persisting neurogenesis in the olfactory bulb. J Comp Neurol 137: 433–458

Baker H, Kawano T, Margolis FL, Joh TH (1983) Transneuronal regulation of tyrosine hydroxylase expression in olfactory bulb of mouse and rat. J Neurosci 3: 69–78

Baker H, Kawano T, Albert V, Joh TH, Reis DJ, Margolis FL (1984) Olfactory bulb dopamine neurons survive deafferentation induced loss of tyrosine hydroxylase. Neuroscience 11: 605–615

Barbaro D, Fisher DE, Strumeyer DH, Fisher H (1978) Developmental changes and dietary histidine manipulation: Effect on rat olfactory bulb and leg muscle components. J Nutr 108: 1348–1354

Brunjes P, Smith-Crafts LK, McCarty R (1985) Unilateral odor deprivation: Effects on the development of olfactory bulb catecholamines and behavior. Dev Brain Res 22: 1–6

Chuah MI, Farbman AI (1983) Olfactory bulb increases marker protein in olfactory receptor cells. J Neurosci 3: 2197–2205

Farbman AI, Margolis FL (1980) Olfactory marker protein during ontogeny: Immunohistochemical localization. Dev Biol 74: 205–215

Ferriero D, Margolis FL (1975) Denervation in the primary olfactory pathway of mice. II. Effects on carnosine and other amine compounds. Brain Res 94: 75-86

Frosch MP, Dichter MA (1984) Physiology and pharmacology of olfactory bulb neurons in dissociated cell culture. Brain Res 290: 321-332

Gonzales-Estrada MT, Freeman WJ (1980) Effects of carnosine on olfactory bulb EEG, evoked potentials and DC potentials. Brain Res 202: 373-386

Graziadei PPC, Monti Graziadei GA (1978) The olfactory system: A model for the study of neurogenesis and axon regeneration in mammals. In: Cotman CW (ed) Neuronal Plasticity, Raven Press, New York, pp 131-153

Grillo M, Margolis FL (1983) Rabbit muscle carnosine synthetase: Purification and generation of mouse antisera. Soc Neurosci Abstr 9: 1134

Halasz N, Shepherd GM (1983) Neurochemistry of the vertebrate olfactory bulb. Neurosci 10: 579-619

Hinds JW (1972) Early neuron differentiation in the mouse olfactory bulb. I. Light microscopy. J Comp Neurol 146: 233-252

Hinds JW, Hinds PL, McNelly NA (1984) An autoradiographic study of the mouse olfactory epithelium: Evidence for long-lived receptors. Anat Rec 210: 375-383

Horinishi H, Grillo M, Margolis FL (1978) Purification and characterization of carnosine synthetase from mouse olfactory bulb. J Neurochem 31: 909-919

Kawano T, Margolis FL (1982) Transsynaptic regulation of olfactory bulb catechol-amines in mice and rats. J Neurochem 39: 342-348

Kream R, Margolis FL (1984) Olfactory Marker Protein: Turnover and transport in normal and regenerating neurons. J Neurosci 4: 868-879

Kream RM, Davis BJ, Kawano T, Margolis FL, Macrides F (1984) Substance P and catecholaminergic expression in neurons of the hamster main olfactory bulb. J Comp Neurol 222: 140-154

Laitinen PH, Huhtinen RL, Hietala OH, Pajunen AEI (1985) Ornithine decarboxyl-ase activity in brain regulated by a specific macromolecule, the antizyme. J Neurochem 44: 1885-1891

MacLeod NK, Straughan DW (1979) Responses of olfactory bulb neurones to the dipeptide carnosine. Exp Brain Res 34: 183-188

Macrides F, Davis BJ (1983) The olfactory bulb. In: Emson P (ed) Biochemical neuroanatomie. Raven Press, New York, pp 391-426

Marasco E, Cornwell-Jones C, Sobrian SK (1979) 6-Hydroxydopamine reduces pre-ference for conspecific but not other familiar odors in rat pups. Pharmacol Biochem Behav 10: 319-323

Margolis FL (1974) Carnosine in the primary olfactory pathway. Science 184: 909-911

Margolis FL (1975) Biochemical markers of the primary olfactory pathway: A model neural system. In: Agranoff BW, Aprison MH (eds) Advances in neuro-chemistry, vol I. Plenum Press, New York, 193-246

Margolis FL (1980a) A marker protein for the olfactory chemoreceptor neuron. In: Bradshaw RA, Schneider D (eds) Proteins of the nervous system. Raven Press, New York, pp 59-84

Margolis FL (1980b) Carnosine: An olfactory neuropeptide. In: Barker JL, Smith T (eds) Role of peptides in neuronal function. Dekker, New York, pp 545-572

Margolis FL (1981) Neurotransmitter biochemistry of the mammalian olfactory bulb. In: Cagan RH, Kare MR (eds) Biochemistry of taste and olfaction. Academic Press, London New York, pp 369-394

Margolis FL, Grillo M (1984) Carnosine, homocarnosine and anserine in vertebrate retinas. Neurochem Intl 6: 207-209

Margolis FL, Grillo M, Kawano T, Farbman AI (1985a) Carnosine synthesis in olfactory tissue during ontogeny: Influence of exogenous ß-alanine. J Neurochem 44: 1459-1464

Margolis FL, Sydor W, Teitelbaum Z, Blancher R, Grillo M, Rogers K, Sun S, Gubler U (1985b) Molecular biological approaches to the olfactory system. Olfactory marker protein as a model. Chem Sens 10: 163-174

Monti Graziadei GA, Stanley RS, Graziadei PPC (1980) The olfactory marker protein in the olfactory system of mouse during development. Neuroscience 5: 1239-1252

Neidle A, Kandera J (1974) Carnosine – an olfactory bulb peptide. Brain Res 80: 359-364

Nicoll RA, Alger BE, Jahr CE (1980) Peptides as putative excitatory neurotransmitters, carnosine, enkephalin, substance P and TRH. Proc R Soc 210: 133-149

Quinn M, Cagan R (1981) Neurochemical studies of the -aminobutyric acid system in the olfactory bulb. In: Cagan RH, Kare MR (eds) Biochemistry of taste and olfaction. Academic Press, London New York pp 395-415

Ribak CE, Vaughn JE, Saito K, Barber R, Roberts E (1977) Glutamate decarboxylase localization in neurons of the olfactory bulb. Brain Res 126: 1-18

Rochel S, Margolis FL (1980) The response of ornithine decarboxylase during neuronal degeneration in olfactory epithelium. J Neurochem 35: 850-860

Simmons PA, Getchell TV (1981) Neurogenesis in olfactory epithelium: loss and recovery of transepithelial voltage transients following olfactory nerve section. J Neurophysiol 45: 516-528

Sobrian SK, Cornwell-Jones C (1977) Neonatal 6-hydroxydopamine alters olfactory development. Behav Biol 21: 329-340

Specht LA, Pickel VM, Joh TH, Reis DJ (1981) Light-microscopic immunocytochemical localization of tyrosine hydroxylase in prenatal rat brain. II. Late ontogeny. J Comp Neurol 199: 255-276

Tonosaki K, Shibuya T (1979) Action of some drugs on gecko olfactory bulb mitral cell responses to odor stimulation. Brain Res 167: 180-184

Wang RT, Halpern M (1982) Neurogenesis in the vomeronasal epithelium of adult garter snakes. I. Degeneration of bipolar neurons and proliferation of indifferentiated cells following experimental vomeronasal axotomy. Brain Res 237:23-39

Electrophysiological Investigations on the Development of the Olfactory System in Laboratory Mice

U Schmidt

Zoological Institute, University of Bonn, Poppelsdorfer Schloß, 5300 Bonn, West Germany

1 Introduction

The importance of the olfactory sense for rodents from the first hours and days of life onwards has been demonstrated in detail (Teicher and Blass 1977; Brown 1982; Cornwell-Jones and Sobrian 1977; Geyer 1981; Pedersen and Blass 1981; Schmidt et al. 1983; Sczerzenie and Hsiao 1977).

However, few electrophysiological investigations (all on rats) have dealt with the function of olfactory receptors or the olfactory bulb during development (Gesteland et al. 1982; Iwahara et al. 1973; Mair and Gesteland 1982; Math and Davrainville 1980; Salas et al. 1969, 1970; Shafa et al. 1981).

In this paper results of ongoing electrophysiological investigations on the postnatal development of the olfactory bulb in Mus musculus are reported.

2 Methodological Remarks

We investigated the bulbar EEG during postnatal development by recording with low resistance tungsten electrodes, permanently implanted in the OB (U Schmidt et al. 1983; Eckert and U Schmidt 1983), and by recording from single units with KCL-filled glass microelectrodes (Schäfer 1983; Schäfer and U Schmidt in press). In the single unit experiments we attempted to distinguish between mitral cells and interneurons

Although the surgery was performed under Xylocain or Evipan narcosis, the recordings were made in unanaesthetized animals, becaus any anaesthetic drastically alters the neural activity of the OB (C Schmidt 1982).

In the EEG the potentials recorded from the OB were filtered (15–110 Hz band-pass), and stored and analysed by a computer (Nicolet MED-80). In the investigations mentioned here the data-processing program carried out a frequency analysis (Fourier transformation; power spectrum density) of the signals; for data reduction the power spectra were integrated over 5 Hz intervals and plotted as histograms. The integral of energy over the frequency range 15–110 Hz served as a measure of the total energy of the OB (for further technical details see Eckert and U Schmidt 1985)

For evaluating the reaction to odourants in the bulbar EEG experiments, we compared the frequency composition of the power spectra before and during odour stimulation. The stimuli used were nest odour, geraniol 10^{-2} vol%, and butyric acid 10^{-2} vol%. In general, the frequency peak (range of oscillations) shifts during stimulation to lower values by 5–19 Hz than when breathing fresh air. The differences in the power spectra were calculated with different statistical methods (Student's t-test; Mann–Whitney U-test; Duncan's multiple range test; L-test after Page; Eckert 1985; Eckert and U Schmidt 1985). In Fig. 6 the amount of incongruity between the spectra is demonstrated by means of the IM-value (incongruity measurement value), based on an analytical method used in electro-encephalography (Matejcek et al. 1984).

3 Results

3.1 The Bulbar EEG During Postnatal Development

The olfactory EEG of adult mice shows a characteristic energy-time structure (U Schmidt 1978): a respiration synchronous slow potential is superimposed by oscillating potentials that begin at the end of inspiration and continue during most of expiration. These frequency modulated oscillations are in the range of 60–100 Hz.

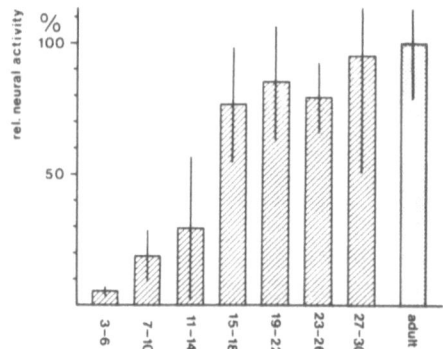

Fig. 1 Development of neural activity in the olfactory bulb in white mice. Abscissa age groups (days; n = 5 – 6 animals / group); ordinate percentage of bulbar activity of the young ones (hatched columns) relative to adult mice (= 100%; dotted column), x̄ ± s. (After U Schmidt et al. 1983)

During the first days of life the typical adult pattern of potentials cannot yet be registered. Up to the end of the second week only irregular, noise-like potentials that increase slowly in energy can be recorded (Fig. 1). Between day 10 and 14 the neural activity increases more rapidly, and at day 15 to 18 about 80% of adult activity is reached (the day of birth is considered as day 1).

The frequency composition of the potentials changes in a characteristic manner. Whereas during the first days only low-frequency components (<40 Hz) are prominent, with increasing age the low frequency region of the power spectra declines gradually and the higher frequency components (>60 Hz) become progressively more pronounced (Fig. 2). In the power spectra two frequency peaks can be distinguished after the second week of life, due to the gradual formation of the oscillatory potentials. The frequency of the oscillations is at first lower, but reaches the frequency range of adult mice between day 18 and 25 (U Schmidt et al. 1983; Eckert and U Schmidt 1985).

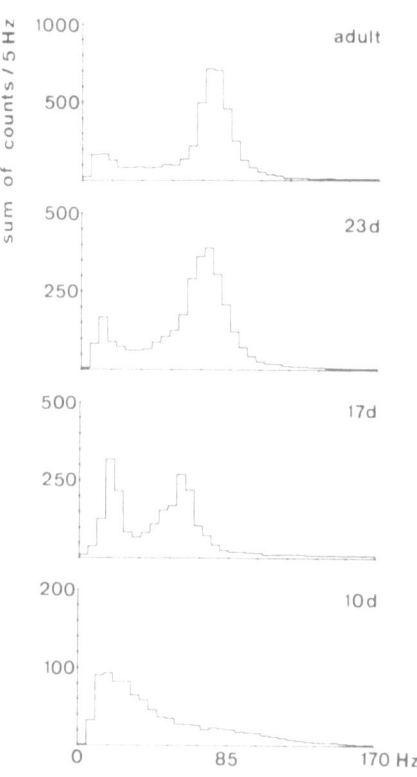

Fig. 2 Frequency content of the bulbar potentials in mice of different age (10, 17, 23 days, adult). The power spectra are integrated over 5-Hz intervals. Abscissa frequency (Hz); ordinate sum of digits / 5 Hz (as a relative measurement of the energy in a given frequency band). Note the different scale at the ordinate

3.2 Single Unit Activity in the Olfactory Bulb During Postnatal Development

In young mice the OB is a very quiet brain region; when a microelectrode is slowly driven through the bulb, only very few spontaneously active neurons can be recorded (Schäfer 1983; Schäfer and U Schmidt in press). Up to the end of the second week, most of these neurons were so weak that they could not be registered long enough to test their reaction to odourants, therefore we did not attempt to investigate mice younger than 10 days. In the age groups 10 to 13 days only 1 to 2 spontaneously active neurons per total bulb penetration were found ($\triangleq$ 0.6 - 1.0 neurons/mm). But at day 14 these values increase rapidly, and at day 15 adult values are reached (ca. 4 units/penetration $\triangleq$ ca. 2 units / mm; Fig. 3). Still more pronounced are the differences when only stable neurons, that could be maintained viable long enough to test odour reactions, are considered. Here the number of neurons per millimeter penetration depth increases from 0.04 at day 10 to 0.1 at day 13, and suddenly to 0.8 at day 15. The time of the sudden increase of neural activity corresponds to the time of increase in energy of the bulbar EEG.

While the number of recordable neurons does not increase after day 15, further neuronal development is especially striking in regard to the spontaneous firing rates. Whereas up to day 15 no neuron could be found with a firing rate > 650 AP/min, by day 25 maximal firing rates reached about 1500 AP/min. These higher frequencies were only found in the stratum glomerulare.

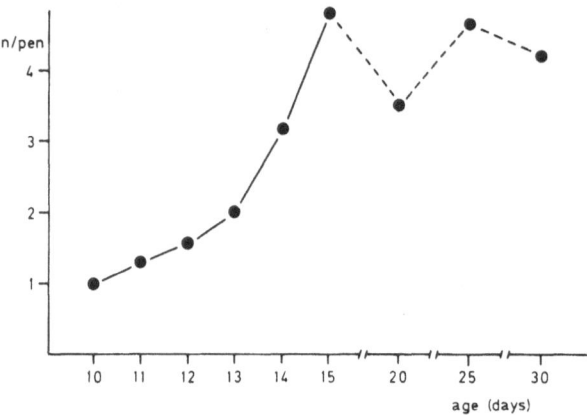

Fig. 3 Number of spontaneously active neurons per bulb penetration. Each day 5 – 6 individuals (28 – 30 penetrations) have been investigated. (After Schäfer 1983)

The single units in young mice show the same patterns of spontaneous activity as in adults. About 60% of the neurons fire continuously with randomly timed spikes, about 27% show irregular spontaneous burst, and about 13% are respiration synchronous.

3.3 Reaction to Odourants

In mice reared under laboratory conditions, slight differences in the power spectra of the potentials recorded from the OB can first be observed at day 15 when stimulated with nest odour; at day 20 significant reactions to all stimuli tested could be recorded.

With single units, responses to odourants could be registered as soon as the neurons were stable enough to be stimulated. As the number of units investigated by us up to day 14 (n = 39) is very small only preliminary trends can be given. The reaction types are the same in young mice as in adults: activations and inhibitions occur to more or less equal extent (20 – 30% of the stimulations), complex reactions are very rare (ca. 6%). Very long responses are fairly common; about 10% of the reactions last longer than 27 s. Figure 4 demonstrates a typical example of a neuron from a 16-day-old mouse that responds with a 30 s activation to a 1 s stimulation with geraniol.

For most of the odourants tested the percentage of responding cells increased slightly with increasing age. Comparing the age groups 11-14 days and 30 days, the percentage increases from 52% to 76% with geraniol, with butyric acid from 38% to 42%, and with grass from 40% to 65%. The percentage of reactions was about the same in young mice (58%) and in adults (56%) only the nest odour.

3.4 Olfactory "Imprinting"

The development of the olfactory system is very much influenced by the olfactory environment in which the pups are raised. Doving and Pinching (1973) report that in rats exposed to cyclo-octanone for 2 weeks to 2 months selective degenerations are found in the OB. This could be explained by a functional alteration in the OB. However up to now the functional significance of these changes is unclear

(Laing 1984). Behavioural data indicate that although rats respond preferentially to odours experienced during early life (Brake 1981; Caza and Spear 1984; Galef and Kaner 1980), the sensitivity to these odours does not change (Laing and Panhuber 1980)

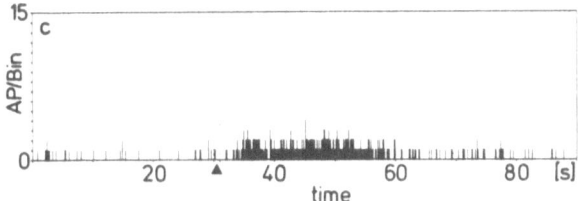

Fig. 4 PST-histogram of a mitral cell responding to geraniol 10^{-2} vol%. The stimulus (duration 1 s) is indicated by a triangle. Ordinate action potentials / bin (bin-width 100 ms)

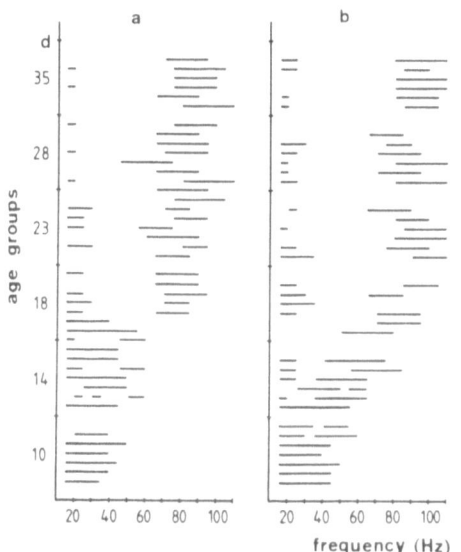

Fig. 5 Development of the main frequency regions of the bulbar EEG in mice. The bars represent 5 - Hz frequency intervals in which the energy exceeds 6.3% of the total energy (frequency range 15-110 Hz). a mice raised under normal laboratory conditions; b mice exposed to geraniol from birth till day of recording

To investigate the effect a strong odour exerts on the electrical activity of the OB, white mice (strain NMRI) were raised from birth in cages odourised with geraniol (Eckert 1985; Eckert and U Schmidt 1985). In different age groups the response to various odourants was studied by means of permanently implanted electrodes (see Section 3.1).

In the first set of experiments, the mice were exposed to geraniol until the day of recordings (group G1-E; experiments at 10, 14, 18, 23, 28 and 35 days post partum). Neural reactions to nest odour and geraniol (10^{-2} vol%) were tested in group G1-E and in normally raised control animals (C1). (for technical details see Eckert 1985)

The neural activity in the controls increased more rapidly than in geraniol exposed mice, but the oscillations appeared earlier in group G1-E. In some day-10 individuals low frequency oscillations could already be recorded, whereas in the controls this stage was not reached until about 4 days later (Fig. 5). The G1-E animals responded to geraniol and nest odour at day 14, but the controls at day 18.

In the controls responses to geraniol were only weak, whereas G1-E-mice responded strongly to the exposure odour at all ages tested (Eckert and U Schmidt 1985).

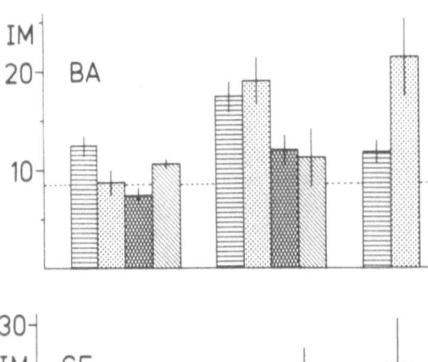
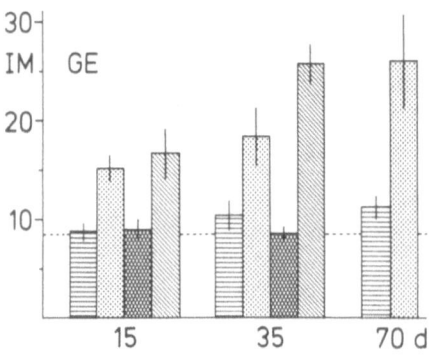

Fig. 6 Neural response to butyric acid **BA** and geraniol **GE** in mice raised under different olfactory conditions. Raised under filtered air **C2** (horizontal hatching); exposed to geraniol from birth till day 14 (G1-14) (punctuated bars); geraniol from birth till day 7 (G1-7) (double hatching); geraniol from day 8 to 13 (G8-13) (oblique hatching). Ordinate IM-value (see Chapter 3.3), $\bar{x} \pm s$; abscissa day of life. (After Eckert 1985)

In the second set of experiments (Eckert 1985) the time of exposure to geraniol was reduced (group G1-14: from birth to day 14; G1-7: from birth to day 7; G8-13: from day 8 to day 13). Except during the exposure period the experimental litters as well as the controls (C2) were kept in cages perfused with filtered air. Group 1-14 and control group C2 were investigated at day 15, 20, 27, 35 and 70, group G1-7 and G8-13 at day 15 and 35.

The neural responses to geraniol in group G1-14 are significantly larger in all age groups than in the controls (U-test, $U \cong 7 \triangleq p < 0.05$). Even at day 70, about 2 months after the last contact with geraniol, these mice show a very marked frequency shift in their bulbar potentials when stimulated with the exposure odour, whereas in C2 only slight responses can be registered.

The results in the groups G1-7 and G8-13 are quite different. Exposure to geraniol during the first seven days of life does not provoke any effect on neural responsiveness, but exposure during the second week leads to the same high reaction values as in G1-14 (Fig. 6).

The results with butyric acid are not quite as consistent. G1-14 shows a slight tendency to give higher values than C2, but there is a large inter-age variation. The higher responses in G1-14 may be due to the fact that butyric acid is an ingredient of the feces and therefore experienced to a higher degree in the exposure cages.

The exposure experiments indicate that in mice the development of the OB's electrophysiological behaviour is crucially dependent on the odours that the animals experience during a distinct phase of their early life. These phenomena resemble imprinting processes, whereby during a critical sensory phase the animal is susceptible to stimuli that lead to stable behaviour patterns. Future behavioural investigations may demonstrate the significance of these olfactory "imprinting".

4 Discussion

In mice the ontogenetic development of the olfactory EEG and of single unit activity in the OB show a striking similarity: The increase in neural activity is very slow up till the 13th day of life, while between day 13 and 15 there is a sudden increase in activity. During this critical period the energy of the EEG increases by about 50%, respiration synchronous oscillations appear in the potentials, and for the first time responses to odourants can be registered; in the single unit studies there is a conspicuous rise in recordable, spontaneous active neurons at this age.

In rats the maturation of the bulb seems to be much more precocious. Iwahara et al. (1973) recorded from day 5 bulbar oscillations ("characteristic rhythmic waves") that increase in frequency until day 30. Mitral cells that were selectively excited by different odourants were found in rat pups already at the day of birth (Mair and Gesteland 1982). The number of spontaneously active neurons starts to increase much earlier in rats (Shafa et al. 1981) than in mice; in 6-day-old rats one unit per bulb penetration could be recorded, the number increases to 3 units at day 10, and by day 16 the adult level is reached (about 4 units / penetration). The frequency of spontaneous active units in the OB of adults seems to be the same in both rodent species, but the course of ontogenetic development is very different.

There is a remarkable discrepancy between the early olfactory performances in mice (pups begin to react to the smell of their nesting material from day 4, U Schmidt et al. 1983) and the late neural responses to odourants. As the accessory system matures earlier than the main system (Pedersen et al. 1983), it could be assumed that the vomeronasal organ is used during the first days for locating biologically important odour sources. However, mice vomeronasal ectomized at day 6, reacted in completely the same manner to nesting material as control animals only one day after surgery (U Schmidt et al. in press). It is likely possible,that a modified glomerular complex, that matures earlier than other regions of the bulb, is responsible for these olfactory performances in rats (Greer et al. 1982) and in mice pups.

In conclusion our electrophysiological data give further indication that the development of the olfactory system is influenced by the odours experienced during the first days of life. Young rodents prefer these odours to novel odour substances (Brake 1981; Cornwell 1975; Galef 1982; Devor and Schneider 1974; Laing and Panhuber 1978). Using the 2-deoxy-D-glucose technique it was shown that the response in the bulb to familiar olfactory cues was enhanced (Coopersmith and Leon 1984). The mechanisms underlying these processes seem to be related to olfactory imprinting as demonstrated with behavioural methods in spiny mice (Porter and Etscorn 1974, 1975, 1976), rats and hamsters (Cornwell-Jones 1979). Both the behavioural studies and the electrophysiological data reveal a sensitive period, when the imprinting effect is most extensive. The long persistence of the preferential neural responses is also a prerequisit for imprinting processes.

Although the remarkable plasticity of the olfactory system during development makes it especially adaptable to environmental changes, the biological significance of these mechanisms remains unknown.

Summary

The postnatal development of the bulbar EEG and of single unit activity in the olfactory bulb was investigated in white mice (NMRI). The data indicate that the neural development of the olfactory bulb is very much influenced by the odorous stimuli an animal experiences during the first days of life. Supporting behavioural and morphological results are presented by Alberts and Coopersmith and Leon in this Vol. The data indicate further that the second week of life is a sensitive period for olfactory imprinting in mice.

Acknowledgements. The investigations were supported by the Deutsche Forschungsgemeinschaft (Schm 322/7).

References

Alberts JR (1986) Postnatal development of olfactory-guided behavior in rodents. In: Breipohl, W Apfelbach R (eds) Ontogeny of olfaction in Vertebrates. Springer

Brake SC (1981) Suckling infant rats learn a preference for a novel olfactory stimulus paired with milk delivery. Science 211: 506-508

Brown RE (1982) Preferences for pre- and post-weanling Long-Evans rats for nest odor. Physiol Behav 29: 865-874

Caza PA, Spear NE (1984) Short-term exposure to an odor increases its subsequent preference in preweanling rats: A descriptive profile of the phenomenon. Dev Psychobiol 17: 407-422

Coopersmith R, Leon M (1984) Enhanced neural response to familiar olfactory cues. Science 225: 849-851

Coopersmith R, Leon M (1986) Neurobehavioral analysis of odor preference development in rodents. In: Breipohl W, Apfelbach R (eds) Ontogeny of olfaction in vertebrates. Springer Verlag

Cornwell CA (1975) Golden hamster pups adapt to complex rearing odors. Behav Biol 14: 175-188

Cornwell-Jones CA (1979) Olfactory sensitive periods in albino rats and golden hamsters. J Comp Physiol Psychol 93: 668-676

Cornwell-Jones CA, Sobrian SK (1977) Development of odor-guided behavior in Wistar and Sprague-Dawley rat pups. Physiol Behav 19: 685-688

Doty RL (1976) Mammalian olfaction, reproductive processes, and behavior. Academic Press, London New York

Devor M, Schneider GE (1974) Attraction to home-cage odor in hamster pups: Specificity and changes with age. Behav Biol 10: 211-221

Doving KB, Pinching AJ (1973) Selective degeneration of neurons in the olfactory bulb following prolonged odor exposure. Brain Res 52: 115-129

Eckert M (1985) Zum Einfluß frühjugendlicher Geruchsbelastungen auf die neuronalen Reizantworten des Bulbus olfactorius bei Labormäusen. Diss, Univ Bonn

Eckert M, Schmidt U (1983) Der Einfluß permanenter Duftreize auf die ontogenetische Entwicklung der neuralen Aktivität des Bulbus olfactorius bei Labormäusen. Verh Dtsch Zool Ges 251

Eckert M, Schmidt U (1985) The influence of permanent odor stimuli on the postnatal development of neural activity in the olfactory bulb of laboratory mice. Dev Brain Res 20: 185-190

Galef BG (1982) Acquisition and waning of exposure-induced attraction to a nonnatural odor in rat pups. Developm Psychobiol 15: 479-490

Galef BG, Kaner HC (1980) Establishment and maintenance of preference for natural and artificial olfactory stimuli in juvenile rats. J Comp Physiol Psychol 94: 588-595

Gesteland RC, Yancey RA, Farbman AI (1982) Development of olfactory receptor neuron selectivity in the rat fetus. Neuroscience 7: 3127-3136

Geyer LA (1981) Ontogeny of ultrasonic and locomotor responses to nest odors in rodents. Am Zool 21: 117-128

Greer CA, Steward WB, Teicher MH, Shepherd GM (1982) Functional development of the olfactory bulb and a unique glomerular complex in the neonatal rat. J Neurosci 2: 1744-1759

Iwahara S, Oishi H, Sano K, Yang KM, Takahashi T (1973) Electrical activity of the olfactory bulb in the postnatal rat. Jpn J Physiol 23: 361-370

Laing DG (1984) The effect of the environmental odours on the sense of smell. In: Animal models on psychopathology. Academic Press, Australia, pp 59-98

Laing DG, Panhuber H (1978) Neural and behavioral changes in rats following continuous exposure to an odor. J Comp Physiol 124: 259-265

Laing DG, Panhuber H (1980) Olfactory sensitivity of rats reared in an odorous or deodorized environment. Physiol Behav 25: 555-558

Leonard CM (1981) Some speculations concerning neurological mechanisms for early olfactory recognition. In: Aslin RN, Alberts JR, Petersen MR (eds) Development of perception, vol I. Academic Press, London New York, pp 383-410

Mair RG, Gesteland RC (1982) Response properties of mitral cells in the olfactory bulb of the neonatal rat. Neurosci 7: 3117-3125

Matejcek M, Neff G, Tjeerdsma H, Krebs E (1984) Pharmaco-EEG-studies with Fluperlapine. Drug Res 34: 114-120

Math F, Davrainville JL (1980) Electrophysiological study on the postnatal development of mitral cell activity in the rat olfactory bulb. Brain Res 190: 243-247

Pedersen PE, Blass EM (1981) Olfactory control over suckling in albino rats. In: Aslin RN, Alberts JR, Petersen MR (eds) Development of perception, vol. I. Academic Press, London New York, pp 359-381

Pedersen PE, Steward WB, Greer CA, Shepherd GM (1983) Evidence for olfactory function in utero. Science 221: 478-480

Porter RH, Etscorn F (1974) Olfactory imprinting resulting from brief exposure in Acomys cahirinus. Nature (London) 250: 732-733

Porter RH, Etscorn F (1975) A primary effect for olfactory imprinting in spiny mice (Acomys cahirinus). Behav Biol 15: 511-517

Porter RH, Etscorn F (1976) A sensitive period for the development of olfactory preference in Acomys cahirinus. Physiol Behav 17: 127-130

Salas M, Guzman-Flores C, Schapiro S (1969) An ontogenetic study of olfactory bulb electrical activity in the rat. Physiol Behav 4: 699-703

Salas M, Schapiro S, Guzman-Flores C (1970) Development of olfactory bulb discrimination between maternal and food odors. Physiol Behav 5: 1261-1264

Schäfer HJ (1983) Elektrophysiologische Untersuchungen zur ontogenetischen Entwicklung der olfaktorischen Sekundarneurone bei der Labormaus. Diss, Univ Bonn

Schäfer HJ, Schmidt U (in press) Ontogenetic development of secundary neurons in the olfactory bulb of laboratory mice. J Comp Physiol

Schmidt C (1982) Behavioural and neurophysiological studies of the olfactory sensitivity in the albino mouse. Z Säugetierkd 47: 162-168

Schmidt U (1978) Evoked-potential measurements of olfactory thresholds of laboratory mice (Mus musculus) to carboxylic acids. Chem Sens Flav 3: 177-182

Schmidt U, Eckert M, Schäfer HJ (1983) Untersuchungen zur ontogenetischen Entwicklung des Geruchssinnes bei der Hausmaus (Mus musculus). Z Säugetierkd 48: 355-362

Schmidt U, Schmidt C, Wysocki CJ (in press) Der Einfluß des Vomeronasalorgans auf das olfaktorisch geleitete Verhalten nestjunger Mäuse. Z Säugetierkunde

Sczerzenie V, Hsiao S (1977) Development of locomotion toward home nesting material in neonatal rats. Developm Psychobiol 10: 315-321

Shafa F, Shineh SN, Bidanjiri A (1981) Development of spontaneous activity in the olfactory bulb neurons of postnatal rats. Brain Res 223: 409-412

Teicher MH, Blass EM (1977) First suckling response of the newborn albino rat: The roles of olfaction and amniotic fluid. Science 198: 635-636

The Effect of Long Duration Postnatal Odour Exposure on the Development of the Rat Olfactory Bulb

H Panhuber

CSIRO, Division of Food Research, P.O. Box 52, North Ryde, NSW, 2113 Australia

1 Introduction

In the last 20 years it has been clearly demonstrated that experience can have an effect on the structure and function of the brain (Harlow and Harlow 1964; Diamond et al. 1964; Rosenzweig 1971). In a recent review, Walsh (1981) cited evidence which showed that exposing young animals to enriched or impoverished environments produced a wide range of anatomical modifications to the brain. These included changes in brain weight and size, number and size of neurons, and the number of synapses and dendritic branches. Studies using exposure to restricted sensory environments have yielded important information about the structure, function and development of parts of the central nervous system (Wiesel and Hubel 1963; Blakemore 1973; Webster and Webster 1977). For example, Blakemore (1973) described how the orientation neurons of the visual cortex of kittens could be tuned to respond predominantly to vertical lines, by exposing visually deprived 4-week-old kittens to an environment of black and white vertical stripes for 6 h. An analogous effect was found in the olfactory system by Doving and Pinching (1973). They found that exposing 14-day-old rats to a single odorant for 2 or more weeks caused mitral cells in specific regions of the olfactory bulb to appear smaller and more deeply stained than in control rats.

Six groups have published results of the effects of long duration odour exposure (Doving and Pinching 1973; Pinching and Doving 1974; Laing and Panhuber 1978; Panhuber 1984; Panhuber et al. 1985; Laing et al. 1985; Jourdan et al. 1980; van As et al. 1980; Dalland and Doving 1981; Cunzeman and Slotnick 1984; Eckert and Schmidt 1985). Of these only the study by Pinching and Doving (1974) unequivocally claimed that exposure to different odours produced different, odour specific patterns of mitral cell alteration. The inconsistency of the results obtained by different laboratories and even in the same laboratory have been attributed to differences in methodology (Laing 1984), and to the lack of robustness of the "exposure-effect" (Cunzeman and Slotnick 1984). In addition, a number of other factors could be involved in producing these inconsistencies. One is that exposure to air-borne chemicals may produce general neurological disorders as well as temporary olfactory deficits, thereby obscuring the purely olfactory effect of the chemical (see reviews by Prockop 1979; Halpern 1982). This does not become a problem if low concentrations of low toxicity chemicals are used. A second factor is that olfactory receptor neurons can be replaced by the differentiation of the basal cells of the epithelium following trauma. In addition, the receptor cells and their axons are generally believed to be continually replaced (Graziadei et al. 1980; Halpern 1982), although this concept has recently been challenged (Hinds et al. 1984; Breipohl et al. 1985). How receptor cell turnover, either following trauma, or as a continual process might effect the results of the odour-exposure experiments is not as yet clear. A third and the most likely factor could be the interaction between the environment and the developing olfactory system. This interaction and its affect on the results of the odour-exposure experiments will be discussed from two points of view, firstly, the importance of the age at which exposure commences, secondly, how previous olfactory experience and genetic variability may explain the poor reproducibility of the maps of altered and normal cells.

2 Age of Starting Exposure: a Possible Critical Developmental Period

When 14-day-old rats are reared in a continuous stream of deodorized air for 2 or 4 months, almost all mitral cells appear densely stained (Nissl staining) and shrunken (Laing and Panhuber 1978). If a pure chemical odorant is added to the deodorized air stream, then only specific regions of mitral cells appear to be altered (Doving and Pinching 1973; Pinching and Doving 1974; Laing and Panhuber 1978; Dalland and Doving 1981). Similar changes are also observed when adult rats are subjected to these environments for two months (Panhuber unpublished observations). The most characteristic feature of altered mitral cells was their dense staining (Fig. 2), which Doving and Piching (1973) likened to the appearance of mitral cells following transneuronal degeneration (Pinching and Powell 1971). However, when rats are exposed from 1 day of age for 2 months to cyclohexanone, no dense staining occurs (Panhuber 1984); this raises the question of whether the first 2 postnatal weeks are critical in the development of the exposure effect.

The rats' olfactory system is at least partly functional at birth (Teicher and Blass 1976; Ruddy and Cheatle 1977; Sczerzenie and Hsiao 1977) and the complex feedback circuitry and centrifugal inhibitory input to the olfactory bulb develops during the first three postnatal weeks (Hinds and Hinds 1976; Grafe 1983; Mair and Gesteland 1982; Bayer 1983; Schwob and Price 1984). Although mitral cells are similar in appearance at birth and in the adult (Singh and Nathaniel 1977), their postnatal electrophysiological activity reflects the input of the receptor cells and is dissimilar to that of the adult (Mair and Gesteland 1982; Math and Davrainville 1980). Such findings indicate that during the first few weeks after birth the olfactory system may only function in a limited, integrative and relay capacity. For example, Eckert and Schmidt (1985) found that the "total neural activity" recorded from the centres of the olfactory bulbs of normal mice only showed a response to the biologically meaningful odour of "nest material" at 18 days, which reached maximum amplitude by 35 days postnatal. No response was obtained to the odour of geraniol at day 18 except with mice that had previously been exposed to that odour from birth. The "total neural activity" recorded from the centre of the olfactory bulbs of 18 day old mice (Eckert and Schmidt 1985), may therefore represent the earliest output of the bulb that is modulated by the developing centrifugal pathways and their associated inhibitory circuitry. Such a view would agree with the findings of Math and Davrainville (1980), who suggest-ed that the excitatory pathways in the rat olfactory bulb mature during the first postnatal week, after which the inhibitory system originating in the granule cells becomes predominant. As mitral cells are the main recipients (via the granule cells) of the centrifugal input to the olfactory bulb (Fig. 1) and in turn connect to the main emotional and appetitive centres of the brain (MacLeod 1971; Shepherd 1974; Price 1977; Macrides et al. 1981), the nature of the initially unmodulated olfactory input to these higher centres may affect their development and in turn the development of the centrifugal input to the olfactory bulb. The olfactory environment after birth may also directly affect the formation of synapses between granule and mitral cells (Fig. 1) which in the mouse peaks during days 12-16 (Hinds and Hinds 1976), about the same time as the granule cell dendrite spines develop in the rat (Greer 1985). In addition the adult pattern of ^{14}C 2-deoxy-glucose (2DG) uptake in the glomeruli of the rat olfactory bulb only becomes established by day 15 (Greer et al. 1982). This suggests that the typical pattern of adult 2DG activity in the glomerular layer depends on the completion of the glomerular wiring, in particular the development of the periglomerular cell dendrites which are almost complete by day 13 (Bayer 1983).

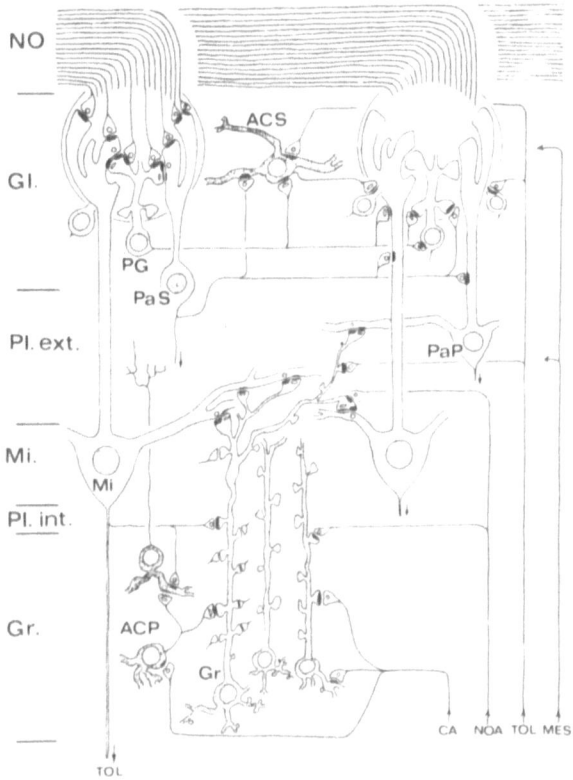

Fig. 1 Schematic diagram illustrating the main neuron types of the olfactory bulb and their synaptic connections (Holley and MacLeod 1977). The different layers of the bulb are indicated on the left of the diagram: **NO** olfactory nerve layer; **Gl** glomerular layer; **Pl. ext.** plexiform layer; **Mi** mitral cell layer; **Pl. int.** plexiform layer; **Gr** granule cells layer. Second order neurons: **Mi** mitral cells; **PaS** superficial tufted cells; **PaP** deep tufted cells. Interneurons: PG periglomerular cells; **Gr** granule cells; **ACS** superficial short axon cells; ACP deep short axon cells. Centrifugal fibres: **CA** from the anterior commissure; **NOA** from the anterior olfactory nucleus; **TOL** fibres from the lateral olfactory tract; **MES** from the mesencephalon. Centripetal fibres: TOL fibres to the lateral olfactory tract

On the basis of the above evidence, the development of the olfactory system of the rat and mouse may be divided into two broad phases. The first is the output phase, where the axons of mitral and tufted cells make connections with higher olfactory centres, this commences before birth and continues through the first 9 days postnatal (Bayer 1983; Schwob and Price 1984). The second phase, which occurs during the first three weeks after birth is the development of the interneuronal circuitry an the centrifugal input to the olfactory bulb (Bayer

1983). When mitral cells are deprived of peripheral excitatory input during the second phase, the higher brain centres may not make the appropriate connections back to the interneurons of the olfactory bulb. Of these interneurons, the granule cells which have been shown to directly suppress mitral cell activity (Nakashima et al. 1978), and the periglomerular cells are the main targets for the centrifugal inhibitory input to the bulb (Price and Powell 1970; Price 1977; Davis and Macrides 1981). The "staining effect" (Fig. 2) observed when exposure to an odour (Pinching and Doving 1974; Laing and Panhuber 1978) or deodorized air (Laing and Panhuber 1978; Panhuber 1984) is commenced after postnatal day 14 could therefore be due to the normal activity of granule cells which had developed before day 14. As no staining effect was observed when exposure was commenced at 1 or 2 days of age (Fig. 2), this suggests that the development of the inhibitory circuitry and possibly the centrifugal input of the olfactory bulb are susceptible to environmental influence during the first 2 weeks of the rats life. In addition, exposure of mice to the predominant odour of geraniol from birth increases the neural activity of the bulb in response to stimulation with that odour and brings forward the onset of a response from day 18 to day 14 postnatal (Eckert and Schmidt 1985).

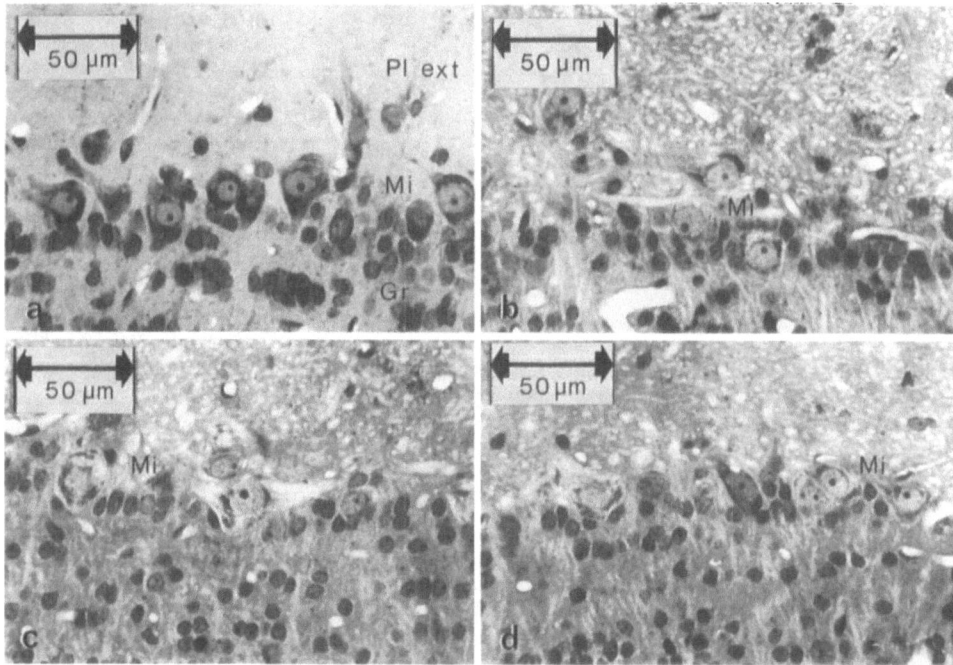

Fig. 2 Light micrographs of a "altered" (deeply stained and shrunken) and b "normal" (cells similar in appearance to those from a normal rat) mitral cells from rats exposed to cyclohexanone from 14 days of age for 2 months. Large mitral cells c and small mitral cells d from rats exposed to cyclohexanone from 1 day of age for 2 months. Epon embedded; 5 µm sectioned; stained with Methylen Blue: Azure II (50:50)

Further evidence that deprivation influences the development of the olfactory bulb comes from studies which have used the technique of unilateral naris closure (Meisami 1976). Again, it has been shown that the first 30 days of the rat's life are critical for the normal development of the olfactory bulb (Brunjes and Borror 1983). Closure of one naris from birth for 30 days significantly reduces the size and various biochemical growth parameters (Meisami 1976; Meisami and Mousavi 1982) of the olfactory bulb. No such changes occurred when one naris was closed at 30 or 60 days of age for 30 days (Brunjes and Borror 1983). The differences in growth between the deprived and the normal bulb only became evident after 10 days of age, and remained stable after 25 days of age (Meisami and Mousavi 1982). Because this period between 10 and 25 days of age coincides with the development of the granule cell dendrite spines (Greer 1985), it is tempting to speculate that the differences between the normal and the deprived bulbs are due to the incomplete development of the centrifugal input to the bulb via the granule cells. Benson et al. (1984) found that there were a significantly fewer mitral to granule, and granule to mitral synapses, on mitral cells from the deprived side than the normal side. They suggested that this meant that olfactory sensation plays an important part in the formation of the inhibitory synapses in the olfactory bulb.

Few observations on changes in the number, size and staining of mitral cells have been made using the technique of nare closure. Of these, Meisami and Safari (1981) found that the number of mitral cells decreased after nare closure at birth, while Benson et al. (1984) reported a reduction in the size of mitral cells but not their number. Mitral cells from the deprived bulbs had a normal appearance (lightly stained cytoplasm), even though they were significantly smaller than those of the non-deprived side (Benson et al. 1984). This observation is in agreement with the outcome of experiments where rats were exposed to deodorized air from 1 day old for 2 months (Panhuber 1984; Laing et al. 1985). In both types of experiment, mitral cells deprived of their peripheral excitatory input from day 1 were smaller than those of control rats, but were not deeply stained.

3 Patterns of Altered Cells Produced by Odour-Exposure

Another aspect of the controversy surrounding the "odour-exposure" effect has been the difficulty in reproducing maps of the patterns of altered and normal cells obtained by exposing 14-day old rats to a single predominant odour for 2 months or more. Pinching and Doving (1974) found that different but overlapping patterns of altered mitral cells occurred when rats were exposed to different odours. These patterns (Fig. 3), however, varied considerably when exposure to some of these odours was repeated in later experiments (Laing and Panhuber 1978; Dalland and Doving 1981). In contrast, Laing and Panhuber (1978) found that the patterns for the different exposure odours were indistinguishable in "blind" subjective studies but histological material from odour-exposed rats could readily be distinguished from that of rats exposed to deodorized air or a rat colony environment.

Slight differences in the fixation procedures, (Panhuber et al. 1985), or the effects of post-mortem trauma (Cammermeyer 1978) may influence the formation of odour-induced patterns of heavily stained and normal cells. An alternative explanation is that the pre-exposure environment during the first two weeks, and/or the genetic make-up of the rats may influence the development of these patterns. Results of electroencephalograms recorded from the olfactory bulb (Freeman 1980) can be taken as support for this explanation. From these recordings it appears that the odour-induced bulbar electroencephalogram is not a function of the odour itself, but is,"...shaped by the initial conditions, the sensory input, the chemical state of the bulb from centrifugal input, and the record of past input embedded in synaptic connections" (Freeman 1983). Similarly, the exposure-induced patterns of altered (densely stained) and normal mitral cells may be a topographic representation of the meaning of an odour to the rat, and not simply a topographic coding of odour quality as suggested earlier (Doving and Pinching 1973; Pinching and Doving 1974; Laing and Panhuber 1978).

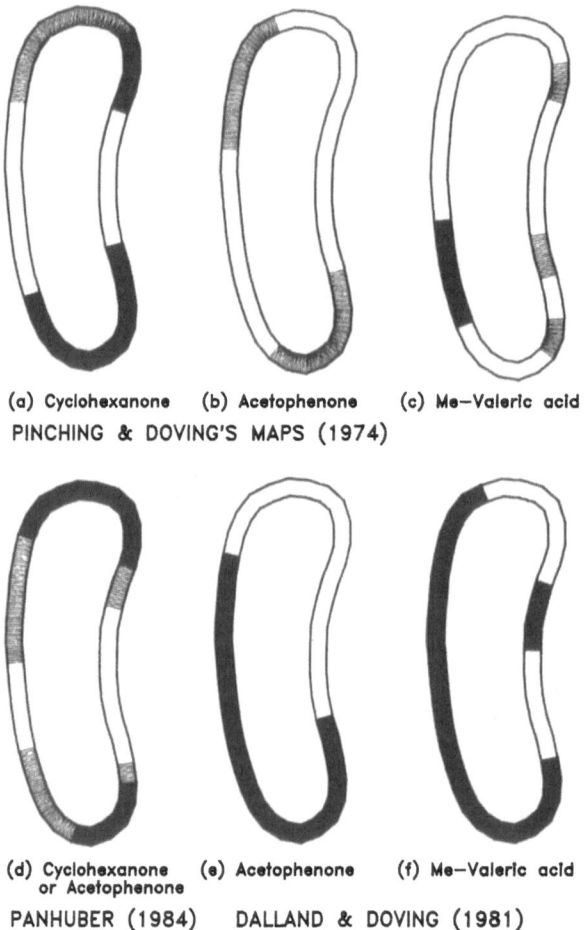

(a) Cyclohexanone   (b) Acetophenone   (c) Me—Valeric acid

PINCHING & DOVING'S MAPS (1974)

(d) Cyclohexanone   (e) Acetophenone   (f) Me—Valeric acid
or Acetophenone

PANHUBER (1984)   DALLAND & DOVING (1981)

Fig. 3 Diagrams of representative coronal sections taken mid-way through the rostro-caudal extent of the olfactory bulb. These diagrams are examples of maps of "altered" (deeply stained and shrunken) and "normal" mitral cells from rats exposed from 14 days of age for 2 months to the three odours as indicated and from different authors as shown. No shading represents regions of "normal" cells; partly shaded regions with a mixture of "normal" and "altered" cells; black regions with all mitral cells altered. Maps are oriented with dorsal to the top of the page and medial to the right

A target for these environmentally induced changes may be the overproduction of granule cell dendritic spines, which reaches a peak in the rat by postnatal day 12 (Greer 1985). The subsequent reduction in spine numbers during the third week, it was suggested (Greer 1985), reflects a stabilization of the reciprocal synaptic contacts between granule cells and the mitral and tufted cells. It is the stabilization of these inhibitory and feedback connections between granule and mitral cells that may be influenced by the continuous exposure of 14 day old rats to one predominant odour for 2 months. Regional differences in the relative strength of the reciprocal, mitral-granule synapses may have therefore given rise to the patterns of altered and normal mitral cells observed when exposure was commenced on postnatal day 14. Commencing the exposure regime on day 1 postpartum could alter the development of this initial overproduction of spines either over the whole bulb or in specific regions, depending on whether the rats had been exposed to deodorized air, or to an odour. In fact, recent work done in collaboration with Breipohl, Rehn and Laing (personal communication) has shown that rats exposed to cyclohexanone from day 1 have fewer mitral/granule cell reciprocal synapses than "normal" rats.

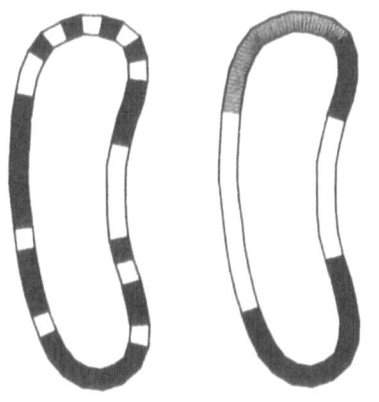

(a) Based on size   (b) Based on staining

CYCLOHEXANONE MAPS

Fig. 4 Coronal representations of the mitral cell layer from the rat olfactory bulb. Details of location and shading are as for Fig. 3. Regions of cells significantly larger than for the same regions of the bulbs of rats exposed to deodorized air are shown in white (a, b), these are equivalent to the regions of "normal" cells in b and d. Maps a and c are from day 1 rats, exposed for 2 months; b and d are maps of "altered" and "normal" cells from rats exposed from day 14 for 2 months (Pinching and Doving 1974)

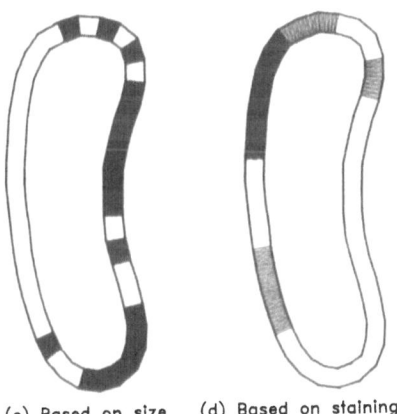

(c) Based on size   (d) Based on staining

AMYL ACETATE MAPS

4 Exposure-Induced Changes in Mitral Cell Size

Because of the difficulties in mapping the patterns of altered mitral cells, an objective method based on cell size was developed (Panhuber et al. 1985). Maps of the differences in mitral cell size between rats exposed to an odour and to deodorized air (Fig. 4a,d), obtained using this method (Panhuber and Laing

1985), were dissimilar to the earlier maps (Fig. 4d,b) produced by Pinching and Doving (1974). This method of measuring cell size also showed that there were regions of mitral cells significantly larger than "normal" (Panhuber and Laing 1985). One possible reason for these dissimilarities is that the earlier maps were based on a subjective assessment of staining and cell size while the later maps were based soley on measurements of cell size. An alternative explanation, however, is that the age at which exposure commenced (day 1-size differences; day 14-staining differences) produced diverse developmental changes (see Sect 3). Taking into consideration that mitral cell activity during the first 2 weeks mainly reflects the activity of the receptor cells (Math and Davrainville 1980; Mair and Gesteland 1982), the patterns of differences in cell size from rats exposed from day 1 (Fig. 4a,c) may represent a topographic coding of odour quality by the mitral cells. Those cells that were stimulated by the exposure odour could develop a normal, or even more extensive interneuron connectivity (cells larger than "normal") than those deprived of stimulation.

5 Behavioural Effects of Exposure to an Odour or Deodorized Air

There is a considerable but confusing literature on the effects that early olfactory environment can have on the subsequent olfactory guided behaviour of an animal. Early exposure to various odours (pure chemicals or conspecific odours) has been shown to increase the olfactory preference for these odours in most rodents (Leon 1975; Marr and Lilliston 1970; Porter and Etscorn 1976; Carter 1972; Cornwell 1975, 1976; Leon et al. 1984). Although exposure to a novel odour influences later preference for that odour, the precise time periods for the acquisition and exhibition of that preference are not clear. For example, Marr and Lilliston (1970), Bronstein and Crockett (1976) and Cornwell-Jones (1979) found that exposure to a novel odour at various times before weaning will produce preferences for that odour when tested up until weaning, but not into adulthood. Hennesy et al. (1977) and Leon et al. (1977) found that pre-weaning exposure can also effect post-weaning preferences. In addition, pre-weaning exposure can also effect adult social behaviour (Marr and Gardner 1965).

One explanation for these contradictory results comes from the work of Galef and Kaner (1980), who showed that the establishment and maintenance of odour preference by young rats is the result of one of two separate processes: simple exposure or associative learning. For example, the use of the young rat's bedding material as an odour source by Cornwell (1976; Cornwell-Jones 1979) could have led to a specific association of the bedding material's smell with the availability of food, an association that was no longer relevant to the rat pups' welfare after weaning. Another developmental change that occurs before weaning which could affect the outcome of the associative type of odour exposure is the change in motivation for suckling. This changes from a reflex reaction to an appetite-controlled behaviour and occurs at 10 to 15 days of age (Nock et al. 1978); a period which coincides with the peak development of the granule/mitral cell reciprocal synapses (Hinds and Hinds 1976; Greer 1985).

Although changes in preference have also been found following exposure to a single odorant or purified air (Laing and Panhuber 1978), the results obtained do not answer the question of whether exposure to an odour or deodorized air significantly alters the ability of exposed rats to smell the exposure odour. To answer

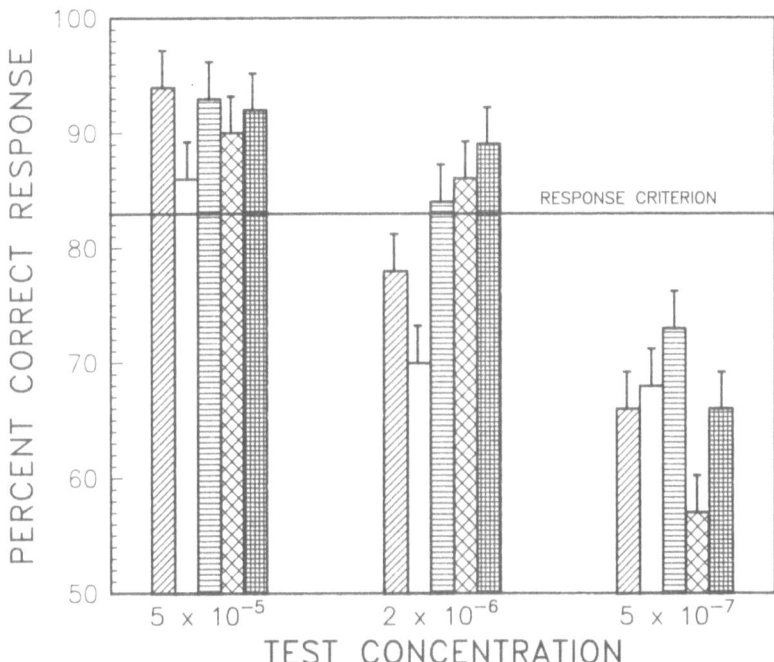

Fig. 5 Histogram of olfactory acuity data for cyclohexanone at three test concentrations (Panhuber 1984). Rats performed at a level significantly better than
chance when they scored better than 83% correct responses (response criterion).
Test concentrations of cyclohexanone are given as fractions of saturated vapour
pressure (SVP). The age at which exposure commenced (AGE) and the treatments
tested are given below the histograms together with a key to the shading. The
results for rats exposed to deodorized air from day 1 and day 14 were similar and
therefore combined. Duration of odour exposure was 2 months. No. of rats refers
to the total number of rats from which the experimental data was combined for
each grouping in the histogram

this question a simple two choice method was devised to measure the rats' ablity
to discriminate between air and deodorized air at decreasing concentrations of an
odour (Laing and Panhuber 1978, 1980). The results obtained using this method
over a number of years and for different groups of rats are presented in Fig. 5.
These combined data show that although rats exposed from day 1, or day 14
exhibited different anatomical changes, their sensitivity to the exposure odour was
the same. Rats exposed to deodorized air performed significantly worse than

odour-exposed or normally reared rats, whereas rats exposed to an odour performed better than did rats exposed to deodorized air, or normally reared rats.

That the olfactory acuity of rats exposed to deodorized air from day 1 or day 14 were indistinguishable suggests that the anatomical differences between these groups had no dramatic functional significance for the rats. In fact, the small absolute differences in olfactory acuity between "normal" rats, rats exposed to one predominant odour, and rats exposed to deodorized air (Fig. 3) indicates that either the anatomical changes caused by such exposure (Fig. 2) have no dramatic effect on the rats' olfactory acuity, or that there is a rapid recovery from the effects of exposure. This functional recovery could occur during the 1 to 2 months needed to train and test the subject rats (Laing and Panhuber 1978).

6 Conclusions

Exposure of young rats to an environment of deodorized air or one containing a single predomonant odour produces two distinct types of changes to the mitral cells.

1. When exposure commences on postnatal day 1, both, increases and decreases in cell size occur, when compared to "normal" rats.

2. If exposure commences on or after postnatal day 14, many or all mitral cells take on an appearance similar to that following transneuronal degeneration; they are densely stained and shrunken.

From other evidence (Sect 2), it would appear that exposure from postnatal day 1 prevents the normal development of the centrifugal and inhibitory input to the mitral cells. The degenerative-like staining observed after rats have been exposed from postnatal day 14 or older, for 2 months, appears to be due to the suppression of mitral cell activity by the granule cells. Maps of the patterns of altered and normal mitral cells obtained from rats exposed to an odour from postnatal day 14 may represent a template, or search image of the odour, based on previous experience (Freeman 1983). The poor reproducibility of these patterns could be a function of their susceptibility to environmental and possibly genetic influence.

Although a considerable amount of evidence has been presented to support the above conclusions, they still need to be rigorously tested. To what extent the centrifugal and inhibitatory cicuitry of other senses and other parts of the central nervous system follow similar developmental scenarios to that outlined here for the olfactory system still needs to be explored. The possibility that the inhibitory pathways of the central nervous system are susceptible to the environmental modification during development however needs to be considered when studying the influence of the environment on the developing senses.

Summary

Continuous exposure of young rats to an odour or deodorized air for 2 months produces two distinct types of effects in the olfactory bulb depending on whether exposure commences on day 1 (increase or decrease in mitral cell size), or day 14 (dense-staining and shrinkage of mitral cells). New evidence will be reviewed which shows that the first 3 weeks of a rats' postnatal life are critical in the development of the olfactory bulbs' interpretative and memory roles. The hypothesis is developed that it is mainly the centrifugal input and the inhibitory circuitry of the olfactory bulb of the rat that are affected by early postnatal experience.

References

As W van, Smit KGJ, Köster EP (1980) Effect of long-term odor exposure on mitral cells of the olfactory bulb in rats. In: Starre H van der (ed) Olfaction and taste, vol VII. London, IRL Press, p 296

Bayer SA (1983) ^{3}H-Thymidine-radiographic study of neurogenesis in the rat olfactory bulb. Exp Brain Res 50: 329-340

Benson TE, Ryugo DK, Hinds JW (1984) Effect of sensory deprivation on the developing mouse olfactory system: a light and electron microscope morphometric analysis. J Neurosci 4: 638-653

Blakemore C (1973) Developmental factors in the formation of feature extracting neurones. In: Schmitt FO, Werden FG (eds) The neurosciences, 3rd Study Program. MIT Press, Cambridge Mass, pp 105-113

Breipohl W, Rehn B, Molyneux GS, Grandt D (1985) Plasticity of neuronal cell replacement in the main olfactory epithelium of the mouse. Abstr XII Int Anat Congr, London

Bronstein PM, Crockett DF (1976) Exposure to the odor of food determines the eating preferences of rat pups. Behav Biol 18: 387-392

Brunjes PC, Borror MJ (1983) Unilateral odor deprivation: different effects due to time of treatment. Brain Res Bull 11: 501-503

Cammermeyer J (1978) Is the solitary dark neuron a manifestation of post-mortem trauma to the brain inadequately fixed by perfusion. Histochemistry 56: 97-115

Carter CS (1972) Effects of olfactory experience on the behaviour of the Guinea-pig. Anim Behav 20: 54-60

Cornwell CA (1975) Golden Hamster pups adapt to complex rearing odors. Behav Biol 14: 175-188

Cornwell CA (1976) Selective olfactory exposure alters social and plant odor preferences of immature hamsters. Behav Biol 17: 131-137

Cornwell-Jones CA (1979) Olfactory sensitive periods in Albino rats and Golden Hamsters. J Comp Physiol Psychol 93: 668-676

Cunzeman PJ, Slotnick BM (1984) Prolonged exposure to odors in the rat: effects on odor detection and on mitral cells. Chem Senses 9: 229-239

Dalland T, Doving KB (1981) Reaction to olfactory stimuli in odor-exposed rats. Behav Neural Biol 32: 79-88

Davis BJ, Macrides F (1981) The organization of centrifugal projections from the anterior olfactory nucleus, central hippocampal rudiment, and piriform cortex to the main olfactory bulb in the hamster: an autoradiographic study. J Comp Neurol 203: 475-493

Diamond MC, Krech D, Rosenzweig MR (1964) The effect of an enriched environment on the histology of the rat cerebral cortex. J Comp Neurol 123: 111-119

Doving KB, Pinching AJ (1973) Selective degeneration of neurones in the olfactory bulb following prolonged odour exposure. Brain Res 52: 115-129

Eckert M, Schmidt U (1985) The influence of permanent odor stimuli on the postnatal development of neural activity in the olfactory bulb of laboratory mice. Dev Brain Res 20: 185-190

Freeman WJ (1980) EEG analysis gives model of neuronal template-matching mechanism for sensory search with olfactory bulb. Biol Cybern 35: 21-37

Freeman WJ (1983) The physiological basis of mental images. Biol Psychol 18: 1107-1124

Galef BG, Kaner HC (1980) Establishment and maintenance of preference for natural and artificial olfactory stimuli in juvenile rats. J Comp Physiol Psychol 94: 588-595

Grafe MR (1983) Developmental factors affecting regeneration in the central nervous system: early but not late formed mitral cells reinnervate olfactory cortex after neonatal tract section. J Neurosci 3: 617-630

Graziadei PPC, Karlan MS, Monti-Graziadei GA, Bernstein JJ (1980) Neurogenesis of sensory neurons in the primate olfactory system after section of the fila olfactory. Brain Res 186: 289-300

Greer CA (1985) A quantitative golgi analysis of granule cell development in the neonatal rat olfactory bulb. Abstract 7th Annu Meet Assoc Chemorecept Sci, April 24-28 Sarasota Florida

Greer CA, Steward WB, Teicher MH, Shepherd GM (1982) Functional development of the olfactory bulb and a unique glomerular complex in the neonatal rat. J Neurosci 2: 1744-1759

Halpern BP (1982) Environmental factors affecting chemoreceptors: an overview. Environ Health Perspect 44: 101-105

Harlow HF, Harlow MK (1964) The effect of rearing conditions of Social Behavior. Bull Menninger Clin 26: 213-224

Hennessy MB, Smotherman WP, Levine S (1977) Early olfactory enrichment enhances later consumption of novel substances. Physiol Behav 19: 481-483

Hinds JW, Hinds PL (1976) Synapse formation in the mouse olfactory bulb. 1. Quantitative studies. J Comp Neurol 169: 15-40

Hinds JW, Hinds PL, McNelly NA (1984) An autoradiographic study of the mouse olfactory epithelium: Evidence for long-lived receptors. Anat Rec 210: 375-383

Holley A, MacLeod P (1977) Transduction et codage des informations olfactive chez les vertebres. J Physiol Paris 73: 725-828

Jourdan F, Holley A, Glaso-Olsen G, Thommesen G, Doving KB (1980) Comparison between the patterns of selective degeneration and marking with ^{14}C 2-deoxy-glucose in the rat olfactory bulb. In: Starre H van der (ed) Olfaction and taste vol VII. IRL Press, London p 295

Laing DG (1984) The effect of environmental odours on the sense of smell. In: Bond N (ed) Animal models in psychopathology. Academic Press, Australia pp 59-98

Laing DG, Panhuber H (1978) Neural and behavioral changes in the rats following continuous exposure to an odor. J Comp Physiol A 124: 259-265

Laing DG, Panhuber H (1980) Olfactory sensitivity of rats reared in an odorous or deodorized environment. Physiol Behav 25: 555-558

Laing DG, Panhuber H, Pittman EA, Willcox ME, Eagleson GK (1985) Prolonged exposure to an odor or deodorized air alters the size of mitral cells in the olfactory bulb. Brain Res, (in press)

Leon M (1975) Dietary control of maternal pheromone on the lactating rat. Physiol Behav 14: 311-319

Leon M, Galef BG, Behse JH (1977) Establishment of pheromonal bonds and diet choice in young rats by odor pre-exposure. Physiol Behav 18: 387-391

Leon M, Coopersmith R, Ulibarri C, Porter RH, Powers JB (1984) Development of olfactory bulb organization in precocial and altricial rodents. Dev Brain Res 12: 45-55

MacLeod P (1971) Structure and function of higher olfactory centers. In: Beidler LM (ed) Handbook of sensory physiology, vol IV. Chemical senses, part 1. Springer, Berlin Heidelberg New York, pp 182-204

Macrides F, Davis BJ, Youngs WM Nadi NS, Margolis FL (1981) Cholinergic and catecholaminergic afferents to the olfactory bulb in the hamster: A neuro-anatomical, biochemical, and histochemical investigation. J Comp Neurol 203: 495-514

Mair RG, Gesteland RC (1982) Response properties of mitral cells in the olfactory bulb of the neonatal rat. Neuroscience 7: 3117-3125

Marr WJ, Gardner LE (1965) Early olfactory experience and later social behavior in the rat: preference, sexual responsiveness, and care of young. J Genet Psychol 107: 167-174

Marr JN, Lilliston LG (1970) Social attachment in rats by odour and age. Behaviour 33: 277-282

Math F, Davrainville JL (1980) Electrophysiological study on the postnatal development of mitral cell activity of the rat olfactory bulb. Brain Res 190: 243-247

Meisami E (1976) Effects of olfactory deprivation of postnatal growth of the rat olfactory bulb utilizing a new method for production of neonatal unilateral anosmia. Brain Res 107: 437-444

Meisami E, Mousavi R (1982) Lasting effects of early olfactory deprivation on the growth, DNA, RNA and protein content, and Na-K-ATPase and AChE activity of the rat olfactory bulb. Dev Brain Res 2: 217-229

Meisami E, Safari L (1981) A quantitative study of the effects of early unilateral olfactory deprivation on the number and distribution of mitral and tufted cells and of glomeruli in the rat olfactory bulb. Brain Res 221: 81-107

Nakashima K, Mori K, Takagi S (1978) Centrifugal influence of olfactory bulb activity in the rabbit. Brain Res 154: 301-316

Nock B, Williams CL, Hall WG (1978) Suckling behavior of the infant rat: modulation by a developing neurotranmitter system. Pharmacol Biochem Behav 8: 277-280

Panhuber H (1984) The effect of environment on the sense of smell. Thesis, Macquarie Univ, Sydney

Panhuber H, Laing DG (1985) Prolonged exposure of young rats to an odour or deodorised air alters the size of mitral cells in the olfactory bulb. Proc Aust Physiol Pharmacol Soc (in press)

Panhuber H, Laing DG, Willcox ME, Eagleson GK, Pitman EA (1985) The distribution of the size and number of mitral cells in the olfactory bulb of the rat. J Anat 140: 297-308

Pinching AJ, Doving KB (1974) Selective degeneration of the rat olfactory bulb following exposure to different odours. Brain Res 82: 195-204

Pinching AJ, Powell TPS (1971) Ultrastructural features of transneuronal cell degeneration in the olfactory system. J Cell Sci 8: 253-287

Porter RH, Etscorn F (1976) A sensitive period for the development of olfactory preference in Acomys cahirinus. Physiol Behav 17: 127-130

Price JL (1977) Structural organization of the olfactory pathways. In: Le Magnen J, MacLeod P (eds) Olfaction and taste, vol VI. IRL Press, London, 1977 pp 87-95

Price JL, Powell TPS (1970) The synaptology of the granule cells of the olfactory bulb. J Cell Sci 7: 125-155

Prockop L (1979) Neurotoxic volatile substances. Neurology 29: 862-865

Rosenzweig MR (1971) Effect of environment on development of brain and behaviour. In: Tobach E, Aronson LR, Shaw F (eds) The biopsychology of development. Academic Press, London New York, pp 303-342

Ruddy JW, Cheatle MD (1977) Odor-aversion learning in neonatal rats. Science 198: 845-846

Schwob JE, Price JL (1984) The development of axonal connections in the central olfactory system of rats. J Comp Neurol 223: 177-202

Sczerzenie V, Hsiao S (1977) Development of locomotion toward home nesting material in neonatal rats. Dev Psychobiol 10: 315-321

Shepherd GM (1974) The synaptic organization of the brain. Oxford Univ Press, New York

Singh DNP, Nathaniel EJH (1977) Postnatal development of mitral cell perikaryon in the olfactory bulb of the rat. A light and ultrastructural study. Anat Rec 189: 413-432

Teicher MH, Blass EM (1976) Suckling in newborn rats: Eliminated by nipple lavage, reinstated by pup saliva. Science 193: 422–425

Walsh RN (1981) Effects of environmental complexity and deprivation on the brain anatomy and histology: a review. Int J Neurosci 12: 33–51

Webster DB, Webster M (1977) Neonatal sound deprivation affects brain-stem auditory nuclei. Arch Otolaryngol 103: 329–397

Wiesel TN, Hubel DH (1963) Effects of visual deprivation on morphology and physiology of cells in the cat's lateral geniculate body. J Neurophysiol 26: 978–993

Transient Postnatal Impacts on the Olfactory Epithelium Proprium Affect the Granule Cell Development in the Mouse Olfactory Bulb

B Rehn and W Breipohl*

Institut für Anatomie, Universitätsklinikum Essen, 4300 Essen, FRG

*Department of Anatomy, University of Queensland, St.Lucia/Brisbane, 4067 Australia

1 Introduction

Olfaction is only partly developed at birth in most altricial mammals (Altman 1969; Alberts 1982; Bayer 1983; Mair and Gesteland 1982; Mair et al. 1982; Schleidt and Hold 1982; Leon et al. 1984; Coopersmith and Leon this Vol; U. Schmidt this Vol). Since some parts seem to be more mature than others (Greer et al. 1982; Schwob et al. this Vol), the question arose as to whether the system as a whole is influenced by postnatal sensory and environmental experiences (cf. Schmidt, Meisami, Lü et al., Panhuber, all this Vol).

Long-term odour exposure and odour deprivation in the early postnatal period have previously been shown to be effective for the conditioning of the olfactory system in adults (Meisami 1976; Laing and Panhuber 1978; Laing 1984). Although theoretically interesting, these experiments add little to our understanding of the plasticity of this part of the nervous system toward transient impacts during early life, which are more likely to occur in a natural environment than under specific long-term experimental conditions. We have previously described a method to produce fairly specific and transient insults to the olfactory periphery in adult animals (C. Schmidt et al. 1980, 1984; Rehn et al. 1981). This study now aims to highlight effects of an early postnatal transitory insult on the normal morphological developmental process in the olfactory bulb and further to discuss the general importance of this kind of investigation for our understanding of the role of environmental and inherent factors on brain function.

2 Material and Method

Abbreviations used:
CB	cell body(ies)
EPL	external plexiform layer
GC	granule cell(s)
IP	inner process(es)
MC	mitral cell
MFM	N-methyl-formimino-methylester
MOB	main olfactory bulb
OEP	olfactory epithelium proprium of the nasal septum
RS	reciprocal synapse(s)
SP	superficial process(es)

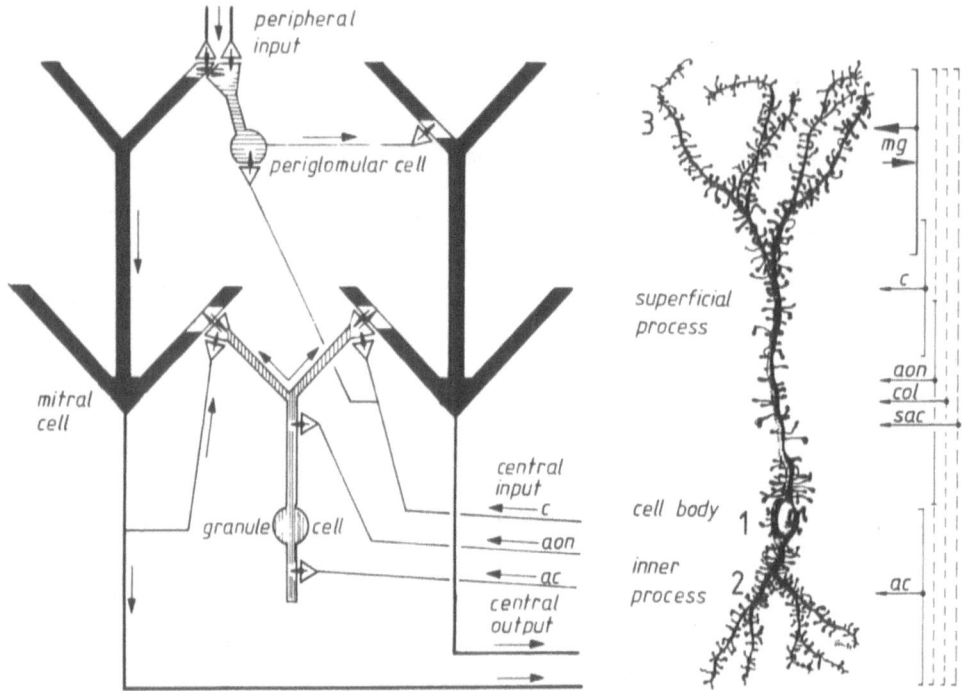

Fig. 1 Schematic drawing of the cell types of the main olfactory bulb, their arrangement and their functional connections. <u>Left</u> Diagram of the spatial arrangement of mitral, periglomular and granule cell and the basic circuit diagram of the olfactory bulb after Shepherd 1979. (◄——— direction of excitation; ◄ synapse; ⚌ reciprocal synapse). <u>Right</u> Schematic drawing of the feature of a typical granule cell. The regions where the number of spines were quantified are numbered. The lines give the distribution of the related synaptic connections of the granule cells according to Price and Powell (1970a,b). (◄— Region of input from **ac** anterior commissure, **aon** nucleus olfactorius anterior, **c** centrifugal fibers and **sac** axons of short axon cells; ⚌ **mg** region of input and output via reciprocal synapses between dendrites of mitral, tufted, and granule cells

A total of 45 NMRI mice (male and female) aged 2, 4, 7, 11, 14, 21, 28, 35 and 120 days (control group) and 4, 7, 14, 21 and 35 days (experimental group) were used. Twenty of these animals were treated with MFM on their first day of life. MFM saturated air was applied to the nose of the animals (0.5 sec; c. 2 to 3 sniffs) by a syringe. All animals were sacrificed by decapitation and the brains (MOB) were fixed by immersion in a solution containing 3.2% glutaraldehyde and 2.6% paraformaldehyde in 0.09M sodium cacodylate buffer (pH 7.3).

For light microscopic examination of the GC, the Golgi impregnation method as described by Fairen et al. (1977) was used. The bulbs were embedded in Agar, cut by hand (coronal sections) with a razor blade into slices about 100 to 200 μm thick which were dehydrated and mounted on slides with DePex. A minimum of 100 completely impregnated GC were evaluated per olfactory bulb regardless of their location within the GC layer. The material was examined at a magnification of 800x. All tissue was coded during the quantification procedure to exclude experimenter bias. Two sets of measure were obtained for the GC. Firstly, all clearly countable GC bodies were classified into three groups according to the number of their visible spines on the CB (cells with 0-2; 3-8; 9 and more spines). Secondly, all spines were counted on different randomly selected GC processes over a length of 50 μm. To compare regional differences, counts of spines were taken from three different regions of the GC: branching segments of the SP, IP and the CB. These three regions approximately corresponded to the areas of different synaptic connections on the GC (Fig. 1).

To check for general effects of MFM on the developing brain, the forebrains of 14 and 21 day old control and MFM animals were investigated in addition to the MOB as described above. Serial horizontal sections of the sensory cortex were cut. Ten sections per animal were randomly selected and on each slide the number of spines was counted on 10 pyramidal cells of layer V of the neocortex (area frontalis), i.e. the number of all visible spines was counted in the middle portion on the apical dendrite over a length of 50 μm. For computing the frequency of spines per 50 μm dendritic length no correction formula was used.

Additional samples from the bulbs (middle portion) of 14 and 21 day old control and MFM treated animals were embedded for transmission electron microscopical investigation, using standard routine procedures. Thin sections were stained with lead citrate and examined with a Philips 400 electron microscope at 60-80kV. From each sample 5 randomly selected areas (10 x 10 μm in size) of the outer half of the EPL beneath the olfactory glomeruli and the periglomerular cells were photographed at a magnification of 15.000x. Synapses were counted and classified from prints at a final magnification of 30.000x. Photographs from all series were mixed and their protocol numbers covered, during analysis.

3 Statistics

All samples were randomly selected and the code of the slides or photographs was unknown to the examiner. The significance of developmental differences and differences between control and experimental group in the number of spines per 50 μm of cell processes were evaluated according to a homogeneity test of Kolmogoroff and Smirnoff as described by Sachs (1982). The different trends in the numerical development of CB without spines was tested by a nonlinear regression. Changes in the number of S or RS per unit area (10 x 10 μm in size) were compared by the unpaired t-test. P-values equal or less than 0.05 were considered as statistically signigficant.

4 Results

In neonatal animals the MFM impact on the olfactory periphery (temporary cessation of mitosis and massive cell death followed by temporary triggered mitosis) appears to be effectively smoothed down after 7 days and is fully repaired by 3-4 weeks (Breipohl et al. 1986). In contrast, adult animals (8 weeks to 6 months) need about 5 to 8 weeks to recover from an MFM insult (C. Schmidt et al. 1984; Rehn et al. 1981). Examination of pyramidal cell maturation in the neocortex revealed no differences between normal and MFM treated animals in the number of spines on the main dendrites (Table 1). The data are within the range Valverde (1967) described for the regio temporalis in mice and indicates that development proceeds normally in the neocortex of MFM treated animals.

Table 1 Number of spines per 50 μm apical length on the main dendrite of pyramidal cells in mouse frontal cortex

Age	MFM treated at postnatal day 1	Control animals
14 d	37.42 +/- 7.3	37.2 +/- 8.6
21 d	48.5 +/- 9	48.6 +/- 7

4.1 Light Microscopy of Golgi-Stained Granule Cells (GC)

Figure 1 gives a simple overview of the cell types within the MOB, according to Shepherd (1979). The morphology and basic circuitry of the mammalian MOB are described in detail elsewhere (Price and Powell 1970a,b; Rall et al. 1966; Shepherd 1972). The GC bodies (CB) are located in a discrete layer (granular cell layer) with their inner processes (internal dendrite) within the granule cell layer and their superficial processes (external dendrite) within the external plexiform layer (EPL). As discussed by Shepherd (1979) the term "process" appears to be more adequate, from a functional viewpoint, than the term dendrite. GC receive input from various regions of the brain (Fig. 1) and possess a wide variety of spines on their dendrites and to a lesser extent on their CB. The term spine is used here for all spine-like processes because the classification and discrimination into spines and gemmules or other protrusions is not possible at the light microscope level. The criteria for the postnatal development of the GC are, for example, changes in the number of spines on the inner and outer processes (IP, SP) and the soma (CB).

4.2 Cell Bodies (Region 1, Fig. 1)

CB with no visible spines on them and others with a different number of spines occur irrespective of the stage of postnatal development of animals. By classifying the population of cells into those with 0-2, 3-8 and 9 and more spines, and estimating the percentage distribution of the three groups, it can be demonstrated that the following developmental changes occur during normal maturation of GC. The number of CB without spines is high at birth, but decreases with further development, reaching a minimum at day 7 and then increases again continuously throughout the life of the animals. In contrast to this, the number of cells with 9 and more spines is low at birth, increases till day 7 and then decreases again. The percentage of cells with 3-8 spines is high at such ages when population shifts from CB without spines to CB with more spines and vice versa (Fig. 2).

Three days after treatment with MFM on postnatal day 1, the numerical spine distribution on the CB resembles that of the 4-day-old control animals. After 4 days the relative number of CB with only 0-2 spines decreases without reaching the low level of the control animals at day 7. From day 7 onward the percentage of CB with 0-2 spines increases more rapidly than in the control group, leading to numerical spine profile patterns at day 21 and 35 in the experimental animals that look more similar to that of 120-day-old controls. (Figs. 2, 3). Nonlinear regression lines were computed from day 7 onwards, and it can be seen that the curves for the normal and experimental animals have not shifted in parallel (Fig. 3). The steeper slope in the experimental group indicates a significantly faster increase in the number of CB with 0-2 spines.

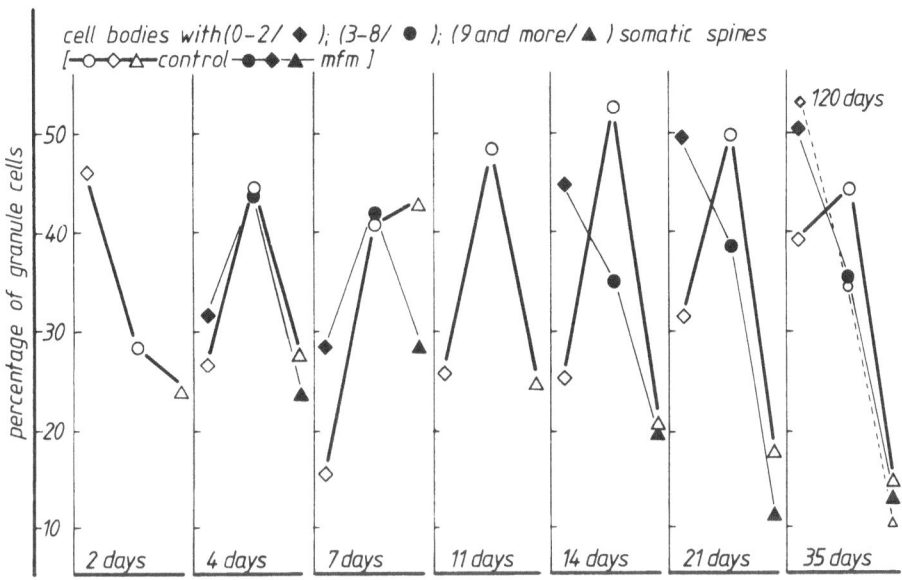

Fig. 2 Age-related percentage distribution of granule cell bodies with different numbers of somatic spines. The total number of cells counted per age group (n.c. 500-1000) is taken as 100%. Cells with 0-2, 3-8 and 9 and more somatic spines are expressed as percentages. Control group (open symbols) ; MFM group (closed symbols). The distribution pattern of 120-day-old animals (broken line, small open symbols) is shown in the 35-day groups

4.3 Inner Process (Region 2, Fig. 1)

Under normal conditions the number of spines on the IP increases during postnatal development, reaching a maximum between the 15th and 21th day of life and then decreases slightly. MFM treatment on postnatal day 1 does not cause any differences in the frequency of spines on the IP with the exception of 4 days postnatal (Figs. 4, 5). Three days after treatment a general reduction in spine frequency can be observed in all GC.

4.4 Superficial Processes (Region 3, Fig. 1)

During normal development the number of the spines on the SP increases postnatally reaching a maximum at day 14 and then decreases (Fig. 6). After MFM treatment, the numerical development of spines on the SP resembles that in the controls up to day 7. Subsequently, numerical spine development slows down asymptotically in the experimental animals reaching at day 14 already values characteristic for 35-day-old control animals (Fig. 6). In Fig. 7, distribution patterns of spines on SP in 1-week and 2-week-old animals of both groups are compared. No obvious differences can be seen in 7-day-old animals while there are significant differences between the 14 day groups. In contrast to the IP, the total reduction in spines on the SP is not due to a general and equal loss of spines for all the GC but merely to a loss on that group of SP with a low frequency of spines.

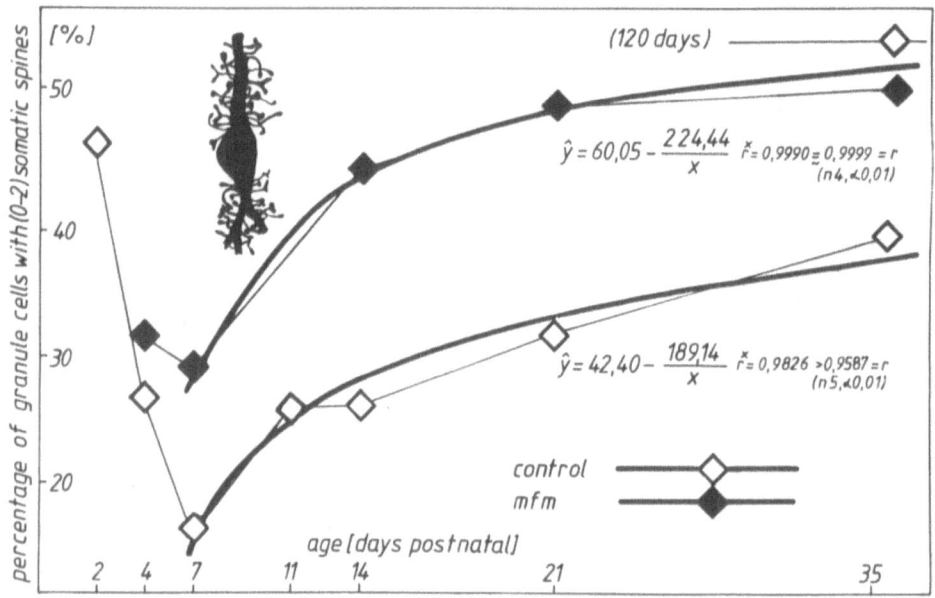

Fig. 3 Age-related distribution of cells with 0-2 somatic spines. Control group re-presented by <u>open symbols</u> and MFM group by <u>closed symbols</u>. <u>Thick lines</u> give computed nonlinear regression. Note that curve of MFM group is not parallel-shifted, but exhibits a different slope

4.5 Electron Microscopy of the External Plexiform Layer

As described above, the SP synapses in the EPL with the lateral dendrites of the mitral and tufted cells via reciprocal synapses (RS). RS within the MOB are known to serve inhibition of MC activity through GC. From previous studies of the ferret MOB development it is known that RS are mainly formed after the decline in spines on the SP becomes evident. The aim of these investigations was to check whether peripheral receptor cell deterioration in mice affects not only spine frequency but also synapse and frequency quality.

Ultrastructural studies were undertaken at peak (14 days) and after peak development (21 days) of spines. Counting the number of synapses per unit area (10 x 10 μm in size) in the outer half of the EPL revealed no significant differences between both animal groups. In both groups, synapses increased to the same level from 14 to 21 days (Fig. 8 top). Only the RS with the synaptic thickenings perpendicular to the plane of the section, could be identified. Despite this methodological deficiency leading only to relative values of RS, the data revealed a significant increase of RS between 14 days and 21 days for the controls and a slight decrease for the MFM-treated animals (Fig. 8 bottom)

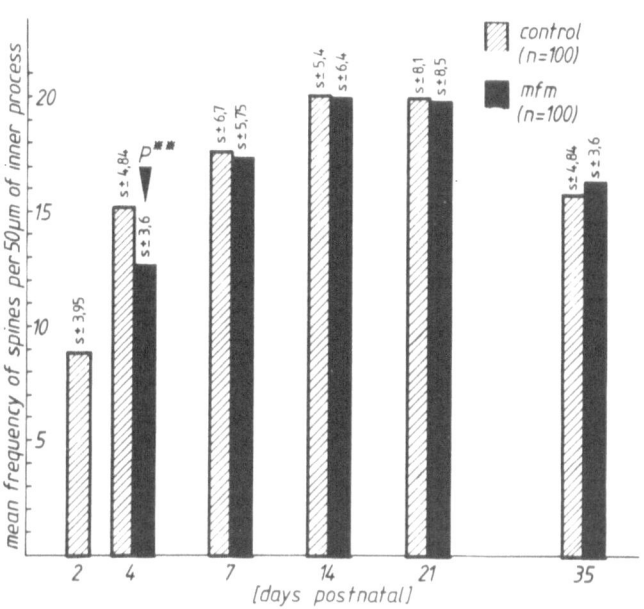

Fig 4   Age-related changes in the mean frequency of spines at the inner process (for orientation see Fig. 1, region 2). Each bar represents the mean spine frequency of 100 randomly selected granule cells. Arrow indicates significant difference between control and MFM group

5 Discussion

Investigations on transient impacts on the developing nervous system are of interest for the understanding of learning and repair mechanisms of the brain during early life. At this phase rapid growth followed by decline can be observed in different subsystems of the central nervous system (Boothe et al. 1979; Brunjes et al. 1982; Duffy and Rakic 1983). The elimination of spines and synaptic contacts after a period of proliferation is a general mechanism, by which the developing nervous system is adapted to its final function (Changeaux and Danchin 1976; Purves and Lichtman 1980). For the olfactory system it has been shown that the sensitive period of imprinting to food odours in ferrets is in phase with a time of overshoot development on the SP of GC in the EPL, and the formation of synapses and RS (Apfelbach et al. 1985; Rehn et al. 1986). Similarly it has been shown for mice that the adult electrophysiological pattern of excitation in response to odorants starts to appear by day 14 (U.Schmidt et al. 1983). As shown above, this is the phase of maximal numerical development of spines on the SP and at the same time a significant increase in number of the inhibitory RS occurs.

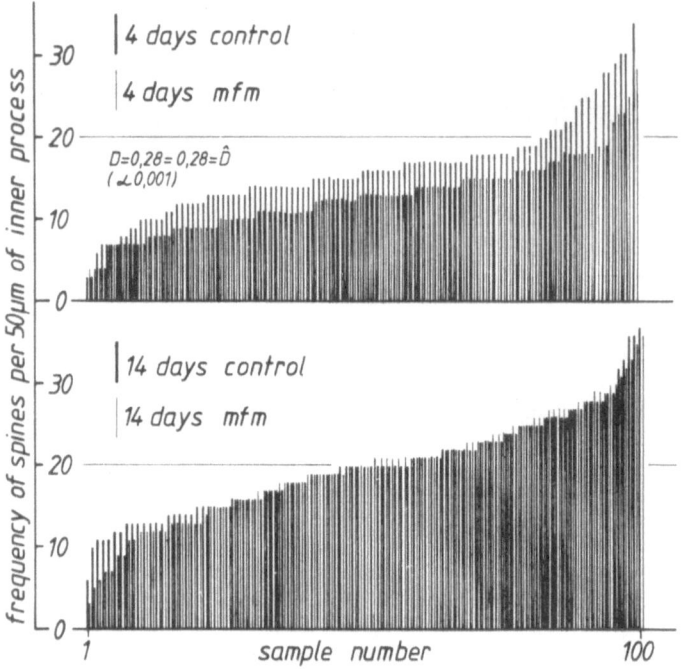

Fig. 5 Comparison of granule cell spine frequency (inner process, region 2) in 4 day old control and MFM treated (upper), and in 14 day old normal and MFM treated mice (lower). For each group, data were taken from 100 randomly selected granule cells and the number of spines on each process plotted (vertical bars) in order of increasing values

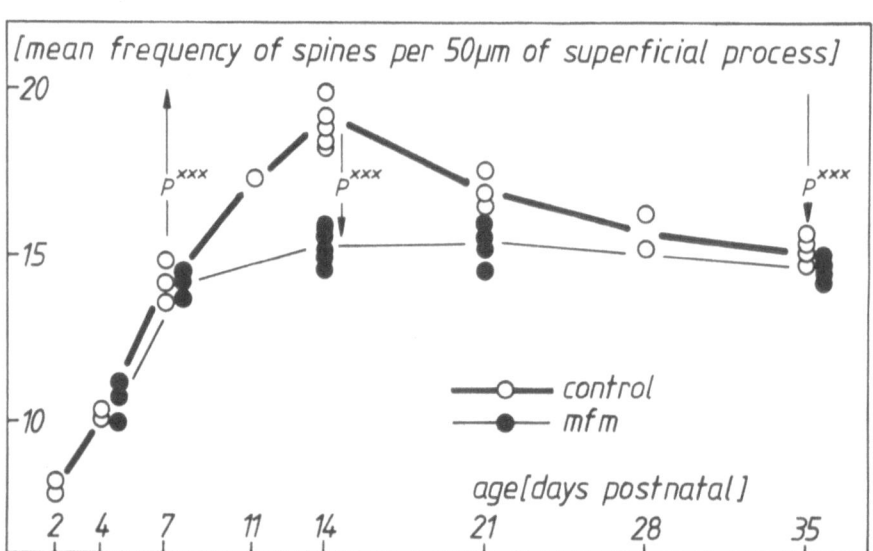

Fig. 6 Age-related changes in the mean frequency of spines of the superficial processes (for orientation see Fig. 1, region 3). Each point represents the mean spine frequency of 100 granule cells SP examined in one animal each. Arrows indicate significant increase and decrease of spine frequencies for the control animals, and between the controls and the experimental groups

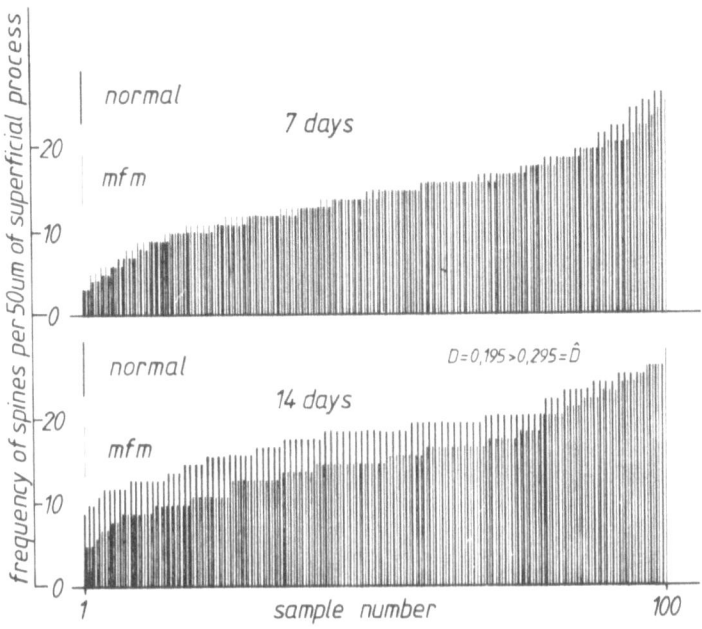

Fig. 7. Comparison of the granule cell spine frequency (superficial process, for orientation see Fig. 1, region 3) in 7-day-old control and MFM-treated (upper), and 14-day-old control and MFM-treated (lower) mice. Otherwise, as for Fig. 5

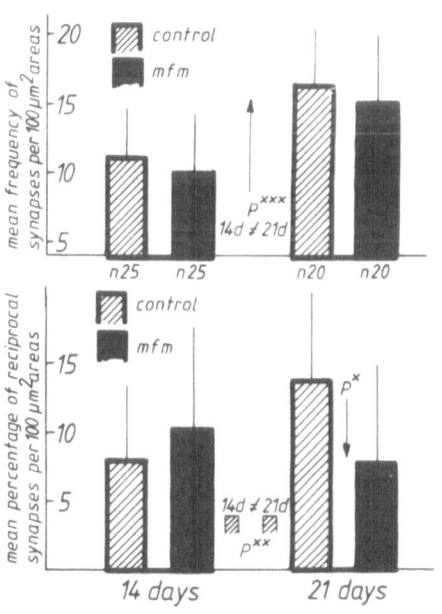

Fig. 8 Synapses and reciprocal synapses in the outer half of the external plexiforn layer in 14- and 21-day-old control and MFM-treated mice. Mean frequency of synapses per 10 x 10 μm area in 14 day old (n = 25 areas) and 21-day-old (n = 20 areas) control and MFM-treated mice (upper). Arrow indicates significant difference between both age groups. Mean percentage of reciprocal synapses per area expressed as percentage of the total number of synapses per area (lower). Arrow indicates significant difference between 21 day old control animals and MFM group

Although the GC receive no direct input from the peripheral olfactory receptor cells, (see Fig. 1) they still react to the temporary sensory deprivation induced by MFM (blockade of olfactory receptor cell function within minutes and consecutively degeneration of sensory cells (C. Schmidt et al. 1980, 1984; Rehn et al. 1981). Reaction is different in different parts of the GC. The IP of experimental animals was immediately affected but recovered after only a few days. In contrast the SP showed a delayed but more pronounced reaction. Spine numbers on the CB were affected within a few days but remained altered for longer periods.

Taken together, it is obvious that mature olfactory functions start to appear when the number of spines on SP or CB begin to decline and RS in the EPL reach a maximum. It is therefore tempting to speculate that the development and the stabilization of this inhibitory system of the MOB seems to be crucial for adult olfactory processing (Shepherd 1979). Otherwise the data presented indicate that the maturation of spines on GC in the mouse MOB differ with respect to their location. CB react with a 4 day delay to MFM treatment and exhibit a permanent reduction in somatic spine numbers (see Fig. 3). On the other hand, the SP react with a 7 day delay but only in a selective group of GC. The most evident changes are the cut off of the normal overshoot development of SP spines and the earlier asymptotic approach toward the lower spine numbers of older animals (see Fig. 6). The modifications on the CB (different slope in numerical spine development) and on the SP (cut off of overshoot phenomena and early asymptotically approach) after MFM may be interpreted as an advanced aging process. It seems that the olfactory periphery modifies not only the general histology (Clairambault 1971; Graziadei and Samanen 1980; Harrison 1958; Lü et al. this Vol) and electrophysiology (Gonzales-Estrada and Freeman 1980) of higher centers, but even the biochemistry (Baker et al. 1983, 1984) and cytology of specific cell types as demonstrated here for the GC (see also Margolis this Vol).

In addition, it has been documented for the primary and secondary olfactory stations that the repair mechanisms after MFM insults differ with age (Rehn et al. 1981; Breipohl et al. 1986). Therefore it seems highly desirable to use these MFM experiments as model to study the age dependency of specific repair mechanisms of different neuronal cell types and at different organisational levels. The above analysis of synapse numbers and synapse types in the EPL is a first step toward this approach.

The fine structural analysis of connections of the olfactory bulb revealed a peculiar synapse type, the reciprocal synapse (RS). RS are not a specific feature of the MOB (Shepherd 1979) and apparently are not an inherently determined type of synaptic contact. Regardless of their location they seem to originate through a close pairing of a asymmetrical and a symmetrical synapses (Jackowski et al. 1978). RS synapses in the EPL of the MOB serve inhibitory control of MC through GC. From our results it follows that the absolute number of RS seem to depend on olfactory sensory experience. This becomes especially clear from the following calculation. If one multiplies the numbers of RS by two because they originate by the association of two synapses then significant differences become evident between MFM and normal animals. The latter shows considerably higher theoretical synapse numbers because of the higher percentage of RS under normal developmental conditions. Thus, it may follow that the modified increase in the inhibitory system in the EPL is the consequence of the early MFM caused imbalance in the OEP (Breipohl et al. 1986) .

During normal development of the olfactory system there are phases which are best suited for an imprinting by environmental inputs (Apfelbach et al. 1985). In the olfactory bulb the phase of maximal spine development on the SP of GC appears to be in phase with the final electrophysiological tuning at this level of the olfactory system (Hinds and Hinds 1976; Mair and Gesteland 1982; Mair et al. 1982; Rehn et al. 1986; Coopersmith and Leon this Vol; U. Schmidt this Vol). To reach this maximal level of competence for olfactory imprinting, the peripheral sensory input may have to guarantee an overshoot development of spines in the MOB. Overshoot developments have been documented for other systems and neuronal cell types as well (Rausch and Scheich 1982; Duffy and Rakic 1983).

Olfactory deafferentation, deprivation and long term odour exposure studies have in common that they all interfere with the normal principles of development in the olfactory system. Our investigations have put a piece in the mosaic to valuate environmental versus inherent factors which determine the specific morphology and function of the vertebrate olfactory system. Though mosaic-like in character the investigations may also contribute to a better understanding of the nervous system in general, since e.g. overshoot phenomena, critical imprinting periods etc. have been reported also for the auditory and visual sensory systems. The different maturational processes on SP, CB and IP of the GC may prove as especially helpful for the models dealing with development of pattern formation and local specifity in the nervous system and brain.

Summary

Postnatal morphological granule cell maturation in the olfactory bulb of NMRI mice is described. Normal development is compared with maturation after early olfactory receptor cell destruction with N-methyl-formimino-methylester.

Though the granule cells receive no direct synaptic input from the sensory periphery they react with specific regional morphological alterations to olfactory receptor cell destruction. The differences in both amounts and time course of the experimentally induced morphological changes are evaluated with respect to postnatal functional maturation of excitatory and inhibitory circuits in the olfactory bulb.

The importance of these results for pattern formation and local specificity in the CNS is discussed.

Acknowledgments. The authors thank Prof. Effenberger (Institut für Organische Chemie, Universität Stuttgart) for his help with the synthesis of N-methyl-formimino-methylester.
Support for the following grants is gratefully acknowledged: DFG-BR 358/5-2; BRF-52; NH & MRC 85/3452.

References

Alberts J (1982) Ontogeny of olfaction: issues and perspectives from contemporary research. In: Breipohl W (ed) Olfaction and Endocrine Regulation. IRL Press Limited, London, pp 161-172

Altman J (1969) Autoradiographic and histological studies of postnatal neurogenesis. IV. Cell proliferation and migration in the anterior forebrain, with special reference to persisting neurogenesis in the olfactory bulb. J Comp Neurol 137: 433-458

Apfelbach R, Weiler E, Rehn B (1985) Is there a neural basis for olfactory food imprinting in ferrets? Naturwissenschaften 72: 106

Baker H, Kawano T, Margolis FL, Joh TH (1983) Transneuronal regulation of tyrosine hydroxylase expression in olfactory bulb of mouse and rat. J Neurosci 3: 69-78

Baker H, Kawano T, Albert V, Joh TH, Reis DJ, Margolis FL (1984) Olfactory bulb dopamine neurons survive deafferentation induced loss of tyrosine hydroxylase. Neurosci 11: 605-615

Bayer SA (1983) ^{3}H-Thymidine-radiographic studies in the rat olfactory bulb. Exp Brain Res 50: 329-340

Breipohl W, Darrelmann C, Rehn B, Tran-Dinh H (1986) Cell death and cell origin in the olfactory epithelium of mice. 21th Ann Conf Anat Soc Australia and New Zealand. J Anat (Lond) in press

Boothe RG, Greenough WT, Lund JS, Wrege KA (1979) A quantitative investigation of spine and dendritic development of neurons in visual cortex (Area 17) of Macaca nemestrina monkeys. J Comp Neurol 186: 473-490

Brunjes PC, Schwark HD, Greenough WT (1982) Olfactory granule cell development in normal and hyperthyroid rats. Devel Brain Res 5: 149-159

Changeaux JP, Danchin A (1976) Selective stabilization of developing synapses as a mechanism for the specification of neuronal networks. Nature 264: 23-30

Clairambault P (1971) Les effets de l'ablation bilaterale de la placode nasale sur la morphogenèse du télencéphale des Anoures. Acta Ebryol Exp 2: 61-92

Duffy CJ, Rakic P (1983) Differentiation of granule cell dendrites in the dentate gyrus of the rhesus monkey: A quantitative Golgi study. J Comp Neurol 214: 224-237

Fairén A, Peters A, Saldanha J (1977) A new procedure for examining Golgi impregnated neurons by light and electron microscopy. J Neurocytol 6: 311-337

Gonzales-Estrada MT, Freeman WJ (1980) Effects of carnosine on olfactory bulb EEG, evoked potentials and DC potentials. Brain Res 202: 373-386

Graziadei PPC, Samanen DW (1980) Ectopic glomerular structures in the olfactory bulb of neonatal and adult mice. Brain Res 187: 467-472

Greer CA, Stewart WB, Teicher MH, Shepherd GM (1982) Functional development of the olfactory bulb and unique glomerular complex in the neonatal rat. J Neurosci 12: 1744-1759

Harrison JL (1958) Some hypoplastic modifications of the telencephalon following unilateral excision of the nasal placode in Rana pipiens embryos. Diss Abstr 19: 605

Hinds JW, Hinds PL (1976) Synapse formation in the mouse olfactory bulb. I. Quantitative studies. J Comp Neurol 169: 14-40

Jackowski A, Parnavelas JG, Lieberman AR (1978) The reciprocal synapse in the external plexiform layer of the mammalian olfactory bulb. Brain Res 159: 17-28

Laing DG (1984) The effect of environmental odours on the sense of smell. In: Animal models in psychopathology. Acad Press, Australia, pp 59-98

Laing DG, Panhuber H (1978) Neural and behavioral changes in rats following continuous exposure to an odour. J Comp Physiol 124: 259-265

Leon M, Coopersmith R, Ulibarri C, Porter RH, Powers P (1984) Development of olfactory bulb organization in precocial and altricial rodents. Devel Brain Res 12: 45-53

Mair RC, Gellman RL, Gesteland RC (1982) Postnatal proliferation and maturation of olfactory bulb neurons in the rat. Neuroscience 7: 3105-3116

Mair RC, Gesteland RC (1982) Response properties of mitral cells in the olfactory bulb of neonatal rat. Neuroscience 7: 3117-3125

Meisami E (1976) Effects of olfactory deprivation on postnatal growth of the rat olfactory bulb utilizing a new method for production of neonatal unilateral anosmia. Brain Res 107: 437-444

Price JL, Powell TPS (1970) The synaptology of the granule cells of the olfactory bulb. J Cell Sci 7. 125-155

Price JL, Powell TPS (1970) An electron microscopic study of the termination of afferent fibres to the olfactory bulb from the cerebral hemispere. J Cell Sci 7: 157-187

Purves D, Lichtman JW (1980) Elimination of synapses in the developing nervous system. Science 210: 153-157

Rall W, Shepherd GM Reese TS, Brightman MW (1966) Dendrodendritic synaptic pathway for inhibition in the olfactory bulb. Exp Neurol 14: 44-56

Rausch G, Scheich H (1982) Dendritic spine loss and enlargement during maturation of the speech control system in the mynah bird (Gracula religiosa). Neuroscience Letters 29: 129-133

Rehn B, Breipohl W, Schmidt C, Schmidt U, Effenberger F (1981) Chemical blockade of olfactory perception by N-methyl-formimino-methylester in albino mice. II. Light microscopical investigations. Chem Sens 6: 317-328

Rehn B, Breipohl W, Mendoza AS, Apfelbach R (1986) Changes in granule cells of the ferret olfactory bulb associated with imprinting on prey odours. Brain Res 373: 114-125

Sachs L (1982) Applied statistics, Handbook of Techniques. Springer Verlag, New York Heidelberg Berlin, pp 291-293

Schleidt M, Hold B (1982) Human odour and identity. In: Breipohl W (ed) Olfaction and Endocrine Regulation. IRL Press Limited, London, pp 161-172

Schmidt C, Schmidt U, Breipohl W (1980) Inhibited olfactory perception in laboratory mice by N-methyl-formimino-methylester. In: van der Starre H (ed) Olfaction and Taste VII. Information Retrieval Ltd., London, p 405

Schmidt U, Eckert M, Schäfer HJ (1983) Untersuchungen zur ontogenetischen Entwicklungen des Geruchssinnes bei der Hausmaus Mus musculus. Z Säugetierkunde 48: 355-362

Schmidt C, Schmidt U, Breipohl W, Effenberger F (1984) The effect of N-methyl-formimino-methylester on the neural olfactory threshold in albino mice. Arch Otorhinolaryngol 239: 25-29

Shepherd GM (1972) Synaptic organization of the mammalian olfactory bulb. Physiol Rev 52: 864–917

Shepherd GM (1979) Synaptic organization of the brain. Ed 2, Oxford University Press, New York

Valverde F (1967) Apical dendritic spines of the visual cortex and light deprivation in the mouse. Exp Brain Res 3: 337–352

Central and Peripheral Influences on Postnatal Growth and Development of the Olfactory Bulb in the Rat

E Meisami

Department of Physiology-Anatomy, University of California, Berkeley, California 94720, USA

1 Introduction

Developmentally, the mammalian olfactory bulb (OB) shows a fairly orderly pattern of biochemical, structural, and functional development, similar to that seen in the cerebral cortex (for reviews see Meisami 1979; Shafa et al. 1980; Meisami and Timiras 1982). In altricial mammals such as the rat, much of this growth and development takes place postnatally, providing an excellent opportunity for experimental manipulation and investigation of the course and causes of neural development. Interestingly, the OB of the newborn rat, notwithstanding many immature features, still shows clear evidence of certain limited functions (Shafa et al. 1980; Leon 1983; Greer et al. 1982; Astic and Saucier 1982).

In the present paper, evidence is presented to show that during early postnatal development, experimental olfactory deprivation by occlusion of a naris, section of the olfactory nerve, transection of the olfactory peduncle (bulbotomy), and complete isolation from the peripheral and central connections interfere markedly with some aspects of OB growth and development, namely gain in weight, cell proliferation, differentiation of plasma and synaptic membrane enzymes (Na-K-ATPase), and cholinergic enzymes, as well as with the number and growth of mitral cells, the principal relay neurons of the OB.

2 Material and Methods

Abbreviations used:
AChE   acetylcholin esterase
CAT   cholinacetyl transferase
MC   mitral cell
OB(s)   olfactory bulb(s)

Albino rats of Wistar strain were used in all the experiments. The operations were carried out unilaterally during days 1-3 after birth. Comparisons were made between the operated and control (contralateral) OB.

3 Growth (Weight Gain)

The results shown in Fig. 1 and Table 1 clearly indicate that postnatal growth of OB as measured by changes in weight is highly dependent on peripheral and central influences. Comparison of the weights of control and operated OBs at day 25 (weaning) when brain development in the normal animal reaches the plateau phase, reveals that neonatal olfactory denervation by axotomy reduces growth by about 20%, while the effect of bulbotomy is twice as severe. When both operations are combined, the deleterious effects are additive (Meisami and Moussavi 1982). However, even under complete isolation from peripheral and central neural connections, the OB is still capable of growth. Thus, while in the normal rat, between birth to day 25, OB weight increases by tenfold, the growth increment in the denervated OB is eightfold, and in the isolated OB fourfold (Meisami and Firoozi 1985).

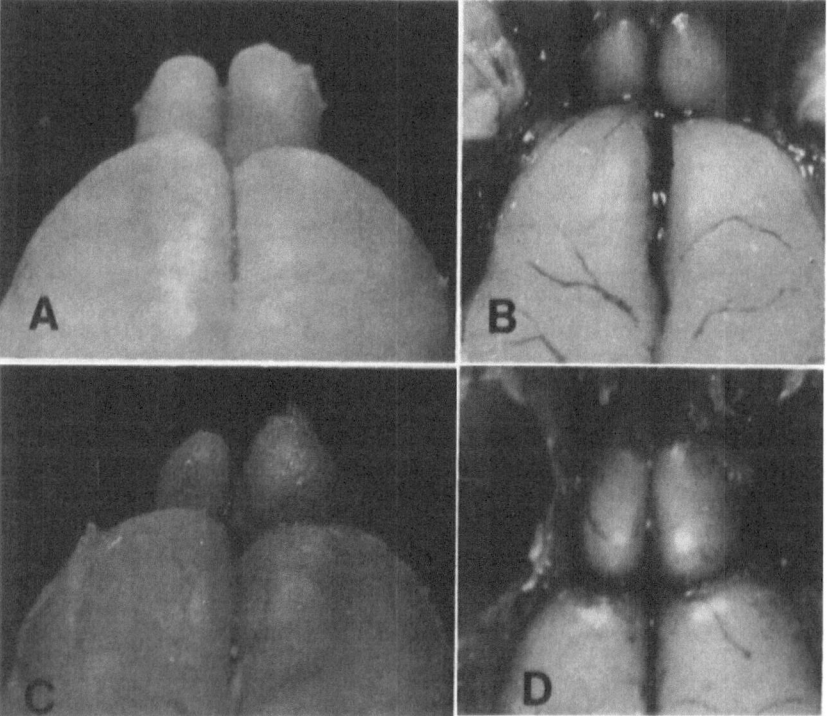

Fig. 1 Appearance of the olfactory bulbs in 30-day-old rats subjected to following operations during the neonatal period (days 2-3); the operated bulb is on the left: **A** transection of the olfactory nerve; **B** transection of the olfactory peduncle; **C** complete isolation from peripheral and central connections by combined denervation and peduncular transection. The left bulb in the animal in **D** was subjected to complete isolation at 30 days of age and inspected for effects at 60 days. Note the lesser severity of effects compared to **C**

Chemical destruction of the olfactory mucosa by treatment with 1% zinc sulfate and the resultant degeneration of olfactory nerve also causes severe retardation in OB

Table 1 Effects of various neonatal unilateral operations (odor deprivation by naris closure, transection of the olfactory nerve, transection of the olfactory peduncle and complete isolation of the bulb) on weight and total DNA content of the control and operated bulbs at day 25

Treatment	Weight (mg / bulb)[a]		Total DNA (µg / bulb)[a]	
	Control bulb	Experimental bulb	Control bulb	Experimental bulb
Odor deprivation	25.2+/-0.9	19.8+/-1.1 (22)*	70.2+/-0.9	56.3+/-2.1 (20)*
Chemical denervation	50.4+/-1.5[b]	35.3+/-1.6[b] (29)	ND	ND
Denervation by axotomy	24.8+/-0.7	20.1+/-0.8 (19)*	91.8+/-3.0	71.5+/-2.1 (22)*
Transection of peduncle	26.9+/-1.1	17.5+/-1.3 (35)*	80.0+/-2.8	48.0+/-2.0 (40)*
Isolation	26.3+/-1.6	12.5+/-0.8 (52)*	85.0+/-3.5	34.0+/-4.0 (60)*

[a] all values are Mean +/- SEM.
[b] values are for a pair of bulbs
* $p < 0.05$
values in parenthesis are per cent reduction compared to control side
N number of animals, at least 10
ND not determined

Newborn values: Weight, 3.1+/-0.2 mg / bulb; DNA, 15.0+/-0.9 µg / bulb.

growth. (Meisami and Manoochehri 1977; Shafa et al. 1980). The magnitude of this effect is about 10% higher than that produced by axotomy (Table 1). This difference may be due to the fact that zinc sulfate-treated animals are usually operated bilaterally, as it is not possible to prevent the leakage of zinc sulfate from one nasal sinus to another. Bilateral anosmia causes the loss of olfactory cues which disturbs nipple search and suckling behavior, resulting in reduced ingestion and undernutrition, a condition detrimental to optimal growth of brain structures including OB (Meisami and Manoochehri 1977; Shafa et al. 1980). Presumably if the influence of undernutrition on OB growth is subtracted from that produced by zinc sulfate treatment, an effect similar in magnitude to that of denervation by axotomy might be obtained.

Even though the olfactory nerve layer constitutes a significant part of OB, the deleterious effects of denervation by axotomy on OB growth cannot be entirely explained by the loss of this layer within the OB. That loss of other factors from the olfactory axons (e.g., trophic factors), may be involved here is suggested when the effects of denervation are compared with those of naris closure. Thus, at day 25, growth of OB homolateral to the closed naris shows a similar reduction compared to that seen after axotomy (Table 1) (Meisami 1976; Meisami and Moussavi 1982). This similarity suggests that olfactory receptor neurons may impart a stimulatory/trophic effect on OB growth which is lost by olfactory deprivation and by denervation.

As shown in Table 1, the reduction in OB growth by bulbotomy is nearly twice as large as that by denervation. This implies that the influences of central connections are quantitatively more critical for OB growth than those of the periphery (Meisami and Firoozi 1985). Transection of OB severs both the centripetal fibers of the lateral olfactory tract and the centrifugal fibers of the more medial bundles. It is unlikely that orthograde degeneration of centrifugal fibers alone could account for the enormous reduction in OB growth after peduncular transection (35%), as the contribution of this fiber system to the weight of OB could not be more than that of the olfactory nerve fibers. The more plausible explanation for the higher severity of the effects of peduncular transection must be due to the impeded flow of the new cells from the lateral ventricular zones to OB (Altman 1969) (see below). The failure of these cells (mainly internal granules and glia) to move into the OB would drastically reduce the growth potential of OB. This interpretation is further strengthened by the data presented below regarding accretion of total DNA and cell proliferation in the operated bulbs.

4 Cell Proliferation

Determination of total DNA provides a reliable estimate for total cell number per brain or brain part in mature as well as the developing animal (see Howard 1973 for refs.). Brain cells, regardless of their size or type (glia or neuron) contain the same amount of DNA (about 7 picogram / cell); the formerly assumed tetraploidy of the large neurons (e.g., Purkinje cells) is now believed to be due to artifacts, as these large cells contain the same amount of DNA as the smaller ones (Cohen et al. 1973). Also in structures like the cerebellum and OB, the large cells, being present in far smaller numbers than the small cells, do not appreciably contribute to the total DNA content.

In the normal OB from birth to day 25 total DNA content increases by about 5 fold. Determination of total DNA in operated OBs revealed a significant reduction compared to the control OB (Table 1). This occurs even though in all experimental OBs, the accretion of DNA continues, albeit at lower and different rates (Table 1). Specifically closure of the naris and the resultant olfactory deprivation produces about 20% reduction in DNA content at day 25. Neonatal denervation has similar effects but neonatally transected OB shows nearly 2 x as much reduction. When bulbotomy is combined with denervation, creating total isolation of OB from both the periphery and the center, total cell loss equals the sum of denervation and transection (Table 1).

These results suggest that although cell acquisition in the postnatal OB is significantly influenced by both peripheral and central factors, it is the central factors that play the most important role. The source of the numerous cells added to OB via the central connections must be the "rostral migratory stream" which delivers new embryonic cells from the tips of the lateral ventricles to the subependymal zone in the olfactory ventricle and from there to the OB proper (Altman 1969). Neonatal transection would produce a lesion in the path of this stream and in the olfactory ventricle, severely impeding the progress of the migratory cells. Thus any new cell acquisition in the transected OB of the postnatal rat (3 x from birth to day 25) must be due to the division of cells already present in the OB at the time of the operation (neonatal period).

The sources of cells lost or not developed in the sensory deprived or denervated OB are not known for certain. Loss or reduction in the rate of acquisition of internal granule cells in these bulbs is very likely. Indeed, recent cell counts by Skeen et al. (1985), have shown that early naris closure in the mouse reduces the number of internal granule cells. Our Golgi studies also support this notion (Meisami and Noushinfar 1985).

5 Biochemical Differentiation: Membrane and Synaptic Enzymes

To determine the effects of central and peripheral connections on the development of specific neurochemical indices of the developing OB, the activity of two important brain enzymes, Na-K-ATPase and acetylcholinesterase (AChE) were determined in the experimental bulbs.

5.1 Na-K-ATPase

This enzyme is localized in the plasma membrane of neurons, particularly in the synaptic membranes (Bertoni and Siegel 1978), and is important both for the operation of the plasma membrane Na-K-pump, as well as for regulation of synaptic mechanisms (Vizi 1979). The activity of brain Na-K-ATPase increases mainly in the postnatal period, coincident with the surge in dendritic and synaptic proliferation. Indeed, the development of this enzyme is linked to the appearance of electrical activity (EEG) (see Meisami 1975; Meisami and Timiras 1982 for reviews). In the normal rat OB, there is no Na-K-ATPase activity at birth. The activity increases gradually in the early postnatal period and markedly so after day 10, reaching adult values by the middle of the second month (Safaii et al. 1986). As seen in Table 2, neonatal olfactory deprivation by naris closure retards the postnatal increase in the activity of this enzyme in the OB, so that the specific activity (activity / mg prot.) is 30% less in the anosmic OB at day 25 (Meisami and Moussavi, 1982). Similar effects are observed in the OB denervated by treatment with zinc sulfate (Meisami and Manoochehri 1977). These results indicate that odor stimulation, activity, and presence of olfactory neurons are essential for optimal development of synaptic membranes in the OB.

Table 2 Effects of various unilateral operations (odor deprivation by naris closure, transection of the olfactory nerve, chemical deafferentation, bulbotomy and complete isolation of the bulb) on total activity of Na-K-ATPase and AChE of the control and operated bulbs at day 25

Treatment	Control bulb	Experimental bulb
	Na-K-ATPase (µmoles Pi / min / bulb)[a]	
Odor-deprivation	0.91 +/- 0.1	0.45 +/- 0.1 (48)*
Chemical denervation	0.72 +/- 0.07[b]	0.33 +/- 0.04 (53)*
	AChE (nmoles ACh / min / bulb)[a]	
Odor-deprivation	160 +/- 13	129 +/- 9 (20)*
Denervation	162 +/- 7	132 +/- 10 (19)*
Bulbotomy	178 +/- 10	88 +/- 12 (50)*
Isolation	152 +/- 12	42 +/- 5 (72)*

[a] all values are Mean +/- SEM; [b] operation carried out bilaterally, but total activity was computed for a single bulb; * $p < 0.05$; N at least 8.
Newborn values: Na-K-ATPase, nill; AChE, 3.6 nmoles Ach / min / bulb.
Values in parenthesis are per cent reduction compared to control

5.2 Acetylcholinesterase (AChE)

AChE is an essential enzyme for regulation of acetylcholine metabolism in the brain (MacIntosh 1981). In the OB, AChE and cholineacetyltransferase (CAT) activities show similar histochemical distribution (Godfrey et al. 1980), indicating that AChE can be used as a marker for cholinergic synapses. Also, Godfrey et al. (1980) and Wenk et al. (1977) have shown that in adult rats cholinergic fibers to OB are mainly centrifugal since destruction of certain forebrain reticular nuclei or lesions of the medial fiber tracts in the olfactory peduncle markedly diminish AChE activity in OB. The cholinergic fibers innervate all OB layers, although their innervation of glomerular and internal plexiform layers is more prodigious (Godfrey et al. 1980).

In the newborn OB, AChE specific activity is about 20% of the adult. The major increase to adult levels occurs between day 10 and 25, most likely due to proliferation and maturation of centrifugal synapses. Day 10 to 25 is the same age period when CAT activity begins to increase in the developing OB (Safaii et al. 1986). The specific activity (per mg prot.) of AChE in the OB homolateral to the

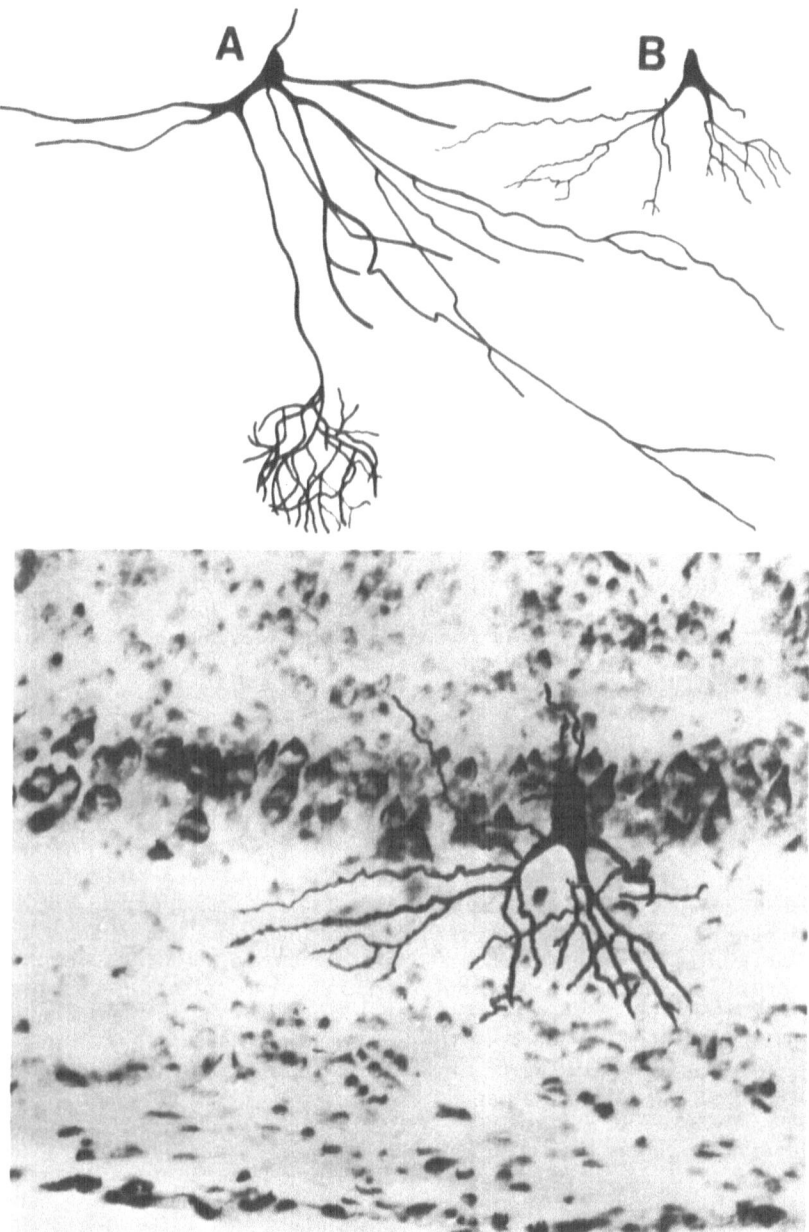

Fig. 2 Top Camera lucida drawing of the mitral cells (rapid Golgi) in the mature
A and newborn B olfactory bulbs. Note the prodigious growth of both apical and
basal dendrites as well at the marked proliferation of the glomerular branching of
the apical dendrite in the first postnatal month. Bottom Camera lucida drawing of
a mitral cell (rapid Golgi) from a newborn is superimposed on the Nissl section
from a bulb at the same age to illustrate possible relations of the newborn mitral
cells to the bulbar layers

closed naris was found to be the same as the contralateral control OB, although total activity (whole OB) was reduced proportional to OB weight (Meisami and Moussavi 1982). Neonatally denervated OB also responds in a similar manner (Meisami and Firoozi 1985). The observed reduction in the denervated OB is likely to be secondary to reduced growth of OB, as olfactory nerve fibers do not possess any AChE activity.

In contrast to the denervated OB, the neonatally transected OB shows markedly reduced AChE activity; isolated OB shows even more marked reductions (Table 2). It is interesting that postnatal growth of AChE in the totally isolated OB is not nil but about a fifth of that occurring in the normal OB. Since there are no known cholinergic cells intrinsic to OB, this latter activity is most likely due to growth of postsynaptic cholinoceptive cells (e.g., internal granule cells, periglomerular cells).

6 Mitral Cells

6.1 Normal Development

In the newborn OB, the only layer which shows signs of structural maturity is that of the mitral cells (MC) (Scheibel and Scheibel 1975; Mair et al. 1982; Meisami 1979). All other layers are still in the process of development. By postnatal day 10, all layers are clearly established but structural differentiation and cytoarchitectural refinements continue up to the end of the first month when OB attains a definitely mature appearance. The MC differ in morphology and size. The functional significance of the different types of these cells is beginning to be investigated (Macrides et al. 1985; Orona et al. 1984).

In Golgi impregnated sections of the newborn OB, some MC exhibit a mature appearance while others show immature features, particularly with respect to the terminal (glomerular) branchings of the apical dendrites and growth and elongation of the basal dendrites (see Fig. 2) (Meisami 1979; Mair et al. 1982). Presumably the growth and elongation of the basal dendrites are accompanied by the formation of increased synaptic contacts between these elements and the superficial dendrites of the internal granule cells. Synaptic structures in the external plexiform layer show prodigious development both in number and in morphology during the postnatal period (Hinds and Hinds 1976a,b). Continued postnatal ultrastructural differentiation in the perikaryon and dendrites of MC has also been demonstrated (Singh and Nathaniel 1977).

A number of functional studies have also indicated that appearance of spontaneous electrical activity and the responsiveness of the MC show significant maturational changes in the postnatal period (Shafa et al.1981; Math and Drainville 1980; Gesteland et al. 1982). Therefore, the mature appearance of some MC in the newborn OB notwithstanding (Scheibel and Scheibel 1975), the above results clearly imply that some of the MC are still in a developing stage and therefore likely to be vulnerable to transneuronal changes induced by olfactory deprivation or denervation.

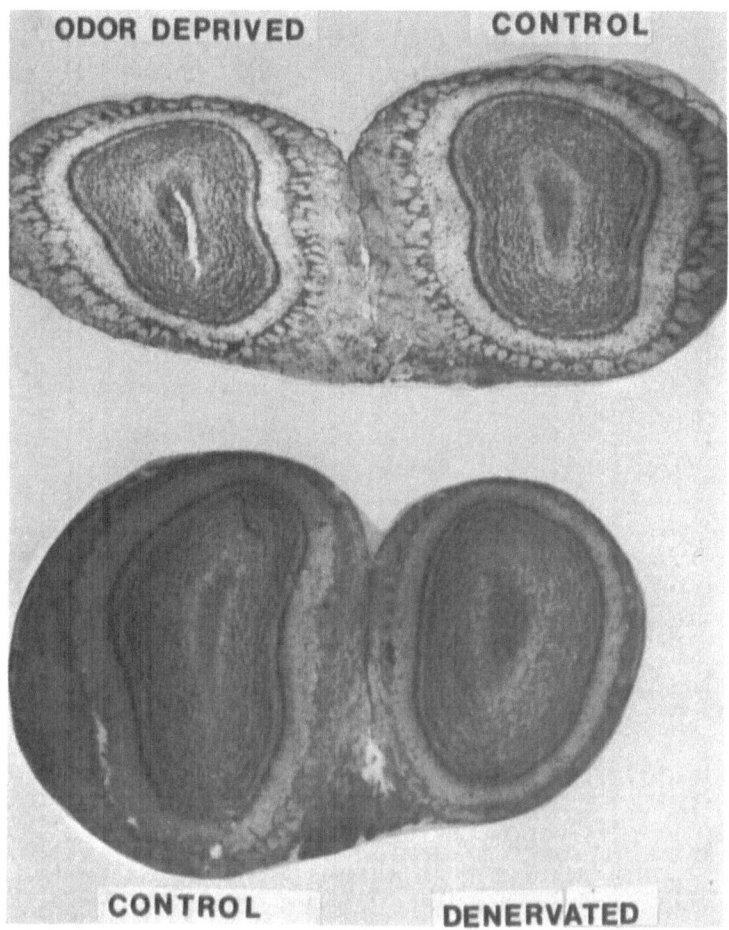

Fig. 3 Top Frontal Nissl (thionin) sections of the control and odor-deprived olfactory bulbs in a 25-day-old rat subjected to neonatal unilateral occlusion of the homolateral nostril. Note that while the odor-deprived bulb is markedly smaller, it shows similar cytoarchitecture, compared to the control bulb. See text for quantitative differences. Bottom frontal Nissl section of the control and denervated olfactory bulbs in a 30-day-old rat subjected to unilateral transection of the olfactory nerve. Note the total loss of the olfactory nerve layers and the marked reduction in the size of the external plexiform and the internal granular layers . The glomerular layer appears also poorly differentiated

6.2 Olfactory Deprivation

As seen in Fig. 3, while neonatal closure of a naris does not alter overall cyto-architecture of OB, it selectively reduces the thickness of external plexiform and internal granular layers (Meisami 1978). Determination of number and distribution of MC in the OB homolateral to the closed naris revealed reductions of 10-30% in total MC number. The reduction occurred throughout the OB, though not to the same extent (Fig. 4, top). Counts of glomeruli revealed no difference between the control and anosmic OB (Meisami and Safari 1981). Since the numbers of MC in

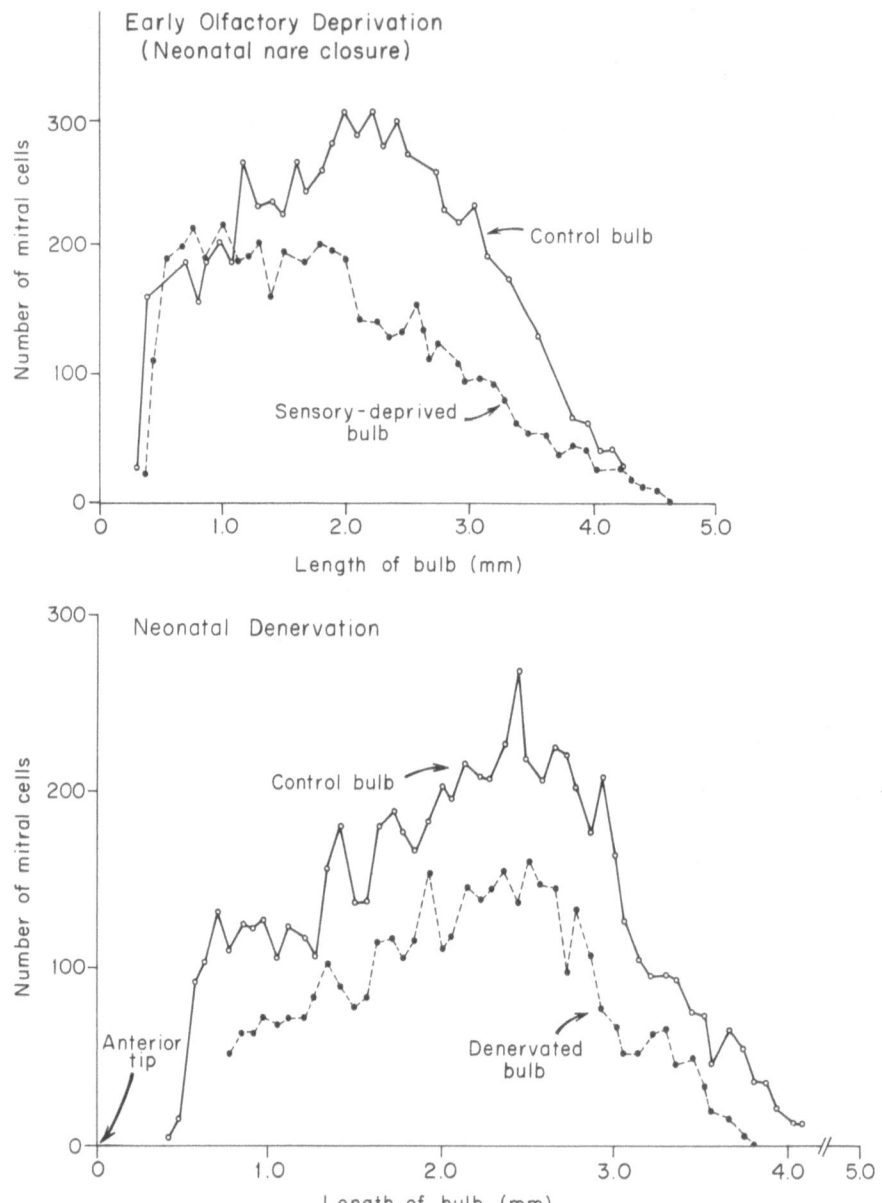

Fig. 4 Distribution profiles of the number of mitral cells along the length of the olfactory bulbs in the control vs. odor deprived bulbs (top) and in the control vs. denervated bulb (bottom). Counts were obtained from serial frontally cut Nissl sections in one-month-old, neonatally operated animals. Presence of at least one nucleolus was used to count the mitral cells. Note that in both cases the mitral cells are less in number throughout the experimental bulbs, although the reduction is not to the same extent in all regions

the neonate and adult OB are the same, it follows that the reduction in the anosmic OB must stem from cell degeneration and death. MC are not the only output cells affected by early sensory deprivation. Number of tufted cells are even more markedly reduced (Meisami and Safari 1981). While the exact nature of functional deficit(s) produced by the loss of mitral and tufted cells is not yet known, the reduction in the number of output channels to higher central structures might diminish the sensitivity of the system.

Loss of MC is evident in Golgi impregnated sections as well. When total number of MC cells impregnated were counted in whole series of thick sections, a marked reduction was found in total MC number in the sensory deprived OB (Meisami and Noushinfar 1986). The length of basal and apical dendrites and diameter of cell bodies of the remaining MC also showed significant reduction (20-30%), compared to the control OB. In addition, the rich and extensive feltwork of dendrites normally found in the external plexiform layer appeared sparse in the sensory deprived OB. Recent ultrastructural and morphometric studies in the mouse (Benson et al. 1984; Stupp et al. 1986) have also found marked reductions in the volumes of various layers of the OB on the close naris side. In addition to marked reduction in the volume of external plexiform layer, synaptic development between the MC and granule cells was altered (Benson et al, 1984); Stupp et al. (1986) have also noted marked reductions in the volume of glomerular layer, particularly on the medial zone of the deprived OB.

The contribution of olfactory deprivation to the proliferation and maturation (including synaptogenesis) of the internal granule cells should not be overlooked. Evidence from our studies indicate that both the number and growth of the granule cells as well as their postnatal spinogenesis are significantly retarded in the olfactory deprived OB (Meisami and Noushinfar 1985). Similar results have been obtained in quantitative Golgi studies of the granule cells in mice by Rehn and Breipohl (this Vol). Also a Nissl study by Skeen et al. (1985), has shown that the number of some internal granule cells in the sensory deprived OB of mouse is significantly reduced.

Due to abundant cytological and physiological interaction between the granule and MC, it is likely that some of the effects seen in the MC may be secondary to those imparted on the granule cells. However, since the MC are directly in contact with the primary olfactory neurons and therefore more likely to be affected by reduction in the activity or other influences of olfactory neurons. That the reduction in number or function of MC may alter the development of higher olfactory centers is indicated by the work of Lü et al. (1984) showing significant loss of cells in the amygdaloid nucleus of unilaterally naris-closed mouse.

6.3 Denervation (Axotomy)

The results of a Nissl study of the denervated OB 30-day-old rats operated at birth by axotomy revealed similarities as well as differences compared to those obtained in the olfactory deprived OB. In the denervated OB, as expected, olfactory nerve layer was completely absent and the size of OB greatly reduced (Fig. 3, bottom). Also total number and size of the glomerular layer was significantly reduced. The latter was in contrast to the results obtained in the sensory deprived OB at day 25 (Fig. 3, top). In addition, similarly to the results obtained in the sensory deprived OB, the thickness of external plexiform and internal granular layers were markedly reduced. When the number of MC were counted, marked reductions (30-40%) in this parameter was found in the denervated OB, compared to the unoperated control. Also distribution profile of the MC in the denervated OB indicated that the loss occurred throughout the OB, though not uniformly (Fig. 4, bottom).

These similarities between the responses of the sensory deprived and the denervated OB are very intriguing, since in one case the entire olfactory nerve is absent while in the other the nerve is present, but presumably functional afferent input is reduced. Yet in both cases, some of the secondary relay cells exhibit vulnerability by showing atrophy or even death. Whether this population of MC is simply less mature and therefore more vulnerable or this is a special sub-population with special properties is not known. The case for vulnerability is a more likely one. MC appear within a 3-day period during embryonic development (E12-E15) (Hinds 1968). Thus at birth some MC must be more mature than others, as indeed shown by the Golgi studies discussed above. It is well known that in other systems, transneuronal cell degeneration and death often occurs following deafferentation during the early developmental period (see Globus 1975 for review).

7 Interpretation and General Discussion

7.1 Influence of Periphery

At the present the real causes of the growth deficits in the developing OB deprived of its afferent input or nerve are not well understood. Neither do we know about the effects of these bulbar deficits on the development of higher olfactory and limbic structures. Similarly little is known about the influences of these neurodevelopmental deficits on the behavioral development of the animal. In the rodents which possess an accessory olfactory system, it would be interesting to know how the homolateral vomeronasal organ and the accessory OB responds to the occlusion of the anterior naris. Is the accessory olfactory system equally affected? Could the connections of the vomeronasal organ to the oral cavity help maintain stimulation of this system? Is there a hypertrophy of the accessory ol-factory system in this condition? Our results have indicated no signs of hyper-trophy in the contralateral main OB (control).

It is noteworthy that the deficiencies observed in the OB in response to neonatal olfactory deprivation are restricted to a critical period of OB growth, because the magnitude of these effects is not increased by prolonging of the duration of deprivation up to a year (Meisami and Moussavi 1982; Meisami and Safari 1981). Furthermore, Brunjes and Borror (1983), have clearly shown that if naris closure begins after postnatal day 30 (when brain and OB growth plateau in rodents), the growth deficits in OB are no longer observed. In this regard, it would be very interesting to investigate the effects of reopening of the nostrils of previously odor-deprived animals to evaluate the ability of the olfactory system to recover and rehabilitate from the earlier damages and deficits.

To understand the causes of reduction in OB growth in olfactory deprivation, we need to know more about the postnatal development of the normal olfactory mucosa and olfactory neurons and about the state of these structures on the deprived side. Preliminary reports by us (Meisami 1984) and by Farbman et al. (1984) indicate that the thickness of the olfactory mucosa on the anosmic side as well as the number of nuclei of olfactory receptor neurons is significantly reduced. What is the cause of this effect? Does stimulus deprivation lead to the loss of olfactory neurons or does it reduce the postnatal formation of these cells?

Structural studies in the mouse indicate that at birth olfactory neurons show an advanced stage of development (Cushieri and Bannister 1975a,b). Similarly, the result of our recent study in the rabbit (Meisami et al. 1986) as well as in the rat and hamster (Meisami 1986), indicate that, qualitatively, the olfactory mucosa of the mammalian neonate has a mature-like cytoarchitecture, although quantitatively its development is far from complete (see below). Electrophysiological studies by Gesteland et al. (1982) indicate that at birth many of the receptor units are functional. The functionally active state of the existing receptor neurons in the neonate is indicated further by increased uptake of radioactive deoxyglucose in the OB, in response to odor stimulation (Astic and Saucier 1982). However, histochemical studies of the olfactory mucosa using antibodies to the olfactory marker protein show that in the rat, the number of neurons expressing this marker protein is small at birth, increasing throughout the postnatal period and reaching a maximum by the age of 30 days, when development of the forebrain structures is basically complete (Monti-Graziadei et al. 1980). This finding indicates that the biochemical differentiation of the primary olfactory neurons in the neonatal period is far from complete.

The last important question is whether the total number of olfactory neurons increases during the postnatal period . Our recent results in the postnatal rabbit (Meisami et al., 1986) as well as in the rat and hamster (Meisami, 1986) indicate that the surface area and thickness of olfactory mucosa as well as the total number of nuclei of olfactory neurons markedly (5x) increases during the postnatal period. Similarly Rehn et al.(1986) have recently shown, in the ferret, evidence of significant increase in the surface density of receptor neurons during the postnatal period .

Thus the critical dependence of OB growth on the activity of olfactory neurons during the early postnatal period may be related in part to the increase in the total number of these neurons, to the increased transport of certain growth factor (olfactory marker protein?) to the bulbar neurons, or simply to the increased excitatory functional activity of the olfactory neurons. Nevertheless, possible effects of olfactory deprivation on blood flow and metabolism of the developing OB should not be overlooked. Indeed, Klosovsky and Kosmarskaya (1963) showed that the OB of puppies denervated early in life, not only is smaller but has markedly reduced network of blood capillaries, particularly in the glomerular and external plexiform layers. Also Singh and Nathaniel (1975) have shown that the development of blood capillaries in the rat OB shows a surge period during the second and third postnatal week, i.e., the same age period during which the effects of early sensory deprivation on OB growth becomes critical (Meisami and Moussavi 1982; Brunjes and Borror 1983).

7.2 Influence of Central Connections

Depriving the developing OB of its central connections by transecting the olfactory peduncle will impart even more deleterious consequences than those seen after sensory deprivation or denervation. This is probably because the normal flow of new cells (granules and glia) form the ependymal zones of the brain ventricles to the main OB as well as the growth and proliferation of centrifugal fibers which occurs prodigiously during the postnatal period, are impeded by peduncular transection. Nevertheless, the effects of loss of centripetal fibers of the mitral and tufted cells on the developing OB may also be important here. Axotomy of the mitral and tufted cells and the loss of their targets may retrogradely impede the development of dendrites and synaptic connections of these cells within the OB.

Similarly, it is not known how the arrival of the centrifugal cholinergic, adrenergic and serotonergic fibers during the postnatal period may influence the development of the OB. The projections of these reticular fiber systems to the OB are very intensive and extensive (Shipley et al. 1985). Selective neonatal lesions of the source nuclei of the serotonergic (raphe) and adrenergic (locus ceruleus) fibers in the brain stem and that of the cholinergic fibers in the forebrain reticular nuclei (e.g., horizontal limb of the diagonal band) may shed light on how these fibers exert influence on the developing OB.

Although we have mostly emphasized the dependency of the developing OB on its peripheral and central connections, the fact that OB has some intrinsic developmental potential even in complete absence of peripheral and central connection, is of considerable theoretical and experimental interest, particularly for those interested in models for brain development.

References

Altman J (1969) Autoradiographic and histological studies of postnatal neurogenesis. IV. Cell proliferation and migration in the anterior fore-brain, with special reference to persisting neurogenesis in the olfactory bulb. J Comp Neurol 137: 433-458

Astic L, Saucier D (1982) Ontogenesis of the functional activity of rat olfactory bulb: autoradiographic study with 2-deoxyglucose method. Dev Brain Res 2: 243-256

Benson TE, Ryugo DK, Hinds FW (1984) Effects of sensory deprivation on the developing mouse olfactory system: A light and electron morphometric analysis. J Neurosci 44. 638-653

Bertoni JM, Siegel GJ (1978) Developemnt of Na^+-K^+-ATPase in rat cerebrum. Correlation with Na^+-dependent phosphorylation and K^+-paranitrophenyl-phosphatase. J Neurochem 31: 1501-1511

Brunjes PC, Borror MJ (1983) Unilateral odor deprivation: differential effects due to time of treatment. Brain Res Bull 11: 501-503

Cohen J, Mares V, Lodin Z (1973) DNA content of purified preparation of mouse Purkinje neurons isolated by a velocity sedimentation techniques. J Neurochem 20: 651-657

Cuschieri A, Bannister LH (1975a) The development of the olfactory mucosa in the mouse: light microscopy. J Anat 119: 277-286

Cuschieri A, Bannister LH (1975b) The development of the olfactory mucosa in the mouse: electron microscopy. J Anat 119: 471-498

Farbman AI, Ritz SM, Brunjes P (1984) Effect of odor deprivation on olfactory epithelium in developing rats. Neurosci Abst 10: 530

Gesteland RC, Yancey RA, Farbman AI (1982) Development of olfactory receptor neuron selectivity in the rat fetus. Neuroscience 7: 3127-3136

Globus A (1975) Brain morphology as a function of presynaptic morphology and activity. In: Riesen AH (ed) The developmental neuropsychology of sensory deprivation. Academic Press, London New York, p 9-92

Godfrey DA, Ross CD, Herrmann D, Matschinsky FM (1980) Distribution and derivation of cholinergic elements in the rat olfactory bulb. Neurosci 5: 273-292

Greer CA, Stewart WB, Teicher MH, Shepherd GM (1982) Functional development of the olfactory bulb and a unique glomerular complex in the neonatal rat. J Neurosci 2: 1744-1759

Hinds JW, (1968) Autoradiographic study of histogenesis in the mouse olfactory bulb. I. Time of origin of neurons and neuroglia. J Comp Neurol 134: 287-304

Hinds JW, Hinds PL (1976a) Synapse formation in the mouse olfactory bulb. II. Morphogenesis. J Comp Neurol 169: 41-62

Hinds JW, Hinds PL (1976b) Synapse formation in the mouse olfactory bulb. I. Quantitative studies. J Comp Neurol 169: 15-40

Howard E (1973) DNA content of rodent brain during maturation and aging. In: Ford DH (ed) Progress in brain research, vol. 40. Elsevier, Amsterdam, pp 91-114

Klosovsky BN, Kosmarskaya EN (1963) Excitatory and inhibitory states of the brain. Natl Sci Found, Washington DC (English translation from russian original)

Leon M (1983) Chemical communication in mother-young interactions. In: Vandenberg JG (ed) Pheromones and reproduction in mammals Raven Press, New York, pp 39-77

Lü Z, Breipohl W, Rehn B, Eckert M, Blank M (1984) Influence of postnatal occlusion of one naris externa on the amygdaloid complex in the NMRI-mouse: A morphometric study. Proc Int Conf Ontogeny of Olfaction in Vertebrates, 9-11 Sept 1984, Tübingen

Macrides F, Schoenfeld TA, Marchband JE, Clancy AN (1985) Evidence for morphologically, neurochemically and functionally heterogenous classes of mitral and tufted cells in the olfactory bulb. Chemical Sens 10: 175-202

MacIntosh FC (1981) Acetylcholine. In: Siegel GJ et al. (eds) Basic neurochemistry, 3rd edn. Little Brown, Boston, pp 183.204

Mair RC, Gellman RL, Gesteland RC (1982) Postnatal proliferation and maturation of olfactory bulb neurons in the rat. Neuroscience 7: 3105-3116

Math F, Davrainville JL (1980) Electrophysiological study on the postnatal development of mitral cell activity in the rat olfactory bulb. Brain Res 190: 243-247

Meisami E (1975) Early sensory influences on regional activity of brain ATPases in developing rats. In: Brazier MAB (ed) Growth and development of the brain, IBRO Monogr Ser, vol. I. Raven Press, New York, pp 51-74

Meisami E (1976) Effects of olfactory deprivation on postnatal growth of the rat olfactory bulb utilizing a new method for production of neonatal unilateral anosmia. Brain Res 107: 437-444

Meisami E (1978) Influence of early anosmia on the developing olfactory bulb. In: Corner MA, Baker RE, van de Poll NE, Swaab DF (eds) Progress in brain research. Maturation of the nervous system, vol 38. Elsevier, Amsterdam, pp 211–230

Meisami E (1979) The developing rat olfactory bulb: Prospects of a new model system for developmental neurobiology. In: Meisami E, Brazier MAB (eds) Neural growth and differentiation. Raven Press, New York, pp 183–206

Meisami E (1984) Early olfactory deprivation and the olfactory epithelium. Pap Chem Sens Day II, Nov 1984. Univ Calif, Berkeley

Meisami E (1986) Postnatal expansion of olfactory mucosal surface and olfactory neuron number in altricial mammals. Proc 5th Int Meet Int Soc Dev Neurosci, Mexico City

Meisami E, Firoozi M (1985) Acetylcholinesterase activity in the developing olfactory bulb: A biochemical study on normal maturation and the influence of peripheral and central connections. Dev Brain Res 21: 115–124

Meisami E, Louie J, Hudson R, Distel H (1986) Marked postnatal increase in the total number of olfactory neurons and surface area of the mucosa in the rabbit. Proc Int Symp Olfaction and Taste, Colorado

Meisami E, Manoochehri S (1977) Effects of early bilateral chemical destruction of olfactory receptors on postnatal growth, Mg-ATPase and Na-K-ATPase activity of olfactory and non-olfactory structures of the rat brain. Brain Res 128: 170–175

Meisami E, Moussavi R (1982) Lasting effects of early olfactory deprivation and the mitral cells of the olfactory bulb. Dev Brain Res 2: 217–229

Meisami E, Noushinfar (1986) Early olfactory deprivation and the mitral cells of the olfactory bulb: A Golgi study. Int J Dev Neurosci (in press) Meisami E, Noushinfar (1985) Effects of early olfactory deprivation on the internal granular cells of the olfactory bulb. Soc Neurosci Abstr 11: 447

Meisami E, Safari L (1981) A quantitative study of the effects of early unilateral olfactory deprivation on the number and distribution of mitral and tufted cells and of glomeruli in the rat olfactory bulb. Brain Res 221: 81–107

Meisami E, Timiras PS (1982) Normal and abnormal biochemical development of the brain after birth. In: Jones CF (ed) The biochemical development of the fetus and neonate. Elsevier, Amsterdam, p 759–822

Monti-Graziadei GA, Graziadei PPC, Stanley RS (1980) The olfactory marker protein in the olfactory system of the mouse during development. Neurosci 5: 1239–1252

Orona E, Rainer EC, Scott JW (1984) Dendritic and axonal organization of mitral and tufted cells in the rat olfactory bulb. J Comp Neurol 226: 346–356

Rehn B, Breipohl W, Mendoza MS, Apfelbach R (1986) Changes in granule cells of the ferret olfactory bulb associated with imprinting on prey odors. Brain Res 373: 114–125

Safaii R, Moussavi RT, Meisami E (1986) Postnatal development of enzymes in the olfactory bulbs of normal and hypothyroid rat. Proc Int Symp Olfaction Taste, Colorado

Scheibel ME, Scheibel AB (1975) Dendrite bundles, central programs and the olfactory bulb. Brain Res 95: 407-421

Shafa F, Meisami E, Moussavi R (1980) Retarding effect of early anosmia on growth of the body, brain, olfactory bulb, and cerebellum and its implications for the development of the olfactory system in the rat. Exp Neurol 67 (1): 215-233

Shafa F, Naghshineh S, Bidangiri A (1981) Development of spontaneous activity in the olfactory bulb neurons of postnatal rat. Brain Res. 223: 409-412

Shipley MT, Halloran FJ, Torre J de la (1985) Surprisingly rich projection from locus ceruleus to the olfactory bulb in the rat. Brain Res 329: 294-299

Singh DNP, Nathaniel EJH (1975) Postnatal development of blood vessels (capillaries) in the rat olfactory bulb: A light and ultrastructural study. Neurosci Lett 1: 203-208

Singh DPN, Nathaniel EJH (1977) Postnatal development of mitral cell perikaryon in the olfactory bulb of the rat. A light and ultrastructural study. Anat Rec 189: 413-432

Skeen LC, Due BR, Douglass FE (1985) Effects of early anosmia on two classes of granule cells in developing olfactory bulbs. Neurosci Lett 54: 301-306

Stupp C, Breipohl W, Rehn B (1986) Effects of early postnatal olfactory deprivation on the structural organization of the olfactory bulb in NMRI mice 24th Ann Meet ASANZ, St.Lucia, Brisbane

Vizi ES (1979) Presynaptic modulation of neurochemical transmission. Prog Neurobiol 12: 181-290

Wenk H, Meyer U, Bigl V (1977) Centrifugal cholinergic connections in the olfactory system of rats. Neurosci 2: 797-800

E. MATURATION OF TERTIARY OLFACTORY CENTERS

Development of Axonal Connections in the Central Olfactory System

JE Schwob, B Friedman* and JL Price

Department of Anatomy and Neurobiology, Washington University School of Medicine, St.Louis, MO 63110

*Section of Neuroanatomy, Yale University School of Medicine, New Haven, CT 06510

1 Introduction

Abbreviations used:

ac,AC	anterior commisure
AOB	accessory olfactory bulb
AON	anterior olfactory nucleus
BM	basomedial amygdaloid nucleus
Co_a	anterior cortical amygdaloid nucleus
Co_p	posterior cortical amygdaloid nucleus
LOT	lateral olfactory tract
Me	medial amygdaloid nucleus
NLOT	nucleus of the lateral olfactory tract
OB	main olfactory bulb
OT	olfactory tubercle
PAC	periamygdaloid cortex
PC	piriform cortex
PC_a	anterior piriform cortex
PC_p	posterior piriform cortex
S	corpus striatum
SEZ	subependymal zone
TT_v	ventral tenia tecta

Elucidation of the ontogeny of olfactory-mediated behavior must include consideration of the development of the central representation of the olfactory system, including the olfactory bulb and the structures of the ventral forebrain which are innervated by the bulb and which are collectively termed the olfactory cortex (including the piriform cortex, anterior olfactory nucleus, olfactory tubercle and lateral entorhinal area) (Schwob and Price 1978).

Studies of the development of the olfactory cortex are made easier by its accessibility and by its simple three layer structure (J.L. O'Leary 1937; White 1965; Price 1973). This consists of a superficial molecular layer (layer I), a layer of superficial pyramidal cells (layer II), and a layer of deep pyramidal cells (layer III). There are two major fiber inputs to the cortex: the first is the projection of the olfactory bulb which axons course in the lateral olfactory tract (LOT) and end in a sharply delimited superficial part of the molecular layer; this terminal zone is called layer Ia (Price 1973). The second major input is the ipsilateral and crossed intracortical fiber systems, termed the associational and commissural fiber systems, respectively, which end more deeply in the molecular layer; their terminal zone is designated layer Ib (Price 1973; Haberly and Price 1978a,b;

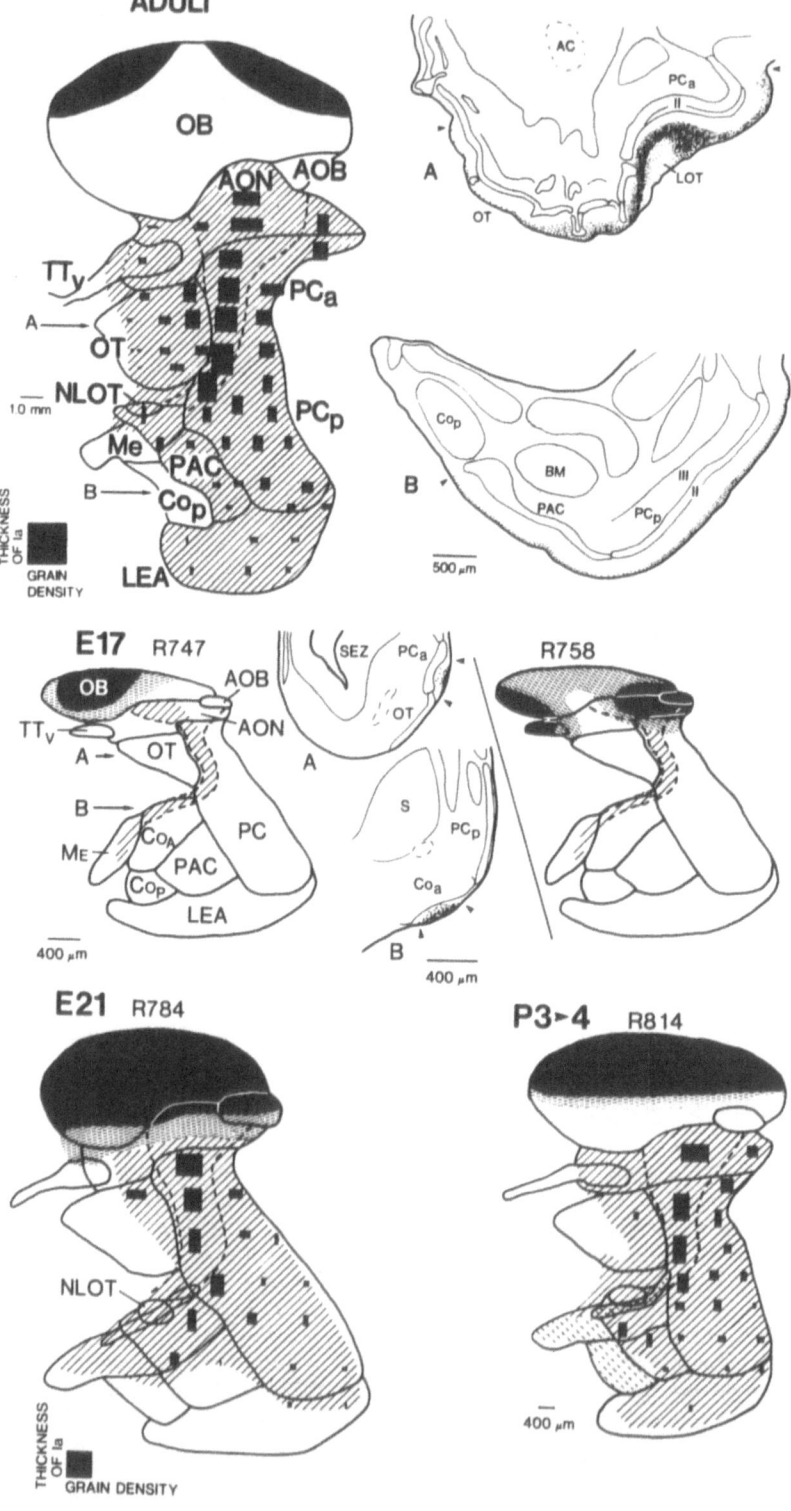

Fig. 1 The projection of the main olfactory bulb across the olfactory cortex in adult, fetal and neonatal rats as demonstrated by the anterograde transport of ^{3}H-leucine from the bulb. **E17** signifies embryonic day 17 (E1 is defined as the first day of gestation). **P3→4** designates that the injection was made on P3 (P0 is the day of birth) followed by postinjection survival for 24 h. The injection site, its center represented in black and its periphery in dense crosshatching, and the distribution of labeled fibers, represented by hatching, are illustrated on unfolded maps of the ventral surface of the brain (medial is to the left and rostral is up) and on coronal sections through the cortex (the arrowheads mark the limits of the projection). The variations in the thickness of the terminal zone of the bulbar fibers (layer Ia) and in the density of the transported label are documented by grain counts through layer Ia at selected cortical locations. The thickness of layer Ia and its average grain density are scaled in proportion to the maximum value recorded in that experiment and are represented as a rectangle of appropriate height and width, respectively.

Luskin and Price 1982). Layers Ia and Ib are precisely complementary to each other (Price 1973; Schwob and Price 1984b).

2 Development of the Projection of the Olfactory Bulb onto the Olfactory Cortex

2.1 The Density of Innervation Varies Across the Surface of the Cortex in the Adult

In rats and other mammals, the olfactory bulb projects via the LOT onto the cortex in a diffuse fashion: the mitral and tufted cells from any small portion of the bulb innervate virtually all of the olfactory cortex (Price 1973; Broadwell 1975; Scalia and Winans 1975; Devor 1976a; Skeen and Hall 1977; Haberly and Price 1977; Scott et al. 1980; Luskin and Price 1982). However, analysis of the projections with either axonal degeneration methods (Devor 1976a) or with axoplasmic transport of ^{3}H-amino acids (Schwob and Price 1978, 1984a) indicates that the thickness of the terminal zone and the density of synaptic innervation covary across the cortex in a systematic way. The thickness of layer Ia and the density of transported label or degenerating axonal terminals are both greatest deep to the LOT and both measures fall (in decreasing order) lateral, medial, and posterior to the LOT (Fig 1). This adult pattern parallels the sequential innervation of the olfactory cortex by the fibers which grow out from the olfactory bulb during embryonic and postnatal development.

2.2 Outgrowth of Axons from the Olfactory Bulb

The initial growth of axons from the olfactory bulb is closely linked with the birth of the mitral cells during the prenatal period E11-15 in mice (Hinds 1968) (embryonic day E1 is defined here as the first day of gestation) and E13-17 in rats (Bayer 1983). The incipient LOT is evident as a coherent fiber bundle positioned against the pia about 2 days after the first mitral cells are born (Hinds 1972). By E14-15 in mice, the LOT fibers form synapses on tangentially oriented cells located superficially in the piriform cortex near the tract (Derer et al. 1977). At these early stages, fibers from the main olfactory bulb are co-extensive with those from the accessory olfactory bulb (Fig 1); both project posteriorly in the LOT to the anterior edge of the amygdala. Since the mitral cells of the accessory olfactory bulb are born a few days in advance of those of the main olfactory bulb (Hinds 1968; Bayer 1983), it is tempting to speculate that the fibers from the accessory olfactory bulb are serving as pioneer axons for the later arriving ones from the main olfactory bulb.

The subsequent stages of the growth of the bulb fibers onto the olfactory cortex parallel and to some extent predict the variation in innervation across the cortex in adult animals. For example, by E17 in rats, fibers begin to leave the LOT and anterogradely transported label from the olfactory bulb can be seen in layer I of the anterior piriform cortex just deep to the tract (Fig 1; Schwob and Price 1978, 1984a). It is precisely this area of the cortex which is most heavily innervated in the adults. Subsequently, over the period E17-21, fibers from the bulb spread caudally and laterally across most of the piriform cortex and medially into the lateral part of the olfactory tubercle. These cortical areas are innervated to a lesser degree than the area deep to the LOT. The posterior edge of the tubercle and far caudal structures such as the lateral enthorhinal area are contacted just after birth, but fibers do not appear in the far medial part of the olfactory tubercle until after the end of the first postnatal week (Schwob and Price 1978, 1984a). In adults, this medial part of the tubercle is the most lightly innervated cortical structure. Distance from the bulb or from the "parent" axons in the LOT is insufficient to account fully for the temporal sequence of innervation. For example, the delay in the innervation of the olfactory tubercle is most likely due to the relatively late birth of these cells and the lateral to medial gradient of neurogenesis in the tubercle (Hinds 1967; Schwob, unpublished observations).

During the prenatal period of axonal growth, synaptogenesis is also ongoing, and a substantial number of both mature and immature synapses are seen in the piriform cortex deep to the LOT at birth (Westrum 1975a,b). These synaptic contacts increase markedly in number during the first and second postnatal week, and the synaptic density and the width of the molecular layer more than double during this time (Westrum 1975a,b). Furthermore, there is electrophysiological evidence that the fibers from the bulb form functional synapses with cortical neurons at birth. In neonates, electrical stimulation of the LOT evokes short latency single and multiple unit activity in the piriform cortex, as well as an extracellular field potential (the A_1 wave) which can be identified with the monosynaptic excitation of the cortex by tract fibers (Schwob et al. 1984).

2.3 Experimental Alterations of the Bulb Fibers

The observations during normal development would suggest that the variation in the density of innervation across the adult cortex is determined by the spatio-temporal sequence of development, but they cannot eliminate an alternative hypothesis which would suggest that the pattern and the timing of axonal branching of individual mitral cells and thus the formation of a certain number of synapses in a particular area of the olfactory cortex may be precisely specified prior to axonal outgrowth. However, this second hypothesis does not easily fit with observations that the pattern of adult innervation is altered by experimental manipulations of the bulb fibers during development. In neonates, both lesions of the LOT and partial ablations of the olfactory bulb provoke an altered distribution of olfactory bulb fibers. As a result of a knife cut in the LOT early in the course of post-natal life axons from the bulb grow over or around the cut but no longer in-nervate the more posterior parts of the cortex (Devor 1976b; Grafe 1983; Small and Leonard 1983; Schwob unpublished results). Areas rostral to the cut are hyperinnervated by the apparently "dammed up" fibers and these include some areas adjacent to the olfactory cortex which do not normally receive a projection from the bulb (Fig 8). Similarly, alteration in the distribution of olfactory bulb fibers is provoked by ablation of the dorsal and anterior parts of the bulb at or within a few days after birth. Since the projection of the bulb is diffuse, partial ablation of the bulb would not be expected to denervate selectively any area of the cortex. However, with destruction of approximately 50% of the bulb there is a substantial and disproportionate reduction in the projection to the lightly innervated areas in the medial olfactory tubercle and the nucleus of the lateral olfactory tract (NLOT), which are normally the last areas to be innervated (Fig 2). There is no retraction of fibers from areas which had been innervated at the time of the lesion such as the entorhinal cortex; a more substantial redistribution in the projection similar to that seen with LOT lesions may require that the ablation be made prior to the time of axonal outgrowth and synaptogenesis.

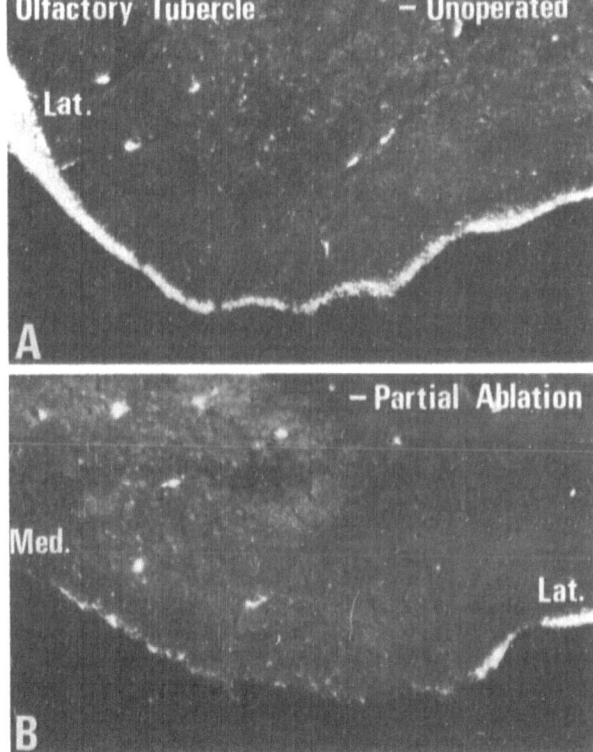

Fig. 2 Darkfield photomicrographs of the olfactory tubercle showing the distribution of anterogradely transported HRP from bilateral injections in the olfactory bulbs of an adult rat subjected to unilateral partial ablation of the bulb at P1 (cf. Fig. 1 -adult- cross section A for a line drawing which is representative of this part of the tubercle). A On the unoperated side, transported label clearly delineates Layer Ia and extends to the medial edge of the tubercle as a distinct zone. B On partially ablated side, the label is light and scattered in medial portions of the tubercle, despite a substantial layer Ia laterally.

In summary, the results from normal animals indicate that the timing of the innervation predicts the density of the innervation. Since attenuation of the olfactory bulb input by LOT lesion or partial bulb ablation reduces the extent of the projection, it is likely that the population of mitral and tufted cells has a fixed and limited intrinsic growth potential (Devor 1976h). There is no evidence that the cortex limits its innervation by the bulb, since areas rostral to LOT lesions can be hyperinnervated by the bulb after the LOT fibers are "dammed up" by the lesion (Devor 1976b; Schwob unpublished observations). We would suggest as a hypothesis for further experimental evaluation that the normal axonal pattern may be explained by the preferential expenditure of this growth potential in the areas innervated earlier in the course of development.

3 Development of Intracortical Associational and Commissural Fibers

The ipsilateral intracortical fiber systems develop to some extent prenatally in rats, but there is also a substantial degree of maturation after birth. Their development can be assessed via axonal transport of ^{3}H-amino acids which have been injected into the cortex (Schwob and Price 1984a), or alternatively, through the use of the Timm stain for heavy metals which selectively labels presynaptic terminals of the intracortical fibers in adults and does not stain fibers from the olfactory bulb (Friedman and Price 1984). The earliest intracortical fibers to develop are those from the posterior piriform cortex which can be labelled by anterograde transport on E19-20 (Schwob and Price 1984a). These fibers extend rostrally for a short distance into layer I of the piriform cortex lateral to the LOT and into the deeper layers of the olfactory tubercle, and caudally into layer I of more posterior parts of the piriform cortex. The rostral and caudal limits of this projection expand over the first few days after birth, and fibers enter layer I of the medial olfactory tubercle at about the same time. Similarly, the ipsilateral fibers from the anterior piriform cortex grow into layer I of the posterior piriform cortex by E20-21 and the entorhinal area by P1. The development of the Timm staining pattern in the cortex (Friedman and Price 1984) correlates well with the analysis of fiber outgrowth by axonal tracing techniques. For example, in the posterior piriform cortex there is Timm staining in layer I at E19-20.

On the other hand, some specific sets of intracortical fibers are markedly delayed in their growth into layer I. These provide an excellent example of apparent competition for synaptic space based on the time of innervation. The anterior commissural fibers from the anterior piriform cortex to the contralateral piriform and entorhinal cortices cross the midline about the time of birth and reach the deeper layer of the cortex by P3, but they do not enter layer I until after P9. In the adult, these fibers occupy the same cortical area and lamina as the much earlier developing ipsilateral associational fibers described above, and it is striking that the density of innervation by the commissural fibers is markedly less than by the ipsilateral fibers. In contrast to this, both the ipsilateral and commissural axonal projections from the anterior olfactory nucleus to the olfactory cortex are delayed and do not grow into layer I until after the end of the first postnatal week, although they are present in the deeper layers of the cortex for several days prior to this. The ipsilateral and contralateral fibers grow into layer I simultaneously on the two sides. Again the two sets of fibers end in the same area and lamina, but here the density of the projection is nearly equivalent on the two sides (Luskin and Price 1983). Thus, in these cases where ipsilateral and commissural fibers innervate the same cortical zone, the relative density of synaptic termination would appear to be determined by the timing of innervation (Schwob and Price 1984a). Similarly a temporal influence on synaptogenesis has been described in the dentate gyrus (Gottlieb and Cowan 1972; D.D.M.O'Leary et al. 1979).

As with the fibers from the olfactory bulb, there is electro-physiological evidence that at least some of the associational fibers are synaptically active at or shortly after birth (Schwob et al. 1984). A component of the field potential response to LOT stimulation which corresponds to the synaptic reactivation of the cortex by the associational fibers (the B_1 wave) can be putatively identified shortly after birth (Fig 9). As in adults, this putative B_1 wave is inhibited and facilitated in parallel with cortical unit activity, and its depth distribution can be distinguished from that of the A_1 wave described above.

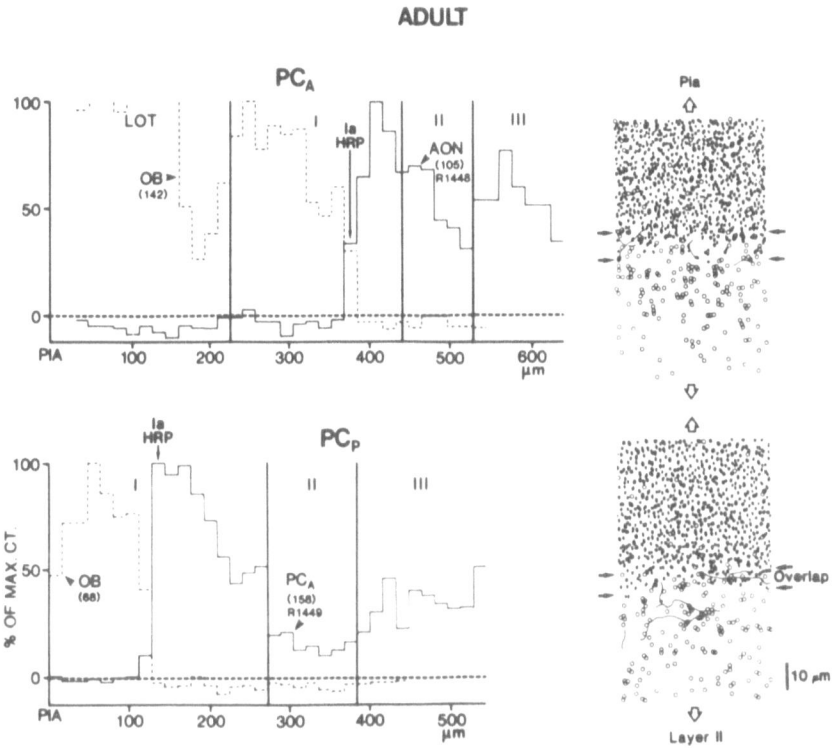

ADULT

Fig. 3 Laminar distribution of labeled bulbar and associational fibers in the anterior and posterior piriform cortex in adult rats. The graphs on the left compare autoradiographic grain counts of transported ^{3}H-leucine from the ol-factory bulb (dashed line) and olfactory cortex (solid line) of two separate animals. The grain counts were adjusted by subtracting background and then normalizing each count to the maximum count in layer I. In the experiments with radioactive labeling of the associational fibers, the deep boundary of the zone oc-cupied by the bulb fibers (designated "la HRP") was also mapped on adjacent sections by the anterograde transport of HRP from the olfactory bulb. The drawings on the right also represent double-labeling experiments and compare, on the same section, the distribution of varicosities on LOT axons diffusely filled with HRP (filled profiles) with the distribution of anterogradely transported radioactive label from the piriform cortex (autoradiographic grains are represented as open circles). These types of experiment indicate that the la-lb boundary is sharp and that the zone where the two types of fibers overlap (parallel sets of arrows) is less than 10 μm wide

4 Development of the Diencephalic Olfactory Cortical Projections

The two major cortical projections to the diencephalon originate from the large neurons of the endopiriform nuclei deep to the piriform cortex and from similar cells situated in the deeper layers of the olfactory tubercle (Price 1977; Price and Slotnick 1983). These send axons to the mediodorsal thalamic nucleus and the lateral hypothalamic area (Powell et al. 1965; Scott and Leonard 1971; Scott and Chafin 1975; Krettek and Price 1977, 1978). By the time of birth, there are fibers from the olfactory cortex which project to the diencephalon. Those to the lateral hypothalamic area have reached the level of the subthalamic nucleus by E18 and the posterior edge of the supramamillary nucleus by E20 (Schwob and Price 1984a). The fibers to the mediodorsal thalamic nucleus have entered the stria medullaris just dorsal to this structure by E20, but a substantial projection into the thalamus is not seen until the time of birth (Schwob and Price 1984a).

5 Development of Afferent Lamination in the Olfactory Cortex

5.1 Normal Development

In adult rats, the boundary in layer I between the fibers from the olfactory bulb (in layer Ia) and the intracortical fibers (in layer Ib) is very sharp, and the zone where these two sets of fibers overlap is less than 10 μm out of a total layer I width of 250-300 μm based on double labeling experiments in adult rats (Fig. 3; Schwob and Price 1984b).

In neonates, the afferent fibers are initially poorly laminated in comparison with the adult but then segregate into complementary zones in layer I (Schwob and Price 1984b). For example, in the anterior olfactory nucleus, anterior piriform cortex and the posterior piriform cortex, the two types of fibers overlap each other throughout layer I at birth, although the bulb fibers are concentrated more superficially and the associational fibers are concentrated more deeply in the molecular layer at this time (Fig. 4). During the remainder of this first week, the zone of overlap in layer I becomes progressively smaller throughout the cortex (Fig. 4). This emergence of precise afferent lamination occurs during the major period of synaptogenesis (Westrum, 1975a, b). Indeed early in the postnatal period there are postsynaptic sites in the piriform cortex which are occupied by more than one presynaptic profile; these shared sites become less common with time (Westrum 1975a). While the presynaptic profiles which contribute to these shared sites have not been identified, their presence and subsequent disappearance are indicators, at the ultrastructural level, of some form of an ongoing synaptic rearrangement early in the postnatal period (Westrum 1975a, b).

These results suggest that some form of interaction between the bulbar and intracortical afferents is responsible for segregating them and setting the boundary in layer I between them. Alternatives to this hypothesis take two forms. The first would suggest that the afferents laminate by virtue of the independent assortment of to types of postsynaptic markers into two spatially segregated zones on the cortical dendrites. Although this assortment of "chemoaffinity" markers (Sperry 1963) might occur during the course of development and thus account for the pattern observed during normal development, it would not be dependent on an

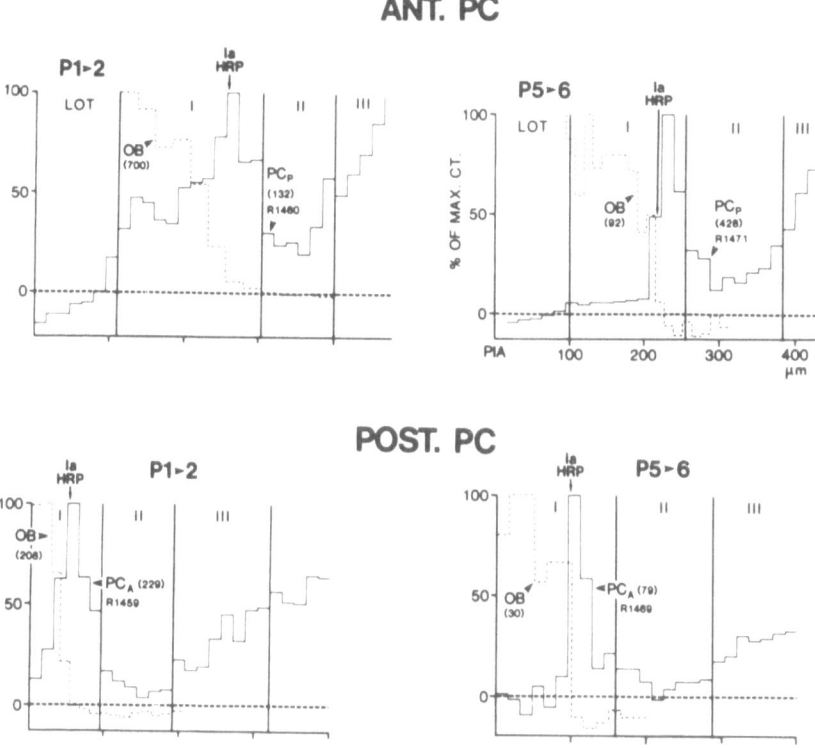

Fig. 4 Laminar distribution of labeled bulbar and associational fibers in the anterior and posterior piriform cortex in neonatal rats. Conventions and analysis as in Fig. 4. In contrast to the adult, the two sets of fibers overlap extensively near the time of birth (**P1→2**) but then segrate into complementary lamina as the Ia–Ib boundary is established during the remainder of the first week (**P5→6**)

interaction between the two sets of fibers. A strong argument against this theory is provided by observations on reeler mutant mice (Devor et al. 1975). This autosomal recessive mutation causes an inversion of cellular lamination in the piriform cortex as well as in other cortical areas (Caviness and Sidman 1972). As a result, the pyramidal cells normally found in layer II are situated deeper in the cortex than usual and are deep to the neurons which would normally occupy layer III and the deeper layers of the cortex. Despite the resulting dendritic displacement, afferents from the olfactory bulb are situated in the superficial part of the molecular layer and are segregated, as in normal animals, from the more deeply placed intracortical fibers (Devor et al. 1975). This result makes it unlikely that afferent lamination is solely produced by factors which are spatially distributed along dendrites as a result of some a priori arrangement independent of the ingrowing fibers. A similar result was obtained in the dentate gyrus of reeler mutant mice (Stanfield et al. 1979; Cowan et al. 1981).

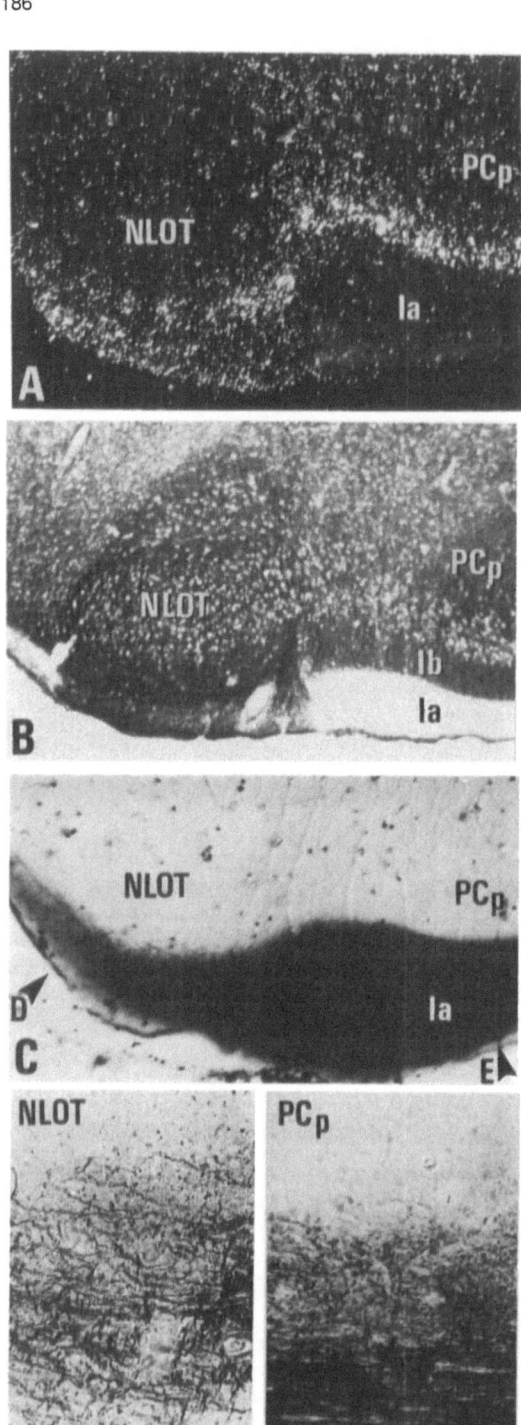

Fig. 5 The distribution of associational and bulbar fibers to the nucleus of the lateral olfactory tract (NLOT) in adult rats. A Darkfield photomicrograph of autoradiographic labelling after injection of ^{3}H-leucine into the anterior piriform cortex. The labelled associational fibers extend to the pia in layer I of the NLOT but not of the adjacent posterior piriform cortex. B The pattern of Timm staining also indicates that the associational fibers extend through layer I of the NLOT but not in the adjacent posterior piriform cortex. C HRP-filled axons from the LOT are present in the middle and superficial part of layer I where they overlap with the associational fibers. This overlap contrasts with the precise complementarity of layer Ia and Ib in the piriform cortex. D, E In higher power photomicrographs (indicated by the arrowheads in C, the LOT axons in layer I of the NLOT are seen to be thicker and lacking in varicosities as compared with those in the posterior piriform cortex

The second alternative hypothesis would suggest that an interaction of the two sets of fibers by themselves (for example, by a type of selective fasciculation of like fibers; Stanfield et al. 1979) could determine the layer Ia-Ib boundary independent of the dendritic field. However, two normal observations tend to rule out this alternative as well (Schwob and Price 1984b). In both cases, segments of olfactory bulb fibers which form few synapses with cortical cells do not segregate from the intracortical fibers. In the first case, a few axons from the olfactory bulb leave layer Ia and traverse layer Ib before re-entering their normal terminal zone. These can be visualized by filling them diffusely with HRP injected directly into the LOT. While they are in layer Ib these aberrant axons are largely free of synaptic boutons, as indicated by the absence of varicosities which have been identified as boutons in many systems (Adams and Warr 1976; Ferster and LeVay 1978; Rainey and Jones 1981). Large numbers of these varicosities are found in layer Ia. The second set of observations was made in the nucleus of the lateral olfactory tract (NLOT). The NLOT is unique among all of the olfactory cortical areas in that olfactory bulb and associational fibers overlap substantially as shown by axonal tracing studies and by Timm staining (Fig. 5) (Schwob and Price 1984b; Friedman and Price 1984). In this case, the fibers from the olfactory bulb are concentrated in the middle of layer I of the nucleus and form very few synapses. This has been shown both by the absence of varicosities along fibers diffusely filled with HRP (Fig. 5) (Schwob and Price 1984b) and by electron microscopical examination (Scalia and Winans 1975; Heimer 1975).

5.2 Effects of Disruption of the Olfactory Bulb Fibers

The observations on normal development suggest that segregation of the afferent fibers depends on an interaction between them and the dendritic field in layer I and that this interaction is most likely based on synapse formation between axons and dendrites. This hypothesis is also suggested by the results of experiments with complete or partial ablations of the olfactory bulb in neonatal animals. Following complete removal of the bulb in early development, the associational fibers extend into the deafferented part of layer I innervating the more superficial aspects of the pyramidal dendrites as shown by axonal tracing techniques or the colored reaction product seen with the Timm stain (Fig. 6; Westrum 1975b; Price et al. 1976, Friedman and Price 1981, 1985). Furthermore, if the ablation is done at birth, layer I and the dendrites which comprise it continue to grow to within 10% of their normal adult size (Friedman and Price 1981, 1985). After such neonatal ablations there would seem to be a competitive type of reinnervation; vacated postsynaptic specializations are rapidly re-occupied by one or sometimes more presynaptic specializations in what may be only a modification of the normal pattern of synaptogenesis (Westrum 1980). Moreover, the process continues until a normal synaptic density is found in layer I on the lesioned side (Friedman and Price 1981). In contrast to this, following bulb ablation in adults there is a marked shrinkage of layer I and a failure of the associational fibers to sprout into the residual layer Ia, although other axons (which have been identified by the granular reaction seen with the Timm staining method and have been seen with the electron microscope) do sprout into this zone (Fig. 6; Caviness et al. 1977; Friedman and Price 1985). Furthermore, this limited synaptic sprouting in layer Ia after adult ablation takes weeks to months (Caviness et al. 1977) while the more extensive sprouting after neonatal ablation occurrs in days (Friedman and Price 1985). After bulb ablations during the second and third postnatal weeks, sprouting by the associational fibers is less vigorous, since the fibers, which do expand to fill layer I, are insufficient to maintain the normal dendritic growth, and the width of layer I grows to only about 70% of the normal side (Friedman and Price 1985). However, it should be noted that the period during which bulbectomy can induce sprouting of associational fibers to fill layer I extents beyond the time when afferent lamination is well established in normal animals (about P7, see above). A similar dissociation between the time when afferents normally become segregated and the end of the period when there is lesion-induced sprouting has been reported for the ocular dominance stripes in the primate visual cortex (LeVay et al. 1980).

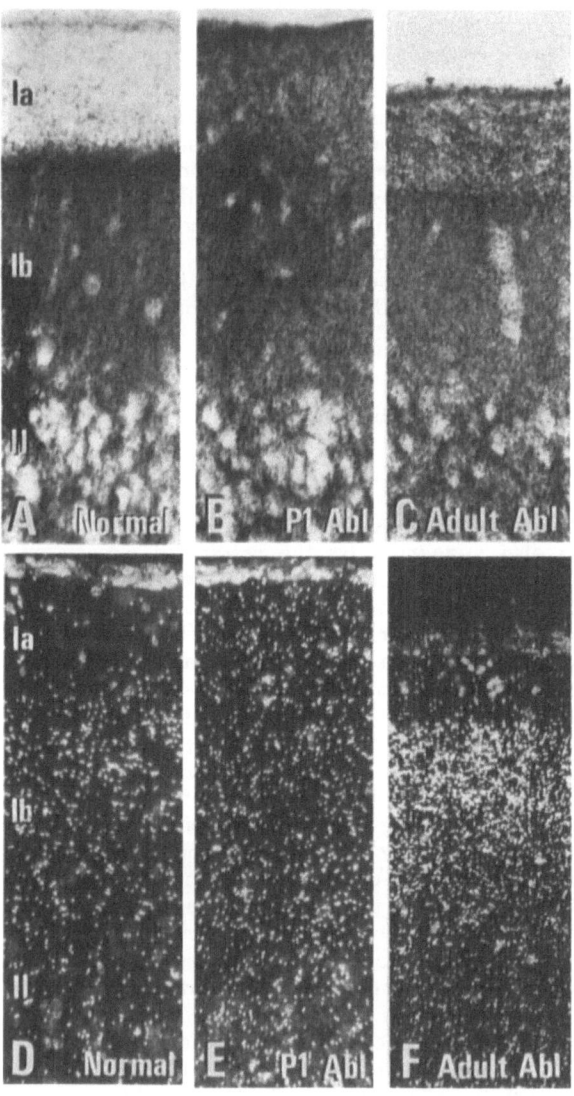

Fig. 6. Distribution of associational fibers in unoperated controls (**A, D**), after complete unilateral ablations of the olfactory bulb at P1 with survival to adulthood (**B, E**) and after complete bulb ablation in adults with long term survival (**C, F**). A–C Timm staining of the posterior piriform cortex. D–F Distribution of anterogradely transported label from rostral piriform cortex on adjacent sections. **B, E** The Timm color reaction which is typical of the associational fibers (Friedman and Price 1984) and transported label in the associational fibers fill layer I after P1 ablations. **C, F** After adult ablation, the associational fibers remain restricted to the deeper part of layer I. The outer molecular layer is shrunken, and there is an increase in the granular type of Timm staining which indicates sprouting of an axonal population other than the associational fibers

Other experimental manipulations such as partial ablations of the olfactory bulb in newborn rats reduce rather than completely eliminate the fibers from the olfactory bulb (see Sect. 2.3). In these experiments, the distribution of both the associa-

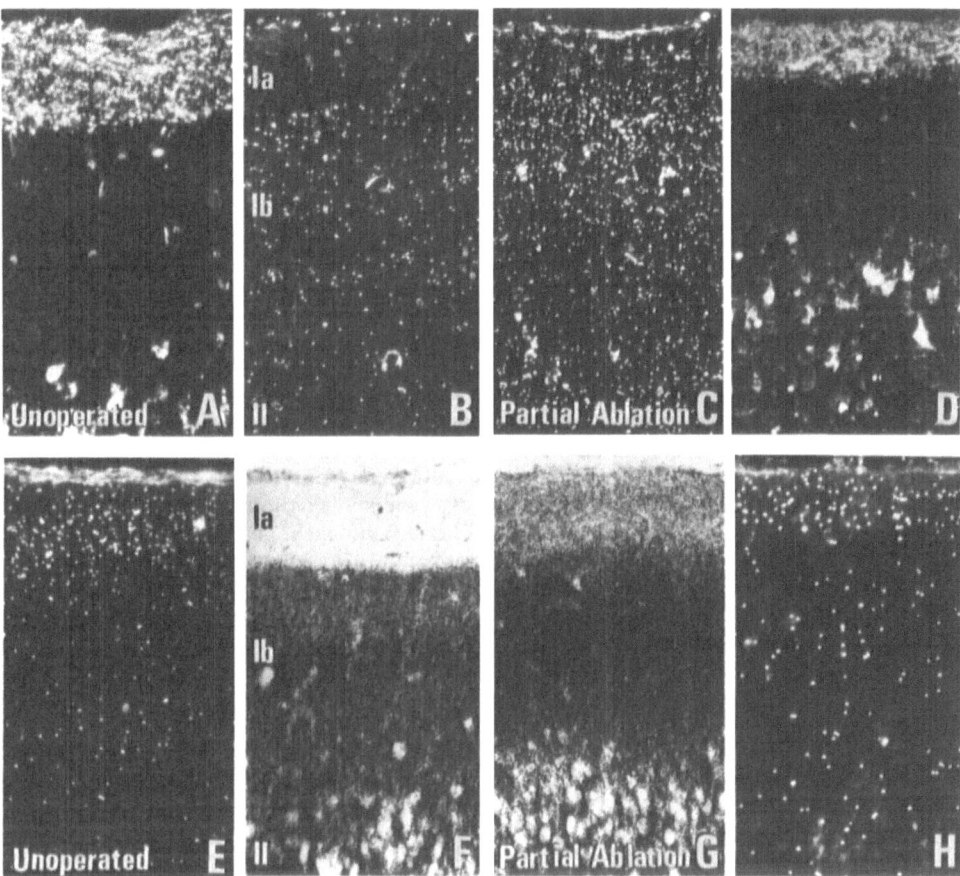

Fig. 7 The distribution of olfactory bulb and intracortical fibers as a result of unilateral partial ablation at P1 assessed after survival to adulthood. Each row of photomicrographs represents a single double labeling experiment and illustrates the posterior piriform cortex on the unoperated (**A,B,E,F**) and partially ablated sides (**C,D,G,H**). The photographs on the flanks in each row show the distribution of olfactory bulb fibers: in the case illustrated in the top row HRP was used as the anterogradely transported marker (**A,D**). In the bottom row ^{3}H-leucine was used 8E,H). The distribution of intracortical fibers was evaluated on adjacent sections: in the experiment shown in the top row, ^{3}H-leucine was used as the anterogradely transported marker from the anterior piriform cortex (**B,C**); in the experiment shown in the bottom row, Timm staining was used to assess the distribution of the intracortical fibers (**F,G**). In both experiments, the bulb and associational fibers form precisely segregated complementary lamina (**Ia** and **Ib**, respectively) on the unoperated side, but the residual fibers from the olfactory bulb overlap the associational fibers in the superficial part of layer I on the partially ablated side

tional and remaining olfactory bulb fibers is altered (Schwob unpublished observations). The zone occupied by the bulb fibers is reduced on the ablated side, and they are not strictly segregated from the associational fibers. This breakdown in the lamination of the afferents is most apparent in the more caudal parts of the posterior piriform cortex and other adjacent olfactory cortical structures where the labelled associational fibers extend throughout the molecular layer, although they are still concentrated more deeply (Fig. 7). With ablations larger than 50% of the bulb, this type of pattern is also observed in more rostral parts of the olfactory cortex, and the breakdown in lamination is more pronounced posteriorly. In some areas, there is a roughly trilaminar arrangement of the fibers in layer I with a central band of olfactory bulb fibers sandwiched between accumulations of associational fibers adjacent to the pia and deeper in the molecular layer. This pattern in the experimental animals is quite similar to the arrangement of the fibers in the NLOT where the bulb and associational fibers overlap in normal rats (see above; cf. Figs. 5, 7) (Friedman and Price 1984; Schwob and Price 1984b).

At first glance, the effects produced by partial ablation are puzzling. During normal development complementary lamination is as well established in areas lightly innervated by the bulb, such as the lateral entorhinal area and the medial olfactory tubercle as in the more densely innervated areas despite the remarkable reduction in the width of layer Ia in these lightly innervated areas as compared to the more rostral piriform cortex deep to the LOT (Fig.1; Schwob and Price 1978, 1984a,b). However, partial ablation of the olfactory bulb at birth produces more than a simple reduction in the number of the innervating bulb fibers, since there would also be rapid degeneration of established synapses in the cortex resulting in multiple vacated post-synaptic sites (Westrum 1975b; Westrum 1980; Bakay and Westrum 1984). The axons from the bulb and their synapses which remain after the lesions may not be sufficiently plastic to detach and then translocate a substantial distance along the dendrite. In the face of the limited growth potential of the bulb fibers (Devor 1976b), this type of translocation would be required in the experimental cases to mimic the more lightly innervated areas in the normal animal where the fewer bulb fibers are concentrated in a narrower zone. As a result, the remaining fibers in the experimental animals would be diluted and unable to exclude fully the associational projection (although they do so partially). This is similar to the situation in the NLOT of normal rats where the bulb fibers make few synapses and where the afferents do not form complementary laminae.

Moreover, the converse condition to a reduction in olfactory bulb innervation is obtained rostral to neonatal LOT section. In these areas which are hyperinnervated by the bulb fibers (possibly as a result of "damming up" of fibers which cannot growth beyond the cut; see Sect. 2.3), an expanded layer Ia is formed which excludes Timm-stained associational fibers (Fig. 8). This is true both in areas which normally receive a very light bulb projection and in areas which receive none at all. This result strongly supports the hypothesis that an interaction of the two sets of fibers is responsible for establishing the Ia-Ib boundary.

6 Development of Interneuronal Circuitry in the Piriform Cortex

In the adult piriform cortex, electrical stimulation of the LOT evokes a strong monosynaptic activation of cortical pyramidal cells manifested by the A_1 wave of the evoked field potential and by short latency activation of cortical units in vivo (Freeman 1968; Biedenbach and Stevens 1969a, b; Haberly 1973a, b); unit activation is then coupled with reactivation of the cortical neurons by the intracortical associational fibers as manifested by the B_1 wave of the field potential (Haberly 1973b; Haberly and Shepherd 1973; Habely and Bower 1984). In addition, a potent inhibitory circuit is engaged by LOT shock which completely blocks unit activation and cortical reactivation by the associational fibers in response to a second LOT shock at a short intershock interval (Freeman 1968; Biedenbach and Stevens 1969a, b; Haberly 1973a,b; Haberly and Shepherd 1973; Haberly and Bower 1984).

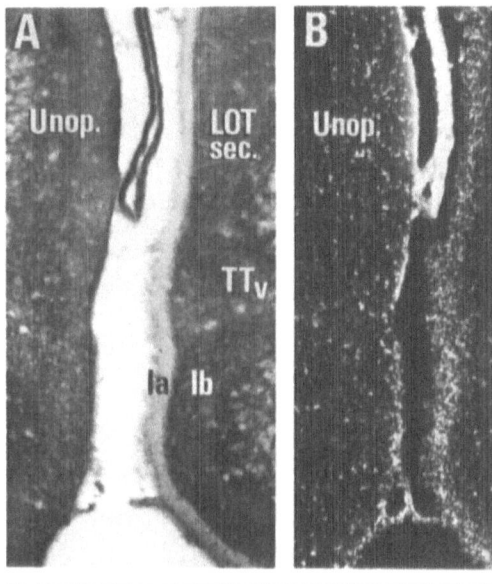

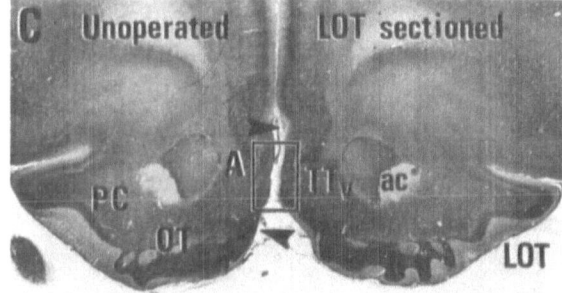

Fig. 8 Distribution of associational and bulbar fibers in an adult rat subjected to a knife cut lesion of the LOT at P1. **A** Timm stained section illustrating the distribution of associational fibers in the vicinity of the ventral tenia tecta on the lesioned and unoperated sides. **B** Adjacent section from the same rat which shows the distribution of labeled fibers following bilateral injections of ^{3}H-leucine in the olfactory bulbs. **C** Low power photomicrograph of the Timm-stained section. The position of **A** and **B** is indicated by the box designated **A**. On the LOT-lesioned side, the bulb projection to the medial surface of the hemisphere is substantially increased in density and tangential extent, and a distinct layer **Ia** which excludes the Timm stained associational fibers is formed here. At this point on the unoperated side where the bulb projection is extremely sparse a Timm-negative layer **Ia** is lacking. The medial extent of layer **Ia** in the Timm-stained section is indicated on the two sides by arrowheads in **C**

In contrast to the potency of the inhibitory circuitry in the adult, at birth inhibition is weak or absent, as documented by paired shock analysis in lightly anesthetized pups in vivo (Schwob et al. 1984). Rather, there is a facilitation of test shock-evoked unit activity which is apparent in recordings of single or multiple unit potentials or of the component of the field potential (termed the S wave) that corresponds to synchronous unit activation (Fig. 9). Paired shocks also facilitate the monosynaptic EPSP as shown by an increase in the A_1 wave, and this synaptic facilitation is most likely responsible for the facilitation of the unit activity at short intershock intervals. A weak inhibitory circuit is established within the first few days after birth, since some inhibition of unit activity becomes apparent with the decay of synaptic facilitation at longer intervals

192

(Fig. 9). Once established, the inhibition lasts for much longer intershock intervals than in adults (Fig. 9). By 2 weeks of age, a conditioning shock of the LOT produces complete or near complete suppression of subsequent evoked unit activity at short intervals, and a more adult-like time course of recovery from the inhibition. This result correlates with the relatively late accumulation of inhibitory-type synapses in the piriform cortex (compared to excitatory synapses) during the second week of life (Westrum 1975a).

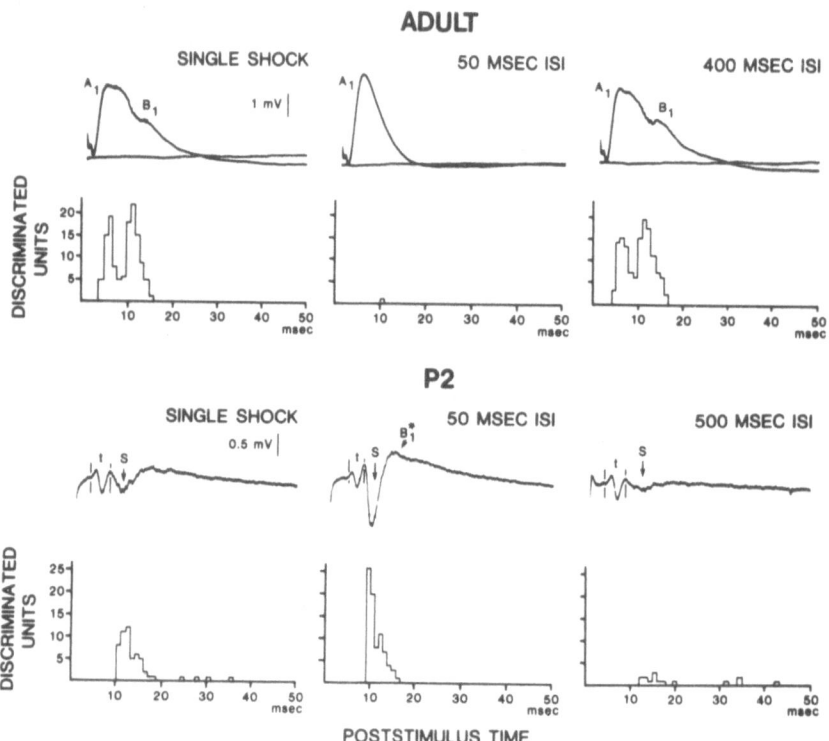

Fig. 9 Field potentials and unit activity in the piriform cortex evoked by single or paired shocks of the LOT in adult rats and at P2. Upper trace in each group is a single sweep of the LOT shock-evoked field potential. Lower graph in each group is a single sweep of the LOT shock-evoked field potential. The lower graph in each group is a poststimulus time histogram of evoked unit activity from multiple sweeps. At P2, S designates the S wave which is the field potential representation of synchronous unit activity and which is simultaneous with evoked unit activity. B_1* designates the presumptive B_1 wave (which in adults corresponds to cortical reactiviation by associational fibers). Note that at P2, evoked unit activity, the S wave and the B_1* wave are facilitated in parallel (when compared with a single shock) by paired shocks with short intershock intervals (ISIs). At P2 inhibition of all three was evident with ISIs longer than 300 ms. Both the adult and P2 records were taken from the vicinity of layer II

Similarly, inhibitory circuitry develops relatively late in the hippocampus (e.g., Mueller et al. 1984) and the lateral geniculate nucleus (Shatz and Kirkwood 1984).

7 Conclusion

Certain features of the central olfactory system are well developed at birth, while others show a significant degree of postnatal development. The olfactory bulb projects widely across the olfactory cortex at birth but has not yet innervated the medial half of the olfactory tubercle and some of the far posterior cortical structures. At this time, there is monosynaptic activation of the piriform cortex. Some of the intracortical associational fibers are also in place at birth and there is evidence of physiological reactivation of the cortex by these fibers after LOT stimulation. The appearance of other intracortical projections is delayed until after the end of the first week; these include the ipsilateral and contralateral projections of the AON and the commissural projection of the piriform cortex. The lamination of afferents in layer I also emerges postnatally. One of the most striking postnatal changes is the virtual absence of functional cortical inhibition in response to LOT shock at birth, and its slow maturation over the first two weeks of life. At birth, outflow fibers from the olfactory cortex can be traced to the mediodorsal thalmic nucleus and the lateral hypothalamic area.

The usefulness of this normative data in the analysis of olfaction in the developing animal has already been illustrated by the studies of Leonard and co-workers on thermotaxis in hamsters. In hamster pups, movement toward warmth is present during the first postnatal week but is subsequently lost (Leonard 1978; Small and Leonard 1983). However, this loss of thermotactic behavior is prevented by ablation of the olfactory bulb during the first week. The normal loss of thermotaxis occurs when pups become responsive to conspecific odors (Devor and Schneider 1974; Crandall and Leonard 1979) and begin to thermoregulate by huddling with littermates (Leonard 1982). It is also temporally associated with innervation of the olfactory tubercle by the bulb (Leonard 1975; Schwob and Price 1978, 1984a9. Indeed, as in normal development, thermotaxis is lost after lesions of the LOT in neonates when the olfactory tubercle is innervated but it is not lost if the tubercle remains denervated (Small and Leonard 1983). There are other examples of olfactory mediated behavior in neonates which must depend on a somewhat different central olfactory substrate. For example, nipple attachment at birth is apparently mediated by olfactory cues (Blass and Teicher 1980) which appear to be processed by a discrete portion of the olfactory bulb which is preferentially activated by nipple washings (Teicher et al. 1980).

The data presented here also raise some question about the role of cortical inhibitory circuitry in olfactory function in the neonate. As early as P2 in rat pups there is evidence for the formation of olfactory mediated conditioned aversions (Rudy and Cheatle 1977) and for behavioral responses to maternal odors (Schapiro and Salas 1970; although see Gregory and Pfaff 1971) and to home nesting material (Sczerzenie and Hsiao 1977); cortical inhibitory circuitry is quite immature at this time. Analysis of the degree of refinement of olfactory discriminations is hampered by the limited behavioral repertoire of the neonatal rat pups, and the acquistion of a fine degree of olfactory discrimination has unfortunately not been assessed.

Summary

The experiments summarized here document the growth and development of the projections of the olfactory bulb, the intracortical and subcortical axonal connections of the olfactory cortex, and the maturation of the physiological response of the piriform cortex to electrical stimulation of the fibers from the bulb. The results obtained during normal development or as a consequence of experimental disruption suggest that factors operating during the course of development determine several important features of axonal connections in the olfactory cortex. The effects of more subtle intervention, such as lesion of the olfactory epithelium or naris occlusion, on the pattern of central olfactory connections, can now be placed within the context of these normative and experimental studies.

Acknowledgments. The preparation of this review was supported by NIH research grants NS09518 and 15070 to JLP and JES was supported by NIH postdoctoral training grant NS07076. The authors thank Joe Hayes for excellent photographic assistance.

References

Adams JC, Warr WB (1976) Origins of axons in the cat's acoustic striae determined by injection of horseradish peroxidase into severed tracts. J Comp Neurol 170: 107–121

Bakay RAE, Westrum LE (1984) Age-related fine structural changes in axons and synapses during deafferentation of the rat piriform cortex: a possible basis for plasticity. J Neurocytol 13: 743–765

Bayer SA (1983) ^{3}H-thymidine-radiographic studies of neurogenesis in the rat olfactory bulb. Exp Brain Res 50: 329–340

Biedenbach MA, Stevens CF (1969a) Electrical activity in cat olfactory cortex produced by synchronous orthodromic volleys. J Neurophysiol 32: 193–203

Biedenbach MA, Stevens CF (1969b) Synaptic organization of cat olfactory cortex as revealed by intracellular recording. J Neurophysiol 32: 204–214

Blass EM, Teicher MH (1980) Suckling. Science 210: 15–22

Broadwell RD (1975) Olfactory relationships of the telencephalon and diencephalon of the rabbit. I. An autoradiographic study of the efferent connections of the main and accessory olfactory bulbs. J Comp Neurol 163: 329–345

Caviness VS Jr, Sidman RL (1972) Olfactory structures of the forebrain in the reeler mutant mouse. J Comp Neurol 145: 85–104

Caviness VS Jr, Korde MG, Williams RS (1977) Cellular events induced in the molecular layer of the piriform cortex by ablation of the olfactory bulb in the mouse. Brain Res 134: 13–34

Cowan WM, Stanfield BB, Amaral DG (1981) Further observations on the development of the dentate gyrus. In: Cowan WM (ed) Studies in developmental neurobiology. Oxford University Press, London, pp 395–435

Crandall JE, Leonard CM (1979) Developmental changes in thermal and olfactory influences on golden hamster pups. Behav Neurol Biol 26: 354–363

Derer P, Caviness VS Jr, Sidman RL (1977) Early cortical histogenesis in the primary olfactory cortex of the mouse. Brain Res 123: 27-40

Devor M (1976a) Fiber trajectories of olfactory bulb efferents in the hamster. J Comp Neurol 166: 31-48

Devor M (1976b) Neuroplasticity in the rearrangement of olfactory tract fibers after neonatal transection in hamsters. J Comp Neurol 166: 49-72

Devor M, Schneider GE (1974) Attraction to home-cage odor in hamster pups: specificity and changes with age. Behav Biol 10: 211-221

Devor M, Caviness VS Jr, Derer p (1975) A normally laminated afferent projection to an abnormally laminated cortex: some olfactory connections in the reeler mouse. J Comp Neurol 164: 471-482

Ferster D, LeVay S (1978) The axonal arborization of lateral geniculate neurons in the striate cortex of the cat. J Comp Neurol 182: 925-944

Freeman WJ (1968) Relations between unit activity and evoked potentials in pre-pyriform cortex of cats. J Neurophysiol 31: 337-348

Friedman B, Price JL (1981) Establishment of normal synaptic density in de-afferented olfactory cortex. Brain Res 223: 146-151

Friedman B, Price JL (1984) Fiber systems in the olfactory bulb and cortex: a study in adult and developing rats, using the Timm method with the light and electron microscope. J Comp Neurol 223: 88-109

Friedman B, Price JL (1985) Plasticity in the olfactory cortex: age-dependent affects of deafferentation. J Comp Neurol (submitted)

Gottlieb DI, Cowan WM (1972) Evidence for a temporal factor in the occupation of available synaptic sites during the development of the dentate gyrus. Brain Res 41: 452-456

Grafe MR (1983) Developmental factors affecting regeneration in the central nervous system: early but not late formed mitral cells reinnervate olfactory cortex after neonatal tract section. J Neurosci 3: 617-630

Gregory EH, Pfaff DW (1971) Development of olfactory-guided behavior in infant rats. Physiol Behav 6: 573-576

Haberly LB (1973a) Unitary analysis of opossum prepyriform cortex. J Neuro-physiol 36: 762-774

Haberly LB (1973b) Summed potentials evoked in opossum prepyriform cortex. J Neurophysiol 36: 775-788

Haberly LB, Bower JM (1984) Analysis of association fiber system in piriform cortex with intracellular recording and staining techniques. J Neurophysiol 51: 90-112

Haberly LB, Price JL (1977) The axonal projection patterns of the mitral and tufted cells of the olfactory bulb in the rat. Brain Res 129: 152-157

Haberly LB, Price JL (1978a) Association and commissural fiber systems of the rat. I. Systems originating in the piriform cortex and adjacent areas. J Comp Neurol 178: 711-740

Habely LB, Price JL (1978b) Association and commissural fiber systems of the olfactory cortex of the rat. II. Systems originating in the olfactory peduncle. J Comp Neurol 181: 781-808

Haberly LB, Shepherd GM (1973) Current-density analysis of summed evoked potentials in opossum prepyriform cortex. J Neurophysiol 36: 789-802

Heimer L (1975) Olfactory projections to the diencephalon. In: Stumpf WE, Grant LD (eds) Anatomical neuroendocrinology. Karger, Basel, pp 30-39

Hinds, JW (1967) Autoradiographic study of histogenesis in the mouse olfactory bulb. Thesis, Harvard University, Cambridge MA

Hinds JW (1968) Autoradiographic study of histogenesis in the mouse olfactory bulb. I. Time of origin of neurons and neuroglia. J Comp Neurol 134: 287-304

Hinds JW (1972) Early neuron differentiation in the mouse olfactory bulb. I. Light microscopy. J Comp Neurol 146: 233-252

Krettek JE, Price JL (1977) Projections from the amygdaloid complex to the cerebral cortex and thalamus in the rat and cat. J Comp Neurol 172: 687-722

Krettek JE, Price JL (1978) Amygdaloid projections to subcortical structures within the basal forebrain and brainstem in the rat and cat. J Comp Neurol 178: 225-254

Leonard CM (1975) Developmental changes in olfactory bulb projections revealed by degeneration argyrophilia. J Comp Neurol 162. 467-486

Leonard CM (1978) Maturational loss of thermotaxis prevented by olfactory lesions in golden hamster pups (Mesocricetus auratus). J Comp Physiol Psychol 92: 1084-1094

Leonard CM (1982) Shifting strategies for behavioral thermoregulation in developing golden hamsters. J Comp Physiol Psychol 96: 234-243

LeVay S, Wiesel TN, Hubel DH (1980) The development of ocular dominance columns in normal and visually deprived monkeys. J Comp Neurol 191: 1-151

Luskin MB, Price JL (1982) The distribution of axonal collaterals from the olfactory bulb and the nucleus of the horizontal limb of the diagonal band to the olfactory cortex demonstrated by double retrograde labelling techniques. J Comp Neurol 209: 249-263

Luskin MB, Price JL (1983) The laminar distribution of intracortical fibers originating in the olfactory cortex of the rat. J Comp Neurol 216: 292-302

Mueller AL, Taube JS, Schwartzkroin PA (1984) Development of hyperpolarizing inhibitory postsynaptic potentials and hyperpolarizing response to aminobutyric acid in rabbit hippocampus studied in vitro. J Neurosci 4: 860-867

O'Leary DDM, Fricke RA, Stanfield BB, Cowan WM (1979) Changes in the associational afferents to the dentate gyrus in the absence of its commissural input. Anat Embryol 156: 283-299

O'Leary JL (1937) Structure of the primary olfactory cortex of the mouse. J Comp Neurol 67: 1-31

Powell TPS, Cowan WM, Raisman G (1965) The central olfactory connexions. J Anat 99: 791-813

Price JL (1973) An autoradiographic study of complementary laminar patterns of termination of afferent fibers to the olfactory cortex. J Comp Neurol 150: 87-108

Price JL (1977) Structural organization of the olfactory pathways. In: LeMagnen J and MacLeod P (eds) Olfaction and Taste, vol VI. Information Retrieval, London, pp 87-95

Price JL, Slotnick BM (1983) Dual olfactory representation in the rat thalamus. An anatomical and electrophysiological study. J Comp Neurol 215: 63-77

Price JL, Moxley GF, Schwob JE (1976) Development and plasticity of complementary afferent fiber systems to the olfactory cortex. Exp Brain Res Suppl 1: 148-154

Rainey WT, Jones EG (1981) Terminations of lemniscal axons in the thalamic ventrobasal complex of the cat labelled by HRP injections into the medial lemniscus. Anat Rec 199: 207A-208A

Rudy JW, Cheatle MD (1977) Odor-aversion learning in neonatal rats. Science 198: 845-846

Scalia F, Winans SS (1975) The differential projections of the olfactory bulb and accessory olfactory bulb in mammals. J Comp Neurol 161: 31-56

Schapiro S, Salas M (1970) Behavioral response of infant rats to maternal odor. Psychol Behav 5: 815-817

Schwob JE, Price JL (1978) The cortical projection of the olfactory bulb: development in fetal and neonatal rats correlated with quantitative variations in adult rats. Brain Res 151: 369-374

Schwob JE, Price JL (1984a) The development of axonal connections in the central olfactory system of rats. J Comp Neurol 223: 177-202

Schwob JE, Price JL (1984b) The development of lamination of afferent fibers to the olfactory cortex in rats, with additional observations in the adult. J Comp Neurol 223: 203-222

Schwob JE, Haberly LB, Price JL (1984) The development of physiological responses of the piriform cortex in rats to stimulation of the lateral olfactory tract. J Comp Neurol 223: 223-237

Scott JW, Chafin BR (1975) Origin of olfactory projections to lateral hypothalamus and nuclei gemini in the rat. Brain Res 88: 64-68

Scott JW, Leonard CM (1971) The olfactory connections of the lateral hypothalamus in the rat, mouse and hamster. J Comp Neurol 141: 331-344

Scott JW, McBride RL, Schneider SP (1980) The organization of projections from the olfactory bulb to the piriform cortex and olfactory tubercle in the rat. J Comp Neurol 154: 519-534

Sczerzenie V, Hsiao S (1977) Development of locomotion toward home nesting material in neonatal rats. Dev Psychobiol 10: 315-321

Shatz CJ, Kirkwood PA (1984) Prenatal development of functional connections in the cat's retinogeniculate pathway. J Neurosci 4: 1378-1397

Skeen LC, Hall WC (1977) Efferent projections of the main and the accessory olfactory bulb in the tree shrew (Tupaia glis). J Comp Neurol 172: 1-36

Small RK, Leonard CM (1983) Early recovery of function after olfactory tract section correlated with reinnervation of olfactory tubercle. Dev Brain Res 7: 25–40

Sperry RW (1963) Chemoaffinity in the orderly growth of nerve fiber patterns and connections. Proc Natl Acad Sci USA 50: 703–710

Stanfield BB, Caviness VS Jr, Cowan WM (1979) The organization of certain afferents to the hippocampus and dentate gyrus in normal and reeler mice. J Comp Neurol 185: 461–484

Teicher MH, Stewart WB, Kauer JS, Shepherd GM (1980) Suckling pheromone stimulation of a modified glomerular region in the developing rat olfactory bulb revealed by the 2-deoxyglucose method. Brain Res 194: 530–535

Westrum LE (1975a) Electron microscopy of synaptic structures in olfactory cortex of early postnatal rats. J Neurocytol 4: 713–732

Westrum LE (1975b) Axonal patterns in olfactory cortex after olfactory bulb removal in newborn rats. Exp Neurol 47: 442–447

Westrum LE (1980) Alterations in axons and synapses of olfactory cortex following olfactory bulb lesions in newborn rats. Anat Embryol 160: 153–172

White LE Jr (1965) Olfactory bulb projections of the rat. Anat Rec 152: 465–480

Impacts of Postnatal Olfactory Deprivation on the Amygdaloid Complex in Mouse

ZB Lü, W Breipohl* and B Rehn**

Department of Cell Research, Beijing Medical College, Beijing, PRC

*Department of Anatomy, University of Queensland, St.Lucia 4067, Australia

**Institut für Anatomie, Universitätsklinikum Essen, 4300 Essen, FRG

1 Introduction

Inherently determined steps of CNS development are manifold. Most obvious are processes like: cell proliferation in specific matrix layers, migration of the daughter cells out of these, and morphological differentiation and maturation of synaptic wiring of the matrix daughter cells after arrival at their final destination (Meller et al. 1966, 1968a,b; Altman 1969; Hinds and Hinds 1976).

All these three steps may be influenced not only by internal parameters (e.g. hormones, vitamins, interaction with targets) but also by environmental impacts. Amongst the latter, early sensory experience is apparently of crucial importance for the function of the mature system (see Breipohl et al., Meisami, Panhuber, Rehn and Breipohl this Vol). In other words, early in life the environment of the animal places its special fingerprints on the developing nervous system (Blakemore and van Sluyters 1975) to guarantee that the adult animal will be able to dif-ferentiate between environments identical or divergent from these early ex-perienced fingerprinting ethological niches.

The aim of the present investigation was to determine whether the morphology of even higher brain centres, which also serve functions other than only olfactory processing (Heimer 1968), is dependent on early postnatal olfactory experience. This report will consider the effect of postnatally performed unilateral olfactory deprivation by closure of the left external naris of the oral-caudal length and number of nerve cells in amygdaloid nuclei in adult NMRI mice.

2 Materials and Methods

Abbreviations used:

MOB	main olfactory bulb
NAB	nucleus amygdaloideus basalis
NAC	nucleus amygdaloideus corticalis centralis
NACA	nucleus amygdaloideus corticalis anterior
NACP	nucleus amygdaloideus corticalis posterior
NAL	nucleus amygdaloideus lateralis
NAM	nucleus amygdaloideus medialis
NLOT	nucleus of the lateral olfactory tract
PC	periamygdaloid cortex
ULO	occlusion of left naris

Nine male NMRI mice were investigated. Six of them received occlusion of the left external naris on postnatal day 1 according to the techniques of Meisami (1976). Animals survived for 8 weeks (3 mice) and 8 months (3 mice) respectively. Controls (3 mice) were 9 weeks old. After anaesthetization with Nembutal the animals were fixed by intracardial perfusion with 2.5% glutaraldehyde and 1.0% paraformaldehyde in 0.12 M cacodylate buffer (pH 7.3). Brains were removed, processed by routine paraffin sectioning, (10 μm in thickness), and stained with toluidine blue. Oral-caudal length of the amygdaloid complex was determined by serial sections. The lengths of the following structures were determined: NAB, NAC, NACA, NACP, NAL, NAM, NLOT, PC. As a further parameter for the estimation of the importance of early sensory experience on brain development the number of nerve cells in various nuclei of the amygdaloid complex were also counted using photographs of serial sections.

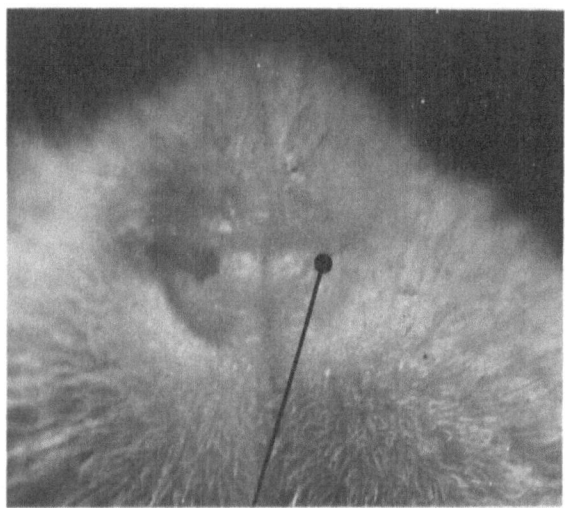

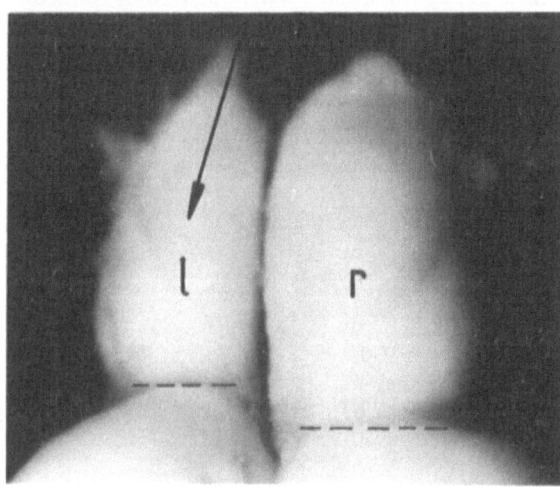

Fig 1 Mouse brain (bottom) 8 weeks after closure of the left external naris (top). The left MOB is remarkably smaller in size and the left forebrain is enlarged

3 Results

Two months after postnatal occlusion of the left external naris there was a macro-scopically visible reduction in the homolateral MOB (Fig. 1). This effect was even more pronounced at 8 months. In addition, the homolateral forebrain showed an oral-caudal enlargement while the contralateral MOB and forebrain remained in the range of untreated control animals.

Light microscope investigations revealed further that not only were the ipsilateral MOB and forebrain heavily affected by a closure of the left external naris, but also nuclei of the amygdaloid complex on that side (Table 1, columns A, C, F).

3.1 Comparison of 8-Week Experimental Group with Controls

When compared with the controls, the oral-caudal length of all amygdaloid nuclei except the NLOT in the 8-week experimental animals were longer (not shorter) on the occluded side (Table 1, column E). On the left side the NLOT was reduced by 15.4% and on the right it was identical to that of the controls. (Table 1, column E). The degree of the enlargement per se, as well as in relation to the contra-lateral side, differed between the various nuclei. In some, the NAC, NACA, NAL and PC the enlargement was smaller on the left side while in other nuclei (NAB, NACP, NAM) the reverse occurred (Table 1, column E).

3.2 Comparison of 8-Month Experimental Group with Control

When comparing the oral-caudal length of nuclei from the 8-month-old group with the 9 week controls (Table 1, column H) the right side was always enlarged with the exception of the right NACP. The right NAC; NACA, NAL, NLOT and PC were especially enlarged. However, it should be noted that this excessive general enlargement of the nuclei could be the consequence of normal postnatal growth. The left side was also clearly enlarged in the NAC, NAL and PC, further indicat-ing ongoing postnatal development of these nuclei between 9 weeks and 8 months. ULO apparently reduced this postnatal growth process in the NAC, NAL and PC, although it can not be excluded at this moment that part of the decline in oral caudal length could also be due to a reduction after initial overshoot development (cf. Rehn et al. 1986). Therefore nuclei on the left side were less developed either because of reduced growth in comparison to the right side (NAC, NACA, NAL, NLOT) or because a true reduction occurred (NAB, NACP). Values for the NAM were almost identical on both sides and only one nucleus (PC) was larger on the left than on the right (Table 1, column H).

3.3 Comparison of 8-Week and 8-Month Experimental Groups

Checking the changes between the experimental 8-week and 8-month groups revealed both reductions and enlargements on both sides during this period (Table 1, column J). Enlargements of approximately 10 and 20% occurred in the NAC and NLOT, respectively on both sides. Postnatal development of the homolateral NLOT appeared to be only temporarily affected by early postnatal ULO at 8 weeks, since normal values were present by 8 months. The right

Table 1 Effects of left ULO on the bilateral oral-caudal lengths of the nuclei in the amygdaloid complex in NMRI-mouse. Columns A, C, F mean length (in µm) of left L and right R nuclei. Columns B, D, G percent L-R differences. The right side is always set as 100%. No standard deviation is given because of the low number of animals (n=3) per group. Percent differences in mean oral-caudal length of amygdaloid nuclei in NMRI-mouse 8 weeks (column E) and 8 months (column H) after left ULO on postnatal day 2 in comparison with control animals (9 weeks. old). Control values are always set as 100%. L = left; R = right. Column J gives the percent differences between values of 8 week old ULO animals (set as 100%) and 8 months ULO animals

	Controls		8 Weeks		8 Weeks/Controls	8 Month		8 Month/Controls	8 Month/8 Weeks
	A	B	C	D	E	F	G	H	J
NAB	L1933 R1966	-1.7	L2000 R2000	+/-0	L +3.5 R +1.7	L1833 R2050	-10.6	L -5.2 R +4.3	L -8.4 R +2.5
NAC	L 600 R 600	+/-0	L 800 R 917	-12.7	L +33.3 R +52.8	L 883 R1000	-11.7	L +47.2 R +66.7	L +10.4 R +9.1
NACA	L1150 R1150	+/-0	L1467 R1550	-5.4	L +27.6 R +34.8	L1200 R1383	-13.0	L +4.3 R +20.3	L -18.2 R -10.8
NACP	L1500 R1533	-2.2	L1550 R1575	-1.6	L +3.3 R +27.0	L1433 R1533	-6.5	L -4.5 R +0.0	L -7.5 R -2.7
NAL	L1466 R1500	-2.3	L1650 R1783	-7.5	L +12.6 R +18.9	L1617 R1783	-9.3	L +10.3 R +18.9	L -2.0 R +0.0
NAM	L1483 R1550	-4.3	L1583 R1650	-4.1	L +6.7 R +6.5	L1600 R1667	-4.0	L +7.9 R +7.5	L +1.1 R +1.0
NLOT	L 650 R 650	+/-0	L 550 R 650	-15.4	L -15.4 R +/-0	L 667 R 783	-14.8	L +2.6 R +20.5	L +21.3 R +20.5
PC	L 833 R 850	-2.0	L1167 R1250	-6.6	L +40.1 R +47.1	L1150 R1133	+1.5	L +38.1 R +33.3	L -1.5 R -9.4

contralateral NLOT, in contrast, showed an overshoot development at 8 weeks which was still present at 8 months.

3.4 Comparison of Left and Right Sides in Controls

Comparing the means for the left and right oral-caudal lengths in control animals revealed comparable values for most nuclei (Table 1, column B). Only the NAM was clearly smaller on the left side. Four other nuclei (NAB, NACP, NAL and PC) showed also moderately smaller left sides. The right side of any given nucleus was never smaller. Whether these data are only due to the small number of animals or indeed reflect another example of brain asymmetry with regard to structures involved in cognitive or sensory processes and systems (Wada et al. 1975; Neri and Aggazani 1984) cannot be decided so far.

3.5 Comparison of Left and Right Sides on Experimental Groups

ULO provoked more pronounced reductions of oral-caudal length in all left nuclei except the NAM and PC at 8 weeks. Eight months after ULO these side differences were still apparent in the NAM, but were reversed in the PC.

3.6 Nerve Cell Number in the NACA and NLOT

The morphometric evaluation of the number of nerve cells 8 months after ULO further revealed in the right NACA and NLOT, nuclei with proven olfactory input, an increase of 78.4% and 11.1% respectively in comparison to 9 week old control animals. Nerve cell numbers in the left NACA increased only by 34.9% and were clearly reduced in the left NLOT (Table 2).

Table 2 Percent differences in mean cell number of NACA and NLOT 8 weeks and 8 months after left ULO in comparison to control animals (9 weeks old). Control values are always set as 100%

| | NACA | | NLOT | |
	Left	Right	Left	Right
8 weeks after ULO	+62.1	+79.1	−33.8	−17.6
8 months after ULO	+34.8	+78.4	−19.9	+11.2

Effects of ULO on the NACA and NLOT became even more obvious after comparison of the side differences in normal amd experimental animals (Table 3).

Table 3 Percent left-right differences in nerve cell numbers in the NACA and NLOT of mouse. The right side is always set as 100%

	NACA	NLOT
Controls (n=3)	+0.2%	+1.0%
8 weeks after ULO (n=3)	-9.3%	-21.5%
8 months after ULO (n=3)	-24.3%	-27.2%

4 Discussion

The present investigation has revealed that unilateral postnatal closure of an external naris leads to remarkable morphological changes in magnitude, duration and direction at both 8 weeks and 9 months later in secondary and tertiary olfactory centres. These changes were unique to the different nuclei.

The effect of a lack of peripheral olfactory input upon the development of higher brain centres has only previously been discussed, to our knowledge, by our own group (Lü et al. 1984) and is extended by this report. Differences in oral-caudal length and cell numbers revealed that the time course of postnatal development varies for the different amygdaloid nucei. As a consequence, their susceptibility towards early postnatal deprivation varied as well. Differences were most evident when the changes in nerve cell numbers were considered, as demonstrated for the NAC and NLOT, two nuclei with proven olfactory input (Heimer 1968).

Brain development in vertebrates, including man, is not complete at birth and the development of various sensory systems seems to be highly dependent on early sensory experience (Marler 1970; Pettigrew and Freeman 1973; Blakemore and van Sluyters 1975; Steffen and van der Loos 1980; Walsh 1981; Wall et al. 1982; Leon et al. 1984). Van der Loos (1985) described the developing cortex cerebri even as being a "slave of the periphery". In addition to the MOB of rodents (Greer et al. 1982; Schmidt this Vol) even higher olfactory centres have not yet reached complete maturity at birth, as was previously demonstrated with the 2-deoxy-d-glucose technique for rat by Astic and Saucier (1982). Thus, these centres can also be regarded as candidates for reduced development under sensory deprivation.

Peripheral influences on the morphological and functional maturation of the olfactory system have been shown by a variety of experimental investigations. For example: and absence of the olfactory placode will cause a failure of MOB development in amphibia (Stout and Graziadei 1980). Postnatal deprivation will still lead to a remarkably dampened maturation of the MOB in rodents (Meisami and Safari 1981). The crucial importance of the sensory periphery for function (Simmons and Getchell 1981) and the formation of normal olfactory structures in the CNS (Land 1973; Land and Shepherd 1974; Graziadei and Samanen 1980; Rehn et al. 1986) has been widely exemplified. On the other hand, it has been well documented that normal olfactory structures in the CNS do not always mean normal olfactory processing (Butler et al. 1984; Samanen and Forbes 1984). Somewhat unexpected in the present investigations was the fact that peripheral olfactory deprivation did provoke quite different changes in both the oral-caudal length and nerve cell numbers in the amygdaloid nuclei. These differences may be the expression of both the relative postnatal maturity or immaturity of the considered nuclei and of the relative importance olfactory sensory stimuli have for them.

An influence of early sensory experience on neuronal plasticity and specificity, as well as animal behaviour, has already been shown for many other systems in the adult brain (Diamond et al. 1964; Harlow and Harlow 1964; Rosenweig 1971; Devor and Wall 1981; Jeanmond et al. 1981; Laing and Panhuber 1978; Panhuber this Vol). The olfactory system has innate general odor preferences (Steiner 1979), sensory preferences and aversions can still be enhanced or even be changed during postnatal life (Cain and Johnson 1979; Pliner 1982; Rozin and Schiller 1980; Rozin and Kennel 1983; Coopersmith and Leon this Vol; Doty this Vol; Panhuber this Vol). The question that arises is at which level in the olfactory system the morphological substrate for such functional alterations should be investigated.

In seeking an answer to this question it must be remembered that final differences in the remaining nerve cell numbers are not only the result of cell origin, but are also highly dependent on the extent of cell maturation and cell death. Maturation can be influenced by deprivation and early postnatal insults in both the olfactory periphery (Rehn et al. 1981; Breipohl et al. this Vol) and the olfactory bulb (Stupp et al. 1986). Although cell death has been documented for a variety of centres in the nervous system (Clarke 1985) it has only briefly been specified in the olfactory system (Breipohl et al. 1985, this Vol). Apparently cell death is not only an inherently determined process, but also highly dependent on environmental factors early in life.

From this it may also be concluded that the pattern of nerve cell origin in some central nervous structures serving the processing of olfactory structures is not only inherently determined but may be highly influenced even by early sensory experience. Therefore, in ongoing investigations of our group we are trying to detail further the effects of inherent factors, for example, hormones (see Breipohl et al. this Vol) early olfactory insults (see Rehn and Breipohl this Vol) and olfactory deprivation (Stupp et al. 1986) on the structural and functional organizations of the developing nervous system in general and the olfactory system especially.

5 Conclusion

In conclusion, our data documented changes in postnatal development of certain tertiary olfactory centers due to unilateral olfactory deprivation.

The observed variation may prove useful parameters for estimating the relative importance of early postnatal olfactory stimulation or deprivation for the morphology and function of tertiary olfactory centres in adults. To investigate the importance of the postnatal developmental stage for the vulnerability of higher brain centres and to estimate their dependence on sensory (environmental) experience, a series of additional studies were started in considerably more animals, and focussed on the structure and plasticity of the primary, secundary and tertiary olfactory centres after olfactory deprivation at different postnatal stages will be reported elsewhere (Breipohl et al. in preparation).

Summary

Impacts of early post-natal olfactory deprivation have been documented morpho-metrically for tertiary olfactory centres in NMRI mice. Parameters considered were oral-caudal length and nerve cell numbers in amygdaloid nuclei. Different brain centres revealed differences with respect to both the time course and the size of the changes. On the basis of these data it is speculated that the developmental stage at which the onset of imbalance occurs between normal inherent or internal factors and sensory experience (here due to unilateral deprivation) will determine the magnitude, duration and direction of the morphological effects on brain centers.

Acknowledgments. The authors gratefully acknowledge the skillful and precise technical assistance of Marian Kremer and Pat Bretherton.

With kind support of grants from Deutsche Forschungsgemeinschaft (Br. 358/5-2), NH&MRC (860587-85/3452) and BRF-32.

References

Altman J (1969) Autoradiographic and histological studies of postnatal neuro-genesis. IV. Cell proliferation and migration in the anterior forebrain, with special reference to persisting neurogenesis in the olfactory bulb. J Comp Neurol 137: 433-458

Astic L, Saucier D (1982) Metabolic mapping of functional activity in the olfactory projections of the rat: ontogenetic study. Dev Brain Res 2: 141-156

Blakemore C, Sluyters RC van (1975) Innate and environmental factors in the development of the kittens visual cortex. J Physiol (London) 248: 663-716

Breipohl W, Rehn B, Molyneux GS, Grandt D (1985) Plasticity of neuronal cell replacement in the main olfactory epithelium of mouse. Abstr 12th Int Anat Congr London

Butler AB, Graziadei PPC, Monti Graziadei GA, Slotnick B (1984) Neonatally bulb-ectomized rats with olfactory-neocortical connections are anosmic. Neurosci Letts 48: 247-254

Cain WS, Johnson F (1979) Lability of odor pleasantness: influence of nerve exposure. Perception 7: 459-465

Clarke PGH (1985) Neuronal death in the development of the vertebrate nervous system. TINS 8: 345-349

Devor M, Wall P (1981) Plasticity in the spinal cord sensory map following peripheral nerve injury in rats. J Neurosci 1: 679-684

Diamond MC, Krech D, Rosenberg MR (1964) The effects of an enriched environment on the histology of the rat cerebral cortex. J Comp Neurol 123: 111-119

Graziadei PPC, Samanen DW (1980) Ectopic glomerula structures in the olfactory bulb of neonatal and adult mice. Brain Res 187: 467-472

Greer CA, Stewart WB, Teicher MH, Shepherd GM (1982) Functional development of the olfactory bulb and a unique glomerular complex in the neonatal rat. J Neurosci 2: 1744-1759

Harlow H, Harlow M (1964) The effect of rearing conditions on social behaviour. Bull Menninger Clin 26: 213-224

Heimer L (1968) Synaptic distribution of centripetal and centrifugal nerve fibers in the olfactory system of the rat. An experimental anatomical study. J Anat 103: 413-432

Hinds JW, Hinds PL (1976) Synapse formation in the mouse olfactory bulb. I. Quantitative studies. J Comp Neurol 169: 15-40

Jeanmond D, Rice FL, Loos H van der (1981) Mouse somatosensory cortex: alterations in the barrelfield following receptor injury at different early postnatal ages. Neuroscience 6: 1503-1535

Laing DG, Panhuber H (1978) Neural and behavioural changes in rats following continuous exposure to an odour. J Comp Physiol 124: 259-265

Land LJ (1973) Localized projections of olfactory nerves to rabbit olfactory bulb. Brain Res 63: 153-166

Land LJ, Shepherd GM (1974) Autoradiographic analysis of olfactory receptor projections in the rabbit. Brain Res 70: 506-510

Leon M, Coopersmith R, Ulibarri C, Porter RH, Powers JB (1984) Development of olfactory bulb organization in precocial and altricial rodents. Dev Brain Res 12: 45-53

Loos H van der (1985) The development of the somatosensory system: the impact of the periphery on the brain. Joint Symp ANS ASANZ, Adelaide

Lü ZB, Breipohl W, Rehn B (1984) Influence of postnatal occlusion of one Naris externa of the amygdaloid complex in the NMRI-mouse (A morphometric study). Symp Ontogeny of the olfactory system in vertebrates, Sept 9th-12th, Tübingen

Marler PA (1970) A comparative approach to vocal learning: song development in white crowned sparrows. J Comp Phys Psychol 71 (2): 1-25

Meisami E (1976) Effects of olfactory deprivation on postnatal growth of the rat olfactory bulb utilizing a new method for production of neonatal unilateral anosmia. Brain Res 107: 437-444

Meisami E, Safari L (1981) A quantitative study of the effects of early unilateral deprivation on the number and distribution of mitral and tufted cells and of glomeruli in the rat olfactory bulb. Brain Res 221: 81-107

Meller K, Breipohl W, Glees P (1966) Early cytological differentiation in the cerebral hemispheres of mice. An electronmicroscopical study. Z Zellforsch 72: 525-533

Meller K, Breipohl W, Glees P (1968a) The cytology of the developing molecular layer of mouse motor cortex. Z Zellforsch 86: 171-183

Meller K, Breipohl W, Glees P (1968b) Synaptic organization of the outer granular layer in the motor cortex in the white mouse during postnatal development. A Golgi-and electronmicroscopical study. Z Zellforsch 92: 217-231

Neri M, Agazzani E (1984) Aging and right-left asymmetry in experimental pain measurements. Pain 19: 43-48

Pettigrew J, Freeman R (1973) Visual experience without lines: effect on developing cortical neurons. Science 182: 599-601

Pliner P (1982) The effects of mere exposure on liking for edible substances. Appetite 3: 283-290

Rehn B, Breipohl W, Schmidt C, Schmidt U, Effenberger F (1981) Chemical blockade of olfactory perception by N-methyl-formimino-methylester in albino mice. Chem Sens 6: 317-328

Rehn B, Breipohl W, Mendoza AS, Apfelbach R (1986) Changes in granule cells of the ferret olfactory bulb associated with imprinting on prey odours. Brain Res 373: 114-125

Rosenweig MR (1971) Effects of environment on development of brain and behaviour. In: Tobach E, Aronson A, Shaw F (eds) The biopsychology of development. Academic Press, London New York, pp 303-342

Rozin P, Kennel K (1983) Acquired preferences for piquant foods by chimpanzees. Appetite 4: 69-77

Rozin P, Schiller D (1980) The nature and acquisition for a preference for chilli pepper by humans. Motiv Emot 4: 77-101

Samanen DW, Forbes WB (1984) Replication and differentiation of olfactory receptor neurons following axotomy in the adult hamster: a morphometric analysis of postnatal neurogenesis. J Comp Neurol 225: 201-211

Simmons P, Getchell TV (1981) Physiological activity of newly differentiated olfactory receptor neurons correlated with morphological recovery from olfactory nerve section in the salamander. J Neurophysiol 45: 529-549

Steffen H, Loos H van der (1980) Early lesions of mouse vibrissal follicles: their influence on dendrite orientation in the cortical barrel field. Exp Brain Res 40: 419-431

Steiner JE (1979) Human facial expressions in response to taste and smell stimulation. Adv Child Dev Behav 13: 257-295

Stout RP, Graziadei PPC (1980) Influence of the olfactory placode on the development of the brain in Xenopus laevis (Daudin). I. Axonal growth and connections of the transplanted olfactory placode. Neuroscience 5: 2175-2186

Stupp C, Breipohl W, Rehn B (1986) Effects of early postnatal olfactory deprivation on the structural organization of the olfactory bulb in NMRI mice. 24th Annual Meeting of ASANZ, St.Lucia, Brisbane

Wada JA, Clarke R, Hamm A (1975) Cerebral hemispheric asymmetry in humans. Arch Neurol (Chic) 32: 239–246

Walker SF (1980) Lateralization of functions in the vertebrate brain: a review. Br J Psychol 71: 329–367

Wall PD, Fitzgerald M, Nussbaumer JC, Van der Loos H, Devor M (1982) Somatotopic maps are disorganized in adult rodents treated neonatally with capsaicin. Nature (London) 295: 691–693

Walsh RN (1981) Effects of environmental complexity and deprivation on the brain anatomy and histology: a review. Int J Neurosci 12: 33–51

Structure and Development of the LHRH-Immunoreactive Nucleus Olfacto-Retinalis in the Cichlid Fish Brain

MD Crapon de Caprona, H Münz, B Fritzsch and B Claas

Fakultät für Biologie, Universität Bielefeld, Postfach 8640, 4800 Bielefeld 1, FRG

1 Introduction

In vertebrates, the release of pituitary hormones involved in sexual behavior is most probably controlled by the gonadotropin releasing hormone (GnRH) system. In mammals, the decapeptid luteinizing hormone releasing hormone (LHRH) is the identified GnRH which guides the release of both gonadotropins LH and FSH. In teleosts, a slightly different LHRH-decapeptide was recently described (for a survey see Crim and Vigna 1983; Peter 1983). It is well known that gonadotropin release is influenced by internal and external factors and depends on the life cycle (Gorbman et al. 1983). Knowledge about the role sensory systems play in endocrine regulation is incomplete (for a survey see Breipohl 1982) but olfaction obviously plays an important role in this context (Kallman et al. 1944; Bronson 1982; Stacey and Kyle 1983; Meisel et al. 1984; Kelche and Aron 1984). In many vertebrates one can distinguish between at least three anatomically different olfactory subsystems:

1. the common or main olfactory system
2. the vomeronasal or accessory system and
3. the nervus terminalis system (NT)

The NT system until recently has received only moderate attention with respect to its possible involvement in reproduction (for a survey see Demski 1984). Therefore, the aim of the present study is to extent earlier investigations (Crapon de Caprona and Fritzsch 1983) and to provide data on the anatomical connection between GnRH systems and olfactory systems, especially the NT system / NOR system, and on the ontogenetic development of this connection in teleost fish.

2 Material and Methods

Abbreviations used:

FMRF-amide	H-Phe-Met-Arg-Phe-NH$_2$ (molluscan cardioexcitatory peptide; Stell et al. 1984)
FSH	follicle-stimulating hormone
GnRH	gonadotropin releasing hormone
HRP	horseradish peroxidase
LH	luteinizing hormone
LHRH	luteinizing hormone releasing hormone

mammalian LHRH	pGlu-His-Trp-Ser-Tyr-Gly-Leu-Arg-Pro-Gly-NH$_2$
	(Jones et al. 1984)
teleosten LHRH	pGlu-His-Trp-Ser-Tyr-Gly-Trp-Leu-Pro-Gly-NH$_2$
	(Sherwood et al. 1983)
NOR	nucleus olfacto-retinalis
NT	nervus terminalis

Teleostean fish were used as experimental animals because they offer three advantages. Firstly, the significance of olfaction in their sexual behaviour is documented (Chien 1973; Crapon de Caprona 1980, 1982; Davis and Kassel 1983; Liley and Stacey 1983). Secondly, many data are available on the anatomy and physiology of the teleostean olfactory system (for a review see Hara 1982). Thirdly, fish presumably possess only one gonadotropin (Gorbman et al. 1983; Peter and Fryer 1983). These facts allow the study of the gonadotropic neuronal system underlying sexual behaviour under more defined conditions than is possible in mammals.

3 Results and Discussion

3.1 Anatomical Connection Between Olfactory and LHRH Systems

Using immunohistochemical methods, LHRH can be traced in various cell groups of the ventromedial forebrain of many vertebrates (for a review see Demski 1984). The cell group consistently containing LHRH in all vertebrate classes is the preoptic area (Demski 1984). Besides this cell group, other cells in the ventromedial forebrain of many gnathostome vertebrates show LHRH-immunoreactive perikarya (Demski 1984). LHRH-immunoreactive cells are also found in the olfactory bulb of sharks (Stell 1986), teleosts (Münz et al. 1982; Stell et al. 1984), amphibians and reptiles (Nozaki and Kobayashi 1980), birds (Jozsa and Mess 1982) and mammals (Schwanzel-Fukuda and Silverman, 1980; Jennes and Stumpf 1980; Phillips et al. 1982). Many, if not all of these cells may belong to the NT (Demski 1984). The NT system is associated with the "vomeronasal" olfactory system (Bojsen-Möller 1974) and is also found in vertebrates with an atrophied olfactory system (Ariens-Kappers et al. 1936; Larsell 1950; Oelschläger and Buhl 1985).

Teleosts possess either an LHRH-immunoreactive NT ganglion or an LHRH-immunoreactive cell group at the olfactory bulb/telencephalic border known as the Nucleus Olfacto-Retinalis (NOR; Münz et al. 1981, 1982; Table 1). The NOR is found only in teleosts in which olfactory bulbs lie directly in front and in contact with the forebrain (Fig. 4) and is the largest LHRH immunoreactive cell group in most teleost brains (Münz et al. 1981, 1982). All other teleosts seem to have a NT-ganglion either in the olfactory bulb (Springer 1983), or attached to the olfactory bulb or within the fila olfactoria (Münz unpublished). The NOR cells possibly establish contacts with fibers ending in the olfactory bulb and in the ventromedial telencephalon and also reach with their processes the preoptic area, the hypothalamus, and the retina (Bartheld et al. 1984; Münz et al. 1982). Since the NOR and the NT show very similar connections and LHRH-positivity, both systems are considered to be homologous (Münz et al. 1982; Crapon de Caprona and Fritzsch 1983; Demski 1984; Springer and Mednick 1985; Stell et al. 1984). These connectivities have been demonstrated either by injection of HRP into the eye (Münz and Claas 1981; Demski and Northcutt 1983), by injection of cobaltous

Table 1 List of the species known to have an NT/NOR system

Order Species	NT/NOR-system LHRH/GnRH[a]	HRP[b]	Reference
Division I			
O. Anguilliformes			
Anguilla anguilla	?	+	Münz and Claas unpubl.
Division II			
O. Osteoglossiformes			
Notopterus chitala	?	+	Münz and Claas unpubl.
Xenomystus nigri	?	+	Münz and Claas unpubl.
Division III			
O. Cypriniformes			
Carassius auratus	+	+	Stell et al. 1984
	?	+	Springer 1983[c]
			Demski and Northcutt 1983
Idus idus	?	+	Münz and Claas unpubl.
O. Siluriformes			
Ictalurus punctatus	+	?	Münz and Claas unpubl.
O. Atheriniformes			
Poecilia sphenops	+	+	Münz et al. unpubl.
Xiphophorus maculatus	+	+	Münz et al. 1981
	+	?	Halpern-Sebold and Schreibman 1983
Xiphophorus helleri	+	+	Münz et al. 1981
Aplocheilus lineatus	?	+	Münz and Claas unpubl.
O. Scorpeniformes			
Sebasticus marmoratus	?	+	Ito et al. 1984
O. Perciformes			
Cichlasoma biocellatum	+	+	Münz et al. 1982
Cichlasoma nigrofasciatum	?	+	Münz et al. 1982
			Crapon de Caprona and Fritzsch 1983
Lepomis macrochirus	+	+	Münz et al. 1982
Haplochromis burtoni	?	+	Münz and Claas unpubl.
	?	+	Crapon de Caprona and Fritzsch 1983
Astronotus ocellatus	?	+	Springer 1983[c]
			Springer and Mednick 1986[c]
Scatophagus argus	?	+	Münz and Claas unpubl.
Macropodus opercularis	?	+	Münz et al. 1982
Mastocembelus opercularis	?	+	Münz and Claas unpubl.
O. Gasterosteiformes			
Gasterosteus aculeatus	?	+	Ekström 1984
O. Tetraodontiformes			
Tetraodon fluviatilis	?	+	Münz and Claas unpubl.
Navodon modestus	?	+	Ito et al. 1984

[a] Antibody reaction. [b] Horseradish peroxidase tracing.
[c] Cobaltous chloride tracing
? = no data. + = method used by author

214

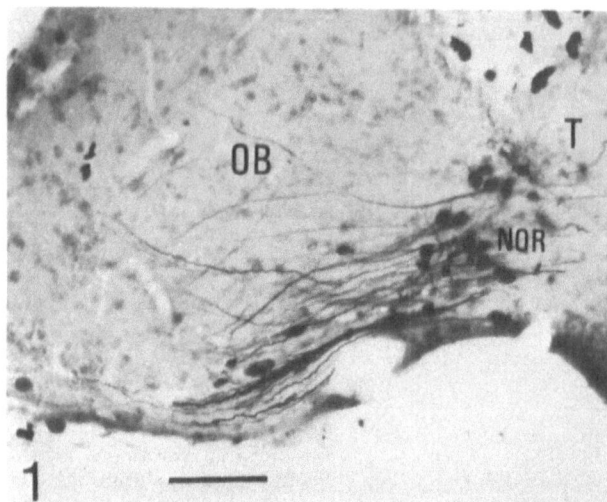

Fig. 1 Horizontal section (50 μm, cryo-section, methylene blue counterstained) through the olfactory bulb **OB** and ventral telencephalon **T** with retrogradly HPR-labeled neurons of the NOR in a cichlid fish. Note the widespread ramification of NOR processes in the olfactory bulb. Bar = 100 μm

chloride into the olfactory mucosa (Springer 1983) or by LHRH-immunostaining (Münz et al. 1982; Halpern-Sebold and Schreibman 1983; Stell et al. 1984). With the two latter approaches, the processes of the NT/NOR cells were shown to reach the retina and ramify within the inner plexiform layer. More recent ultrastructural data confirm the suggestion of Münz et al. (1982) that these fibers end on amacrine and other retinal cells (Ball and Stell 1983). Concerning the probable function of the NOR/NT fibers in the retina, one should remark that teleostean and mammalian LHRH can act as neuromodulators (Jan et al. 1979; Jones et al. 1984) with postsynaptic effects in the time range of minutes. It was shown that injections of teleostean, but not of mammalian, LHRH into the fish retina have excitatory effects on the color channel (Stell et al. 1984). LHRH co-exists with another putative neuromodulator in the NT of the goldfish (FMRF-amide; Stell et al. 1984) and sharks (Stell 1986). In teleosts, but not in other vertebrates the NT and NOR interconnect anatomically and probably also functionally the olfactory with the optical system and both with the gonadotropic centers of the thalamus/nucleus preopticus (Demski 1984; Stell 1986).

3.2 Structure of the Adult Teleostean NOR

The NOR (Fig. 1) of adult cichlids consists of 50 to 100 neurons (Münz et al. 1982; Crapon de Caprona and Fritzsch 1983) which can be divided into large multipolar and smaller bipolar neurons (Figs. 1, 2):

(1) The first (25 μm or more in diameter) have a large lobulated nucleus and three to six nucleoli. The nucleus is in an eccentric position. Infoldings of the nuclear envelope protrude in the direction of the nucleoli where they may branch (Fig. 2b). (2) the second (15 μm in diameter) have pear-shaped perikarya. The

nucleus contains only one, rarely two nucleoli and it is surrounded by a smooth nuclear envelope without infoldings (Fig. 2b).

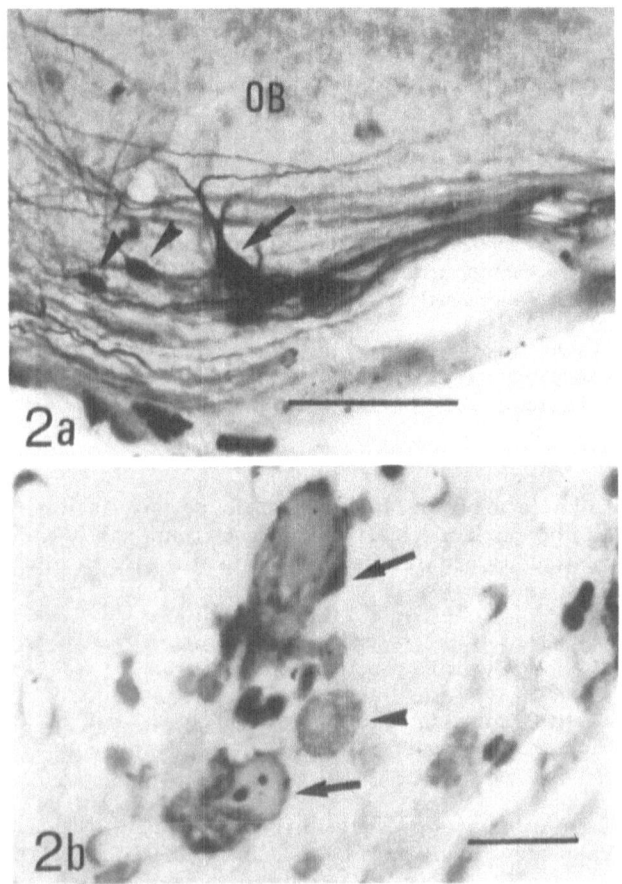

Fig. 2 **a, b** Higher magnification of HRP-labeled NOR neurons (**a** horizontal cryo-section, 50 µm, methylene blue counterstained; **b** epoxy-resin cross-section, 4 µm, toluidin blue counterstained) showing the cytoarchitectonic differences between the large cells (arrows) and the small cells (arrowheads) in an adult male Haplochromis burtoni. Notice the differences in shape **a,b** and the nuclei of these two cell types **b**. The large multipolar neurons have a large lobulation nucleus in an eccentric position and 3 to 6 nucleoli. Infoldings of the nuclear envelope protrude in the direction of the nucleoli where they may branch. The small mainly bipolar neurons have pear-shaped perikarya, the nucleus contains one, rarely two, nucleoli and is surrounded by a smooth nuclear envelope without infoldings. Bars= **a** 100µm, **b** 25µm

There are about eight times more small neurons than large ones. No sexual differences occur in the number of these cells, but the large cells tend to be larger in the larger animals, i.e. usually the males. Large and small HRP-labeled neurons show long dendrites to large areas of the olfactory bulb (Figs. 1, 2a). Similar results but somewhat larger numbers (about 200 cells) were obtained in a study of a cichlid NOR using cobalt chloride as a tracer (Springer and Mednick 1986). This numerical variability may be species-specific (Münz et al. 1982) or depend on the methods used. Two cell types and comparable cell numbers were also found in the goldfish NT (Springer 1983) a fact which supports the hypothesis that the NOR and the NT ganglion are homologous.

3.3 Ontogenic Development of the NOR

During ontogeny, the LHRH-immunoreactive cells of the NT/NOR system develop prior to all other LHRH-immunoreactive centers in teleosts (Halpern-Sebold and Schreibmann 1983) and mammals (Schwanzel-Fukuda et al. 1981). In the teleostean fish Xiphophorus maculatus their maturation is established around puberty (Halpern-Sebold and Schreibmann 1983).

We studied the developing NOR of two cichlid fish species: Haplochromis burtoni (Hb), an African mouth-brooder and Cichlasoma nigrofasciatum (Cn), a South-American substrate-spawner. Hb shows a conspicuous sexual dimorphism in coloration. The gravid female produces an odor which stimulates courtship in conspecific males (Crapon de Caprona 1980). Cn males and females differ more in size than coloration. In Hb the fry is free swimming around day 18, in Cn already around day 10. Hb and Cn have a different reproductive behavior.

The development of the NOR system was studied by injecting larvae at different ages in the eye with HRP. Details of the injection and processing methods are described elsewhere (Crapon de Caprona and Fritzsch 1983).

We addressed the following questions:

1. Does the NOR develop according to a proliferation/cell death pattern as it was found in the retinopetal isthmo-optic nucleus (ION) of birds (Cowan and Clarke 1976) or does it develop with continuous cell proliferation as is the case in other parts of the teleost nervous system (Easter 1983)?

2. Is the ontogeny of the NOR different if the larvae develop under different environmental constraints (mouth-brooders versus substrate-spawners)?

3. Does the development of the NOR coincide with puberty in these fish as in other teleosts (Halpern-Sebold and Schreibman 1983), i.e., can this phenomenon be generalized to all teleosts?

Undifferentiated and differentiated cells are present in the developing NOR at hatching in both species. The cells are already in a subpial position in the lateral wall of the telencephalon. For this reason it was not possible to show either the origin or the migration route of these cells. At day 5 some NOR cells are already differentiated in both species and show an euchromatic nucleus and more cytoplasm than any other cell of the forebrain. These cells are in continuity with a group of undifferentiated cells which border the developing olfactory bulb. Only the differentiated cells with euchromatic nuclei can be labeled with HRP. The diameter of NOR cells increases with age. Between day 9 and 10 postspawning, the first dendrite-like processes can be seen. At day 11, the whole group of neurons forming the NOR still shows both undifferentiated and differentiated cells. It is only at day 30 that the two adult types of neurons can be distinguished. The ratio small/large cells seems to be comparable to adults (about 7:1).

Although the NOR neurons develop very early compared to other parts of the forebrain, their final position in the ventro-medial part of the forebrain is gradually achieved as a consequence of the growth and migration patterns of other cells in the forebrain, i.e. the forebrain everts during ontogeny (Northcutt and Braford 1980). The numbers of NOR cells which show labelling after injection of HRP into one eye show a gradual increase with age (Fig. 3).

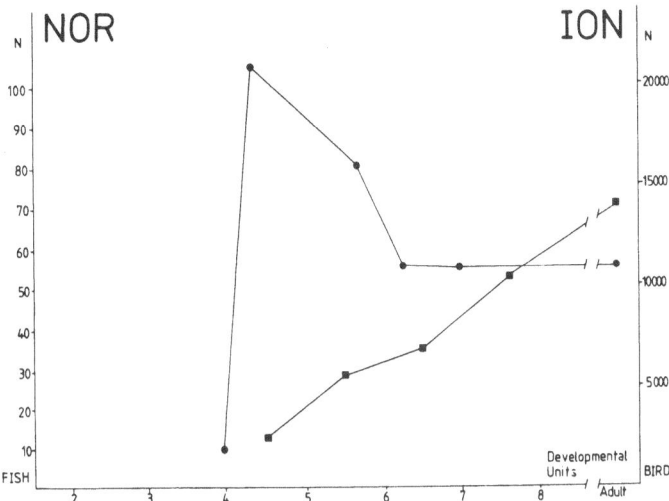

Fig. 3 Changes in the cell number of the bird ION and the cichlid NOR during comparable developmental stages (indicated as developmental units). The mean values are indicated by dots (**ION**) or by squares (**NOR**). THe number of neurons in the ION is maximal immediately after the first neurons can be labelled with HRP from the eye. In later stages it is reduced to about 50%. In contrast, cell numbers in the NOR continuously increase until adult numbers are reached. (After Crapon de Caprona and Fritzsch 1983)

Concerning question (1) and within the limits of the method we used, we cannot demonstrate an excess of cells projecting to the retina followed by a subsequent reduction, as shown with similar methods in the bird's ION (Cowan and Clarke 1976). We therefore conclude that no conspicuous cell death occurs in the NOR (Fig. 3). This conclusion clearly relates to the very different ontogenetic processes in birds and teleosts. Whereas in birds almost all neurons are produced within a limited ontogenetic period, teleosts obviously show a relatively long proliferation of neurons until adulthood, at least in the retina and some parts of the brain (Easter 1983). However, we cannot exclude that as in the anuran isthmic nucleus (Udin and Fisher 1985), a very small continuous celldeath and replacement by newly formed neurons occurs in the NOR.

Concerning question (2), no differences were found in the numerical development of the NOR between substrate-spawners and mouth-brooders. Likewise, the NOR is very similar in the adults of both species. Thus, the structure and development of the NOR do not reflect the large differences in reproductive behaviour and larval developmental speed of the two species.

Concerning question (3), the cytoarchitectonic differentiation between large and small neurons coincides with the first signs of adult male coloration in the young mouth-brooders (the territorial lacrymal black bar; Crapon de Caprona and Fritzsch 1983). Full sexual maturity, which we define as the first spawning, occurs around 4-5 months after hatching in both species, much later than the differentiation of the NOR. In mammals the NT is the first group of cells to develop immunoreactivity (Schwanzel-Fukuda et al. 1981). In the platyfish (Xiphophorus), the NOR is the first LHRH-immunoreactive center to mature. Moreover, a sequential appearance of LHRH-immunoreactive neuronal groups seems to be timed with specific changes of puberty (Halpern-Sebold and Schreibman 1983). We currently test the possibility that, as in the platyfish, the NOR of cichlids may become LHRH-immunoreactive at puberty, i.e., between 30 days and 4 months of age, after the large and small neurons have differentiated.

4 Conclusion

Our results raise three questions which are listed as a conclusion:

1. How do olfaction and vision influence each other during reproduction and how do they influence the hormanal state of the animal?

2. Is the NOR involved in the regulation of reproductive cycles and if so how do optical and in particular olfactory cues act on this system?

3. What is the influence of the NOR on the development and differentiation of the brain?

With respect to (1), LHRH-immunoreactive processes have been described as pre-synaptic elements in the hamster accessory olfactory bulb. This has led to the suggestion that LHRH may play a neuromodulatory role in the hamster olfactory information processing (Phillips et al. 1982). In contrast, in fish, LHRH-immuno-reactive processes have been described only as postsynaptic elements. Different inputs probably impinge upon the NT/NOR cells:

a) Axons of olfactory receptor cells and of olfactory bulb neurons most likely convey olfactory information onto the NOR cells (Münz et al. 1982). In addition, the NT/NOR cells may be able to obtain olfactory information directly in the mucosa (Demski and Northcutt, 1983).

b) NT/NOR cells send processes into the retina (Münz et al. 1982; Stell et al. 1984). Via these processes they may receive visual information and modulate visual input as well (Demski and Northcutt, 1983)

c) Axons from the telencephalon (Northcutt and Davis 1983) conveying presumably optic, gustatory, and lateral-line information may reach the NOR cells.

d) Axons from the locus coeruleus bringing information from the reticular formation are reported to project to the NOR cells (Fernald and Finger 1984)

All this input can be channeled via the NT/NOR fibers to the preoptic area (Davis and Kassel 1983; Demski 1984). Thus, besides the direct access of the NT/NOR to olfactory and visual cues, other inputs probably converge via forebrain efferents on the long processes of the NT/NOR cells. Because of these multifunctional inputs the NT/NOR is perhaps an essential part of the neuronal network neces-sary for the sophisticated regulation of external and internal stimulation in sexual behaviour.

With respect to (2), Demski and Northcutt (1983) were able to produce sperm release in the goldfish by electrical stimulation of the olfactory nerve and the optic nerve. This release was most likely achieved by direct stimulation of the NT (the NOR homolog in the goldfish). Studies using partial forebrain ablations support the assumption that the NT/NOR is important in regulating sexual behav-

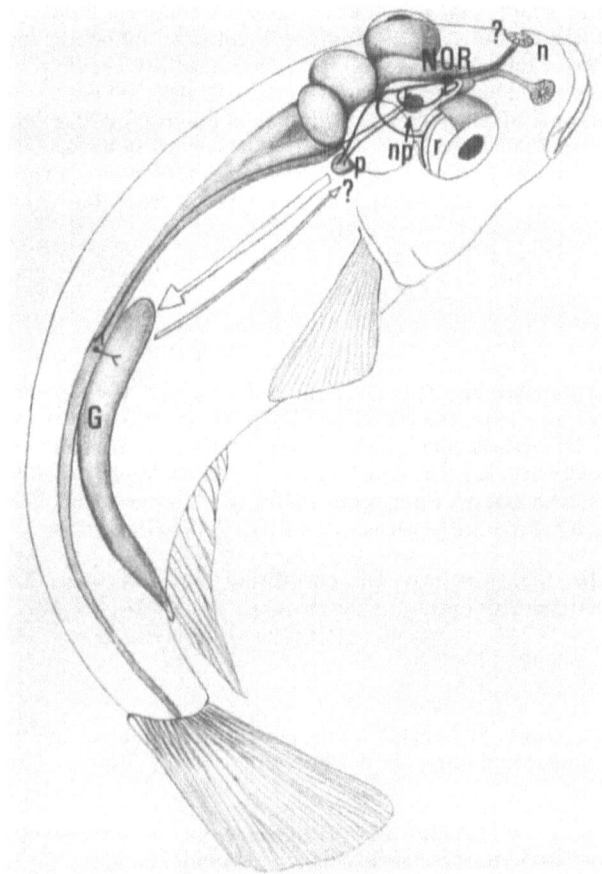

Fig. 4 Drawing of a fish showing the position of the NOR (black) with the known connections to the retina **r**, the nucleus preopticus **np**, and the contralateral NOR. The likely connections with the nasal mucosa **n**, the hypothalamus and the pituitary **P** are indicated with a question mark. Notice that the nucleus preopticus has well-established connections to the hypothalamus and to the sperm release center in the spinal cord. The gonads **G** and the gonadotropic axis of the fish brain (nucleus preopticus – hypothalamus – pituitary) form a feedback system of regulation. Data are predominantly derived from studies on cichlid fish and goldfish

iour. Noble (1939) has already shown that telencephalon-ablated cichlids are infertile and that their gonades degenerate. Data on other fish species indicate, however, that telencephalic ablation is not invariably antigonadotropic (Davis and Kassel 1983). The simplest explanation of the results in cichlids would be that the destruction of the LHRH-immunoreactive cells in the NOR and the preoptic area leads to the observed degeneration of the gonads. The results of Demski and co-workers (Demski and Northcutt 1983; Demski and Dulka 1984) on sperm release in the goldfish and of Stacey and Kyle (1983) and Koyama et al. (1984) on the effect of partial telencephalic ablations seem to support this hypothesis. The possible connections between NOR and the regulation of gonadal activity is depicted in Fig. 4.

Whereas pheromone-like odors have been postulated for goldfish (Partridge et al. 1976; Stacey and Kyle 1983) and cichlids (Crapon de Caprona 1980), little is known in that respect in most other fish families (Liley and Stacey 1983). In this context it is noteworthy that the olfactory mucosa varies in size and structure among teleosts as much as among tetrapods (Yamamoto 1982). These varying anatomical features indicate that olfaction may play a different role during sexual interaction and ontogenetic maturation in different fish species.

With respect to (3), ontogenetic studies are important in that they may describe the developmental sequence of the various brain centers involved in gonadotropic regulations. A coincidence between development of LHRH-immunopositive centers of the forebrain and puberty was described in the platyfish (Halpern-Sebold and Schreibman 1983). We have shown that in cichlids the establishing of the adult cytoarchitectonic patterns of the NOR coincides with the onset of puberty. The onset of LHRH-immunoreactivity as well as of the FMRF-immunoreactivity, shown to be co-localized in the NT (Stell et al. 1984), in the cichlid NOR has to be established in further studies.

Summary

Data are presented on the adult structure, the possible function and the development of the LHRH-immunoreactive nervus terminalis (NT)/nucleus olfacto-retinalis (NOR) system in cichlid fish. The results are compared with those on other vertebrates and the importance of olfaction for reproduction in vertebrates in general and in teleosts in particular is discussed. We suggest that the influence of olfaction on reproduction is mediated by the LHRH-immunoreactive NT/NOR system.

Acknowledgments. We thank Dr. D. Graham for correcting our English. This study was supported by the Deutsche Forschungsgemeinschaft, Im 1/14, Fr 572.

References

Ariens-Kappers CU, Huber GC, Crosby EC (1936) The comparative anatomy of the nervous system of vertebrates, including man. Reprinted 1960, Hafner, New York

Ball AK, Stell WK (1983) Structure of LHRH- and FMRFamide-like immunoreactive fibers in the goldfish retina. Invest Ophthalmol Vis Sci, Suppl 23: 66

Bartheld CS von, Meyer DL, Fiebig E, Ebbesson SOE (1984) Central connections of the olfactory bulb in the goldfish, Carassius auratus. Cell Tissue Res 238: 475-487

Bojsen-Möller F (1974) Demonstration of terminal, olfactory, trigeminal and perivascular nerves in the rat nasal septum. J Comp Neurol 159: 245-256

Breipohl W (1982) Olfaction and endocrine regulation. IRL Press, London

Bronson FH (1982) Pheromonal influences on endocrine regulation of reproduction. In: Breipohl W (ed) Olfaction and endocrine regulation. IRL Press, London, pp 103-113

Chien AK (1973) Reproductive behavior of the angel fish Pterophyllum scalare (Pisces, Cichlidae). II. Influence of male stimuli upon the spawning rate of females. Anim Behav 21: 457-463

Cowan WM, Clarke PHG (1976) The development of the isthmo-optic nucleus. Brain Behav Evol 13: 345-375

Crapon de Caprona MD (1980) Olfactory communication in a cichlid fish, Haplochromis burtoni. Z Tierpsychol 52: 113-134

Crapon de Caprona MD (1982) The influence of early experience on preferences for optical and chemical cues produced by both sexes in the cichlid fish Haplochromis burtoni (Astatotilapia burtoni, Greenwood 1979). Z Tierpsychol 58: 329-361

Crapon de Caprona MD, Fritzsch B (1983) The development of the retinopetal nucleus olfacto-retinalis of two cichlid fish as revealed by horseradish peroxidase. Dev Brain Res 11: 281-301

Crim JW, Vigna SR (1983) Brain, gut and skin hormones in lower vertebrates. Am Zool 23: 621-638

Davis RE, Kassel J (1983) Behavioral functions of the teleostean telencephalon. In: Davis RE, Northcutt RG (eds) Fish neurobiology. Univ Michigan Press, Ann Arbor, pp 237-264

Demski LS (1984) The evolution of neuroanatomical substrates of reproductive behavior: Sex steroid and LHRH-specific pathways including the terminal nerve. Am Zool 24: 809-830

Demski LS, Dulka JG (1984) Functional-anatomical studies on sperm release evoked by electrical stimulation of the olfactory tract in goldfish. Brain Res 291: 241-247

Demski LS, Northcutt RG (1983) The terminal nerve: A new chemosensory system in vertebrates? Science 220: 435-437

Easter SS (1983) Postnatal neurogenesis and changing connections. Trends Neurosci 6: 53-56

Ekström P (1984) Central neural connections of the Pineal organ and retina in the teleost Gasterosteus aculeatus L. J Comp Neurol 226: 321-335

Fernald RD, Finger TE (1984) Catecholaminergic neurons of locus coeruleus project to the ganglion cells of the nervus terminalis (NT) in goldfish. Soc Neurosci, Abstr 10: 50

Gorbman A, Dickhoff WW, Vigna SR, Clark NB, Ralph CL (1983) Comparative endocrinology. Wiley, New York

Halpern-Sebold LR, Schreibman MP (1983) Ontogeny of centers containing luteinizing hormone-releasing hormone in the brain of platyfish (Xiphophorus maculatus) as determined by immunocytochemistry. Cell Tissue Res 229: 75-84

Hara TJ (1982) Chemoreception in fishes. Development in aquatic and fisheries science, vol VIII. Elsevier, Amsterdam Oxford New York

Ito H, Vanegas H, Murakami T, Morita Y (1984) Diameters and terminal patterns of retinofugal axons in their target areas: an HRP study in two teleosts (Sebasticus and Navodon) J Comp Neurol 230: 179-197

Jan YN, Jan LY, Kuffler SW (1979) A peptide as a possible transmitter in sympathic ganglia of the frog. Proc Natl Acad Sci USA 76: 1501-1505

Jennes L, Stumpf WE (1980) LHRH-systems in the brain of the golden hamster. Cell Tissue Res 209: 239-256

Jones SW, Adams PR, Brownstein MJ, Riviers JE (1984) Teleost luteinizing hormone-releasing hormone: Action on bullfrog sympathetic ganglia is consistent with role as neurotransmitter. J Neurosci 4: 420-429

Josza R, Mess B (1982) Immunohistochemical localization of the luteinizing hormone releasing hormone (LHRH)-containing structures in the central nervous system of the domestic fowl. Cell Tissue Res 227: 451-458

Kallman FJ, Schoenfeld WA, Barrera SE (1944) The genetic aspect of primary eunuchoidism. Am J Ment Defic 48: 203-236

Kelche C, Aron C (1984) Olfactory cues and accessory olfactory bulb lesion: effects on sexual behavior in the cyclic female rat. Physiol Behav 33: 45-48

Koyama Y, Satou M, Oka Y, Ueda K (1984) Involvement of the telencephlic hemispheres and the preoptic area in sexual behavior of the male goldfish Carassius auratus: A brain-lesion study. Behav Neurol Biol 40: 70-86

Larsell O (1950) The nervus terminalis. Ann Oto Rhinol Laryngol 59: 414-457

Liley R, Stacey E (1983) Hormones, pheromones, and reproductive behavior in fish. In: Hoar WS, Randall DJ, Donaldson EM (eds) Fish physiology, vol IXB. Behavior and fertility control. Academic Press, London New York, pp 1-63

Meisel RL, Sachs BD, Lumia AR (1984) Olfactory bulb control of sexual function. In: Early brain damage, vol II. Academic Press, London New York, pp 253-268

Münz H, Claas B (1981) Centrifugal innervation of the retina in cichlid and poecilid fishes. A horseradish peroxidase study. Neurosci Lett 22: 223-226

Münz H, Stumpf WE, Jennes L (1981) Lh-RH systems in the brain of platyfish. Brain Res 221: 1-13

Münz H, Claas B, Stumpf WE, Jennes L (1982) Centrifugal innervation of the retina by luteinizing hormone releasing hormone (LHRH)-immunoreactive telencephalic neurons in teleostean fishes. Cell Tissue Res 222: 313-323

Noble GK (1939) Effect of lesions of the corpus striatum on the breeding behavior of cichlid fishes. Anat Rec 70: 58

Northcutt RG, Braford MR (1980) New observations on the organization and evolution of the telencephalon of actinopterygian fishes. In: Ebbeson SOE (ed) Comparative neurology of the telencephalon. Plenum Press, New York, pp 41-98

Northcutt RG, Davis RE (1983) Telencephalic organization in ray-finned fishes. In: Davis RE, Northcutt RG (eds) Fish neurobiology. Univ Michigan Press, Ann Arbor, pp 203-236

Nozaki M, Kobayashi H (1980) LH-RH-like substance in the brain of lower vertebrates. In: Farner DS, Lederis K (eds) Neurosecretion: Molecules, cells systems. Plenum Press, New York, pp 452-453

Oelschläger HA, Buhl EH (1985) Development and rudimentation of the peripheral olfactory system in the harbor porpoise Phocoena phocoena (Mammalia: Cetacea). J Morphol 184: 351-360

Partridge BL, Liley NR, Stacey NE (1976) The role of pheromones in the sexual behaviour of the goldfish. Anim Behav 24: 291-299

Peter RE (1983) Evolution of neurohormonal regulation of reproduction in lower vertebrates. Am Zool 23: 685-695

Peter RE, Fryer JN (1983) Endocrine functions of the hypothalamus of actino-pterygians. In: Davis RE, Northcutt RG (eds) Fish neurobiology. Univ Michigan Press, Ann Arbor, pp 165–202

Schwanzel-Fukuda M, Silverman AJ (1980) The nervus terminalis of the guinea pig: A new luteinizing hormone-releasing hormone (LHRH) neuronal system. J Comp Neurol 191: 213–225

Phillips HS, Ho BT, Linner JG (1982) Ultrastructural localization of LH–RH-immunoreactive synapses in the hamster accessory olfactory bulb. Brain Res 246: 193–204

Schwanzel-Fukuda M, Robinson JA, Silverman AJ (1981) The fetal development of the luteinizing hormone-releasing hormone (LHRH) neuronal systems of the guinea pig brain. Brain Res Bull 7: 293–315

Sherwood N, Eiden M, Brownstein M, Spiess J, Rivier J, Vale W (1983) Charac-terization of a teleost gonadotropin-releasing hormone. Proc Natl Acad Sci USA 80: 2794–2798

Springer AD (1983) Centrifugal innervation of goldfish retina from ganglion cells of the nervus terminalis. J Comp Neurol 214: 404–415

Springer AD, Mednick AS (1986) Retinofugal and retinopetal projections in the cichlid fish Astronotus ocellatus. Brain Res (in press)

Stacey NE, Kyle AL (1983) Effects of olfactory tract lesions on sexual and feeding behavior in the goldfish. Physiol and Behav 30: 621–628

Stell WK (1986) Luteinizing hormone-releasing hormone (LHRH)- and pancreatic polypeptide (PP)-immunoreactive neurons in the terminal nerve of the spiny dogfish, Squalus acanthias. Anat Res (in press)

Stell WK, Walker SE, Chohan KS, Ball AK (1984) The goldfish nervus terminalis: A luteinizing hormone-releasing hormone and molluscan cardioexcitatory peptide immunoreactive olfactoretinal pathway. Proc Natl Acad Sci USA 81: 940–944

Udin SB, Fisher MD (1985) The development of the nucleus isthmi in Xenopus laevis. I.Cell genesis and the formation of connections with the tectum. J Comp Neurol 232: 25–35

Yamamoto M (1982) Comparative morphology of the peripheral olfactory organ in teleosts. In: Hara TJ (ed) Chemoreception in fishes. Elsevier, Amsterdam Oxford New York, p 39

F. POSTNATAL DEVELOPMENT OF OLFACTORY-GUIDED BEHAVIOR

Postnatal Development of Olfactory-Guided Behavior in Rodents

JR Alberts

Indiana University, Department of Psychology, Bloomington
In 47405, USA

1 Introduction

Olfaction is among the first sensory systems in vertebrates to become functional during ontogenesis (Alberts 1984). Even in the most altricial species, i.e., those that produce extremely immature newborns, the sense of smell is operable at birth and is often vital to the early survival of the young (Alberts 1976, 1981, 1985). From birth to sexual maturity, most vertebrates encounter an expanding array of social and environmental challenges. The roles of olfaction in survival and reproductive success broaden accordingly. In rodents, for example, the sense of smell is a critical avenue for information affecting exploration, aggression, feeding, sexual maturation, estrous cycles and sexual behavior. Similarly, olfaction regulates maternal behavior, dominance relations, as well as recognition of kin, non-kin, mates, prey, predators, and territory (see Alberts 1976, 1981; Cheal 1975; Doty 1976; Schulz and Tapp 1973).

The purpose of the present chapter is to discuss and review selected aspects of the postnatal ontogeny of olfactory-guided behavior in rodents: the relationship between the development of olfactory perception and behavior, anatomical, psychophysical, and behavioral analytic perspectives on the development of olfactory guided behavior in general, and social huddling by rats in particular. These perspectives emphasize odor-warmth associations as a necessary and sufficient precursor to species-typical social odor preferences.

2 Problems Relevant to the Development of Olfactory-Guided Behavior

To appreciate fully the development of olfactory-guided behavior, it is necessary to integrate information on: the morphological development of the olfactory apparatus, age-related changes in perceptual processing of olfactory information, the development of behavioral repertoire, functional and experiential (stimulative) influences, and reciprocal interactions among these variables. As indicated earlier, I will emphasize in the present chapter an approach to understanding experiential aspects of olfactory and behavioral development.

2.1 Predetermined and Probabilistic Processes in Sensory and Behavioral Development

The vertebrate olfactory system is organized around two bilaterally symmetrical masses of paleocortex, the olfactory bulbs. The bulbs are composed of three types of neuron: mitral, tufted, and granule cells. During the process of neurogenesis, the mitral cells arise first, followed by the tufted cells, and then the granule cells. Other classification schemes are also applicable to the same neuronal populations. Some workers refer to microneurons and macroneurons, to emphasize cells that arise, generally, during prenatal and postnatal life, respectively (Altman 1967; see also Jacobson 1974). Mitosis, migration and differentiation of mitral and tufted cells occurs during the prenatal period, which is typical for most macroneurons of the CNS (Hinds 1968a,b).

The granule cells are the smallest neurons of the olfactory bulb. These cells comprise the largest volumetric area of the bulb, and are mostly situated as interneurons to perform integrative functions. Most spectacular, however, is that birth and maturation of the granule cells occurs almost entirely during the postnatal period, i.e., later than in nearly all other regions of the vertebrate brain.

This picture of anatomical development suggests a view of ontogenetic organization relevant to understanding olfactory/behavioral development. The early-arising neural networks may subverve the preadaptations required for immediate adjustments to the postnatal world. For instance, the newborn rat has a congenital response to specific olfactory cues that activate its search for nipples (Teicher and Blass 1978). There is evidence for a prenatally active "glomerular complex" (Greer et al. 1982) that could mediate such preparedness. It has also been suggested that regional activity in the early-maturity accessory olfactory system might be responsible for in utero olfactory detection (Alberts 1976; Pedersen et al. 1983). These cells and the sensory afferents in the nasal cavity could be attuned to stimuli in the amniotic fluid (Pedersen and Blass 1981). The mechanism(s) of in utero induction of stimulus specialization (Alberts 1981) remain unelucidated.

The proliferation of granule cells during early postnatal life exposes them to the full constellation of environmental stimuli present in the complex, extra-uterine world. Such neurons are presumed to be highly sensitive to modification by stimulation (Jacobson 1974). This may be the substrate on which olfactory-guided behavior is shaped and channeled by postnatal experience. Gottlieb (1973) has elaborated on the classical embryological concepts of predetermined and probabilistic epigenesis, particularly with regard to the development of species-typical perception. These concepts appear to be applicable and useful in regard to neural substrates of the olfactory system.

2.2 Development of Olfactory Perception

Corresponding to the anatomical changes seen in the olfactory system during postnatal development are changes in sensory-perceptual function (the "operating characteristics" of the sense of smell). These functional developments are the parameters of the sensory-perceptual function. In the rat pup, olfactory sensitivity to both natural and synthetic stimuli increases during the first 3 weeks of life (Alberts and May 1980b). The ability to make olfactory discriminations is present in the infant (Johanson and Hall 1979) and improves steadily (for a review see Alberts 1981). Moreover, the neuromuscular apparatus required for sustained, modulated stimulus sampling, i.e., sniffing, also develops dramatically during the first few postnatal weeks (Alberts and May 1980a). It is clear that the status of olfactory-guided behavior can be limited by each of these processes during the postnatal development of olfactory processing.

2.3 Patterns in Behavioral Development

Expansion of the animal's behavioral repertoire is a fundamental kind of developmental change that provides new behavioral capacities for the expression of olfactory influence. Rosenblatt (1976) has made the interesting observation that most emergent behaviors in mammalian young undergo a sequence of sensory controls: first thermotactile and then olfactory control. The transition between sensory controls of behavior occurs at different developmental points for different behaviors within the same organism. It is the sequence rather than the emergence of olfactory control, per se, which is the ontogenetic regularity. Olfactory guidance often yields to guidance by visual input. Little is known about the functional transitions from control by one modality to another, and this could be a fruitful problem for analysis.

3 Filial Huddling by Rat Pups: Olfactory-Guided Contact Behavior

In the remainder of this paper, some of our recent findings are described concerning the Norway rat's olfactory-guided contact behavior, or huddling. The Norway rat, like most rodents, is born into a social group, the huddle, where it lives in direct contact with its littermates and mother. Even among adult rats, huddling behavior plays an important role in the species' social behavior (Barnett 1963).

The cues that control huddling vary at different stages of early life. Until postnatal day 10, huddling by rat pups is guided predominantly by thermal cues (Alberts 1978a; Alberts and Brunjes 1978). By postnatal day 15, however, huddling by rat pups is guided predominantly by olfactory stimuli. Species-odors attract and maintain the pups' huddling (Alberts and Brunjes 1978). We refer to this olfactory-guided huddling as "filial huddling" because the determinative stimulus, rat odors, represents the animal's affiliation to its species.

We have studied intensively the organization and control of huddling by rat pups (e.g., Alberts 1978a,b ; Alberts and Brunjes 1978), using standardized tests and well-controlled huddling stimuli. Our typical test is a preference test, in which a pup is presented with two motionless targets with which it can huddle. These stimuli are usually identical in size, temperature and texture, but vary in olfactory characteristics. We observe 4h of continuous behavior via time-lapse videography, and calculate preference ratios for the various stimuli (see cited papers for procedural details).

3.1 Predetermined Onset and Probabilistic Assignment of Filial Attractions

In our standardized huddling test, the rat's filial preference for the odors of its species emerged at 2 weeks of age. This observation reveals little about the underlying, developmental determinants. A species-typical preference could be a predetermined expression of a genetic "program" that results in an attraction to a specific, predetermined cue. Stimulus and response are both specified by the rat's social milieu. Alternatively, the same preference could be <u>acquired</u> from the animal's <u>experience</u>. Rats may display a species-typical attraction because they experience a species-typical environment during their development.

We reasoned that if the rat's normal preference is acquired by its experiences in the nest, then we might be able to "assign" its preference by altering its olfactory experiences. To this end, we superimposed each day a novel, synthetic scent on the ventrum of a rat dam and otherwise allowed her to rear her pups normally. Control females were handled similarly, but were treated with water instead of a fragrance. The standardized huddling preference test was administered on either day 5, 10, 15, or 20. The huddling test consisted of a choice between contact with the flank of a "normal", unadulterated rat versus that of another anesthetized rat anointed with the scent borne by the pup's dam. A variety of controls for odorant and the specificity of the pups' preferences were included.

The most important result of the investigation was that on day 15, pups displayed filial huddling preferences for the target bearing the scent associated with their mother. Thus, for normally reared pups the preference was for normal rat odors, whereas the offspring of the scented dams preferred the synthetic scent to the normal, species-typical stimulus (Brunjes and Alberts 1979). Onset of filial huddling was unchanged; prior to day 15 the pups did not show reliable preferences. Onset of filial responses may be predetermined. The stimulus to which the pup displays its affiliation, however, is not rigidly fixed, and appears to derive from a "probabilistic" process (cf., Gottlieb, 1976). The plasticity in this system is reflected in our ability to reassign the animal's filial attraction (Brunjes and Alberts 1979).

3.2 Equipotentiality of Odors in the Development of Filial Attractions

Although we found evidence for plasticity in the development of the rat's filial preferences, it remains possible that there is a bias inherent in the developing olfactory system that would predispose the organism to form attractions to the natural, species odors. We examined the question of "stimulus specialization" (Alberts 1985). To pursue this question, we conducted a study in which rat pups were reared equivalently and alternately by two mothers (Addison and Alberts unpublished). One dam was anointed with a synthetic olfactant (two different scents were used in different replicates of this experiment, but the outcomes were identical, so the results were collapsed). The second mother's olfactory characteristics were not altered. Thus, for pups in this experiment maternal care was associated with two different olfactory experiences: by day 15 (around the onset of olfactory-guided filial huddling) pups had 7 days of maternal care from a scented dam and 7 days of maternal care from a olfactorily normal dam.

A standardized, 4-h huddling preference test was administered on day 15. Pups (n=73 from 10 litters) that had received equal amounts of contact from a species-typical and perfumed mothers demonstrated equal preferences for the two olfactory signals.

The result of this investigation support our previous contention that the rat's olfactory preferences are acquired (Brunjes and Alberts 1979) and suggest that the developing olfactory system responds equivalently to species-typical and nonbiological stimuli (see also Alberts and May 1980a).

4 Kinds of Experience That Induce Olfactory Preferences

The next stage of analysis concerns the kind of experience that are necessary and sufficient for the establishment of an olfactory-guided behavioral preference. At least two types of experiential processes must be evaluated. The first, familiarization, refers to a general, nonassociative form of experience that acts on an organism to produce a relatively long-lasting, specific effect on responsivity to particular cues. Familiarization is believed to occur simply on the basis of "perceptual registration" (Sluckin, 1964), that is, in the absence of any contiguous events or associative forces. Sluckin (1964) and others (Leon et al. 1977) explicitly view familiarization as a form of learning, calling it "exposure learning" or simply, "nonassociative learning".

We tested whether mere exposure to an odor is sufficient to render that odor attractive in a huddling test. Pups were exposed to an arbitrary odor for 4 h/day from day 1 to day 14 and otherwise lived in a standard maternity cage with their own unadulterated dam. On day 15, we conducted a standardized test of huddling preference. Two soft, furry flanks of acrylic fur were presented to individual pups in a small arena (15 cm diam). Time spent in contact with each huddling target during a 4-h test was measured. We found that mere exposure to an odor for 4-h/day was sufficient to produce a relative preference for that odor.

4.1 Olfactory Preference Induced by a Dam Versus Mere Exposure

Next, it was crucial to determine whether these two kinds of experience, each sufficient to produce a robust huddling preference, are equivalent. We incorporated both kinds of experience into a single experiment. Litters of rat pups were removed from their home nest for 4h each day, from day 1-14, and were exposed on alternate days to each of two different odor-experience pairings. Thus, on day 15, the pups had had equal experience with the two odorants. By manipulating the conditions of odor exposure (e.g., on dam, mere exposure) we measured the strength of odor preference established by the two kinds of experience.

When either of the two test odors was presented on the ventrum of a lactating foster dam, the odor was strongly preferred to the odor of exposure. Exposure learning is sufficient to alter the pups' behavioral responses, but it does not induce the same kind or degree of attraction as that derived from olfactory experience contiguous with some aspect(s) of maternal care. Similarly, Galef and Kaner (1980) reported that odor preferences established by simple exposure conditions are not as long-lasting as odor preferences established by contact with a foster mother.

These data suggest that the rat's species identity is based on an arbitrary olfactory emblem, and that preference for the emblem is induced by an explicit association of odor and a stimulative experience in the nest e.g., stimuli presented by a foster mother. We have made progress in isolating the crucial factors that induce these olfactory preferences.

4.2 Does Suckling Induce Filial Attachments?

As in the previous study, one olfactant was an "odor of exposure", presented in the atmosphere of a barren plastic chamber. Another odor was presented on the ventrum of a foster dam that rigorously cared for the pups but did not present nipples for suckling (Rosenblatt 1967). Following two weeks of alternating daily exposure to the odor-stimulation pairings, the odor associated with non-nutritive maternal care were vastly preferred to the odor of mere exposure, indicating that the rewards of suckling are not necessary for pups to form a filial preference.

4.3 Induction of Filial Preference, by Lactating and Non-lactating Mothers

Mother's milk is not necessary for the formation of a filial preference; however, the milk may influence filial attraction in an additive manner. In the next experiment, then, pups were given equivalent exposure to two odors on alternate days. Odor A was borne on a lactating foster dam. On alternate days the litter was tended for 4 h by a perfumed, nonlactating foster dam, rendered maternal by standard procedures (Rosenblatt 1967) and anointed daily with a different odor (Odor B). Thus, both odors were associated with maternal care; but only one odor was associated with nipples to suckle and the receipt of milk. The results indicated that there was no difference in the subsequent affiliative attraction to odors A and B. The rewards of suckling, although powerful in other situations (Kenny and Blass 1977), made no additive contribution to the pups' filial huddling preferences.

4.4 Induction of a Filial Preference by a Warm, Surrogate Mother

The results of these analyses point clearly to an important contribution by maternal care to the induction of an odor preference in rat pups. This contribution, however, is not transmitted through the suckling interactions. We have recently become aware that thermal aspects of the rodent nest exert powerful influence on the mother and offspring (Alberts 1978a; Cosnier 1965; Leon et al. 1978; Rosenblatt 1976). We therefore examined whether odor preferences could be induced in young pups by association with sources of heat.

Pups received daily alternating exposures to inanimate, odorized cylinders or "surrogate mothers". One cylinder was left at room temperature (22°C) and the other warmed to the surface temperature of an adult rat (36°C). The results indicated that the odor borne on the warm cylinder was strongly preferred in the standard huddling test.

4.5 Olfactory Preferences Induced by Warm Surrogates and Lactating Dams

We again used the within-subject, alternating odor procedure to determine the relative strength of perceptual preferences produced by experience with a warm tube versus those induced by a lactating dam. The results of this experiment indicated that the magnitude of the filial preferences derived from the warm cylinder are equivalent to those induced by experience with a lactating mother. We have concluded, therefore, that the perceptual preference mediating olfactory-guided huddling in rat pups is induced by non-nutritive, thermotactile stimulation (Alberts and May 1984). It is noteworthy that we have found other, independent indices of odor associations in rat pups created by thermal reinforcements. For instance, we have also discovered that neonatal rat pups rapidly acquire odor-temperature associations after brief (10 min) exposures to a test odor paired with severe cold (10°C). Using a measure of cardiac responsivity to odor cues, we found that infant rats had formed conditioned aversive associations to odors on the basis of the odor-temperature pairings (Martin and Alberts 1982).

These types of analysis, we believe, are essential to the refinement of our understanding of the developmental sources of olfactory-guided behavior. In animals it is more than a sensory problem, it is a behavioral issue as well. Our analyses point to the importance of odor-temperature pairings.

Acknowledgments. Preparation of this chapter was supported by Grant MH-28355 and Research Scientist Development Award MH-00222, both from the National Institute of Mental Health to the author.

References

Alberts JR (1976) Olfactory contributions to behavioral development in rodents. In: Doty RL (ed) Mammalian olfaction, reproductive processes and behavior. Academic Press, London New York, p 67

Alberts JR (1978a) Huddling by rat pups: Multisensory control of contact behavior. J Comp Physiol Psychol 92: 220-230

Alberts JR (1978b) Huddling by rat pups: Group behavioral mechanisms of temperature regulation and energy conservation. J Comp Physiol Psychol 92: 231-245

Alberts JR (1981) Ontogeny of olfaction: Reciprocal roles of sensation and behavior in the development of perception. In: Aslin RN, Alberts JR, Petersen MR (eds) Development of perception: Psychobiological perspectives, vol 1. Academic Press, London New York, p 321

Alberts JR (1984) Sensory-perceptual development in the Norway Rat: A view toward comparative studies. In: Kail R, Spear NE (eds) Comparative perspectives on the development of memory. Lawrence Erlbaum Assoc, Hillsdale, p 65

Alberts JR (1985) Ontogeny of social recognition: An essay on mechanism and metaphor in behavioral development. In: Gollin ES (ed) Comparative aspects of adaptive skills: Evolutionary imlications. Erlbaum Assoc, Hillsdale, p 65

Alberts JR, Brunjes PC (1978) Ontogeny of thermal and olfactory determinants of huddling in the rat. J Comp Physiol Psychol 92: 897-906

Alberts JR, May B (1980a) Development of nasal respiration and sniffing in the rat. Physiol Behav 24: 957-963

Alberts JR, May B (1980b) Ontogeny of olfaction: Development of the rat's sensitivity to urine and amyl acetate. Physiol Behav 24: 965-970

Alberts JR, May B (1984) Nonnutritive, thermotactile induction of filial huddling in rat pups. Dev Psychobiol 17: 161-181

Altman J (1967) Postnatal growth and differentiation of the mammalian brain, with implications for a morphological theory of memory. In: Quarton G, Melnechuk T, Schmitt FO (1967) The neurosciences: A study program. Rockefeller Univ Press, New York

Barnett SA (1963) The rat: A study in behavior. Methuen, London

Brunjes PC, Alberts JR (1979) Olfactory stimulation induces filial huddling preferences in rat pups. J Comp Physiol Psychol 93: 548-555

Cheal M (1975) Social olfaction: A review of the ontogeny of olfactory influences on vertebrate behavior. Behav Biol 15: 1-25

Cosnier J (1965) Le comportement gregaire du rat d'elevage. Thesis, Univ Lyon

Doty RL (ed) (1976) Mammalian olfaction, reproductive processes and behavior. Academic Press, London New York

Galef BG, Kaner HC (1980) Establishment and maintenance of preference for natural and artificial olfactory stimuli in juvenile rats. J Comp Physiol Psychol 94: 588-596

Gottlieb G (1973) Introduction of behavioral embryology. In: Gottlieb G (ed) Studies in the development of behavior and the nervous system, vol 1. Academic Press, London New York, p 3

Gottlieb G (1976) The role of experience in the development of behavior and the nervous system. In: Gottlieb G (ed) Studies in the development of behavior and the nervous system, vol III. Academic Press, London New York, p 25

Greer CA, Steward WB, Teicher MH, Shepherd GM (1982) Functional development of the olfactory bulb and a unique glomerular complex in the neonatal rat. J Neurosci 12: 1744-1759

Hinds JW (1968a) Autoradiographic study of histogenesis in the mouse olfactory bulb: I. Time of origin of neurons and neuroglia. J Comp Neurol 134: 287-304

Hinds JW (1968b) Autoradiographic study of histogenesis in the mouse olfactory bulb: II. Cell proliferation and migration. J Comp Neurol 134: 305-322

Jacobson M (1974) A plentitude of neurons. In: Gottlieb G (ed) Studies in the development of behavior and the nervous system, vol II. Academic Press, London New York, p 151

Johanson IB, Hall WG (1979) Appetive learning in 1-day-old rat pups. Science 205: 419-421

Kenny JT, Blass EM (1977) Suckling as an incentive to instrumental learning in preweanling rats. Science 197: 898-899

Leon M, Galef BG, Behse JH (1977) Establishment of pheromonal bonds and diet choice in young rats by odor pre-exposure. Physiol Behav 18: 387-391

Leon M, Croskerry PG, Smith GK (1978) Thermal control of mother-infant contact in rats. Physiol Behav 2: 793-811

Martin LT, Alberts JR (1982) Associative learning in neonatal rats revealed by heartrate response patterns. J Comp Physiol Psychol 96:668-675

Pedersen PE, Blass EM (1981) Olfactory control over suckling in albino rats. In: Aslin RN, Alberts JR, Petersen MR (eds) Development of perception: Psychobiological perspectives, vol 1. Academic Press, London New York, p 359

Pedersen PE, Stewart WB, Greer CA, Shepherd GM (1983) Evidence for olfactory function in utero. Science 221: 478-480

Rosenblatt JS (1967) Nonhormonal basis of maternal behavior in the rat. Science 156: 1512-1514

Rosenblatt JS (1976) Stages in the early development of selected species of non-primate mammals. In: Bateson PPG, Hinds RA (eds) Growing points in ethology. Cambridge Univ Press, Cambridge, p 345

Schulz EF, Tapp JT (1973) Olfactory control of behavior in rodents. Psychol Bull 79: 21-44

Sluckin W (1964) Imprinting and early learning. Aldine, Chicago

Teicher MH, Blass EM (1978) The role of olfaction and amniotic fluid in the first suckling response of newborn albino rats. Science 198: 635-636

Neurobehavioral Analysis of Odor Preference Development   in Rodents

R Coopersmith and M Leon

Department of Psychobiology, University of California, Irvine
Ca 92717 USA

1 Introduction

At birth, altricial rodents are blind, deaf, and have limited motor control. The
olfactory system, however, is functional even during their prenatal life, when
pups are capable of learning about odors (Pedersen and Blass 1982; Stickrod et
al. 1982). Olfaction continues to be the dominant sensory modality for young
rodents until they are weaned. During the first 2 weeks postpartum, the young
use the odor of the maternal nipples to determine an appropriate target to attach
for suckling (Pedersen and Blass 1981 for a review). By the third week of life,
rat pups can stray from the maternal nest, where they must return for
nourishment and warmth until they are weaned.

In this chapter we will describe a mechanism whereby odor familiarity, induced by
early olfactory experience, functions as a bond to keep pre-weanlings in the
proximity of their mother. We will also present a neural correlate of early odor
familiarity and discuss how olfactory bulb organization during the first 2 weeks of
life may allow the development of a special olfactory responsiveness to behaviorally
relevant odors.

2 Development of Olfactory Preferences

2.1 Maternal Chemoattractants

Leon and Moltz (1971) found that 16-day-old rat pups reliably approach the odor
of a lactating female in preference to that of a nonlactating female. The olfactory
nature of the attractant was demonstrated when pups approached a box previously
occupied by a lactating female, and when they no longer approached her when
placed upwind of her.

Mothers begin to emit the pheromone in quantities sufficient to attract pups from
a distance by 14 days postpartum and continue to emit it until the end of the
fourth postpartum week (Leon and Moltz 1972). Pups begin to show an attraction
to maternal odors by 14 days, and continue to be attracted to the odor through
the fourth week of life (Leon and Moltz 1972). The emission of the maternal
attractant is therefore synchronized with the development of pups' responsiveness
to it.

2.2 Emission and Synthesis of Maternal Attractants

The source of the odor is the anal excreta of lactating females (Leon 1974). Anal excreta were collected from dams at different lactational ages; the same temporal pattern of attactiveness that was observed for the mothers was found for the anal excreta.

A specific component of the anal excreta, cecotroph, was found to be the substance capable of attracting pups. Cecotroph is a semi-solid substance produced in the cecum, a large structure at the junction of the small and large intestine (Harder 1949). Rich in bacterial population, it is excreted with the feces and immediately eaten by adult rats (Harder 1949). Lactating females, however, excrete a large amount of cecotroph, which is not consumed (Leon 1974). Pups approached the cecotroph of lactating females, and also approached cecal contents taken from either mothers or virgin females. These data indicate that the differential attractiveness of mothers is due to their higher probability of emitting the attractive substance, rather than to a difference in synthesis.

The attractant is synthesized by cecal bacteria in all rats, regardless of their reproductive state (Leon 1974). Antibiotics suppress the cecal odor, as does a nutritionally adequate diet with sucrose as its sole carbohydrate by depriving the bacteria of their cecal substrate.

2.3 The Role of Experience

Different diets select for different cecal bacterial populations, and thereby cause mothers on different diets to produce different odors (Leon 1975). The pups therefore develop their attraction to the mother postnatally and have a preference for the odor produced by mothers eating the same diet as their own (Leon 1975).

Under natural circumstances, rat pups receive early experience principally with their mother's odor. The mechanism by which the odor is produced, the manner in which pups become familiar with their own mother's odor, and the ontogeny of the approach response appear to be species-typical phenomena. Under laboratory conditions, however, exposure to arbitrarily selected odors such as peppermint odor, even when separated from the mother, similarly induces a prefential approach response in young Norway rats.

3 Neural Correlates of Olfactory Preferences

3.1 Comparative Development

Pups of other rodent species show similar attraction to maternal odors, and the ontogeny of the approach response varies with the general developmental rate of the animal. Spiny mice (Acomys cahirinus), a precocial murid species, are attracted to a diet-dependent olfactory cue emitted by lactating dams beginning on the first day postpartum (Porter and Doane 1976, 1977). The maternal attractant is also diet-dependent and pups can be induced to approach artificial odors (Porter and Doane 1976; Porter and Etscorn 1976). Mongolian gerbils have altricial young, that are reliably attracted to maternal nest odors, beginning at 3 weeks of age, a week later than are rat pups (Gerling and Yahr 1982; Yahr and Anderson-Mitchell 1983).

If pup attraction to familiar odors is linked to a specific phase of olfactory system development, then one might expect a similar pattern of neural and behavioral maturation in these three species. Indeed, the onset of attraction to maternal odors is correlated to a particular stage of olfactory bulb development (Leon et al. 1984). In each species, pups begin to be attracted to maternal odor at the time when the gross histological organization of their olfactory bulbs first appeared to reach its adult level. This stage was characterized by the thinning of the mitral cell body layer to a single row of cells, and the narrowing of the internal plexiform layer relative to the external plexiform layer.

Perhaps odors experienced during an immature period of olfactory bulb development are processed differently from odors experienced later, in such a way as to alter the subsequent response to the now-familiar odor by the mature olfactory system. Much olfactory bulb neural development occurs postnatally (Altman 1969; Bayer 1983) and the output neurons of the neonatal bulb fire with a differnt pattern than at maturity (Mair and Gesteland 1982).

3.2 Differential Olfactory Bulb Reponsiveness

Enhanced Neural Responsiveness. If odors experienced during early development change the subsequent response to the odor, then the neural response to familiar odor should differ from that to an unfamiliar odor. To test this possibility, we used ^{14}C-2-deoxyglucose (2DG) autoradiography, which allows one to measure relative neural activity simultaneously throughout the brain (Sokolof et al. 1977). Neurons which are more active take up more glucose; when radiolabeled 2DG is administered in tracer doses, isotope uptake and subsequent entrapment in a cell is proportional to the glucose uptake of a neuron (Sokolof et al. 1977). Autoradiographs prepared from tissue sections can then be analyzed to measure relative neural activity in different brain areas.

Nineteen-day-old rat pups show an enhanced olfactory bulb neural response to a familiar odor. Pups were given daily 10-min exposures either to peppermint odor or to clean air for the first 18 days of life. On day 19, all pups were given 2DG and then exposed to peppermint odor for 45 min. Odor-familiar pups had 64% higher 2DG uptake in three complexes of glomeruli, located on the lateral aspect of the bulb, 1.5 - 2.2 mm from its rostral pole (Coopersmith and Leon 1984).

Respiratory Analysis. One explanation for the differential uptake was that the odor-familiar pups increased respiration of the odor during the 2DG test. The enhanced neural response would then simply reflect the higher stimulus intensity, rather than an enhanced response to the same stimulus. We therefore measured respiration rate of odor-familiar and odor-unfamiliar pups to peppermint odor on day 19. No difference was found between the groups (Cooopersmith and Leon 1984), even when the respiratory frequency distribution was considered (Coopersmith and Leon, submitted). The enhanced neural response, therefore, is clearly not a result of an increased respiration of the familiar odor.

Odor Specificity. An alternative explanation for the enhanced neural response is that the daily peppermint experience induced a nonspecific enhancement of responses to all odors. Daily exposure to a different odor then would produce the same enhanced response to a peppermint odor on day 19 as did daily peppermint experience. Therefore, we gave pups daily 10-min exposure either to the odor of peppermint or to the odor of cyclohexanone. On day 19, all pups were given peppermint as the test odor during 2DG administration. The peppermint-familiar pups showed significantly higher uptake in the previously identified peppermint-responsive glomeruli than did the cyclohexanone-experienced animals (Coopersmith, Lee and Leon 1984). There were no differences in respiration frequency pattern between the groups. Thus, the enhanced neural response appears to be odor-specific.

We also examined the neural response of cyclohexanone-familiar and peppermint-familiar pups to a test stimulus of cyclohexanone. As expected, all pups showed a pattern of 2DG uptake different from pups tested with peppermint. However, a medial group of glomeruli showed higher activity in the cyclohexanone-familiar pups than in the peppermint-familiar pups. Thus, the response of cyclohexanone-familiar pups to a subsequent cyclohexanone stimulus was analogous to that of peppermint-familiar pups to peppermint odor. Although the spatial pattern of uptake within the glomerular layer was dependent on the odor presented during the 2DG test, the enhanced uptake was dependent on previous experience with that particular odor.

Context Specificity. Pups were made familiar with the odors under conditions that mimicked those produced by rat mothers and reliably induced an approach reponse to the odors (Coopersmith and Leon 1984). However, pups may become familiar with an odor under a variety of circumstances. As early at 2 days of age, rat pups can learn to avoid odors previously associated with toxicosis (Rudy and Cheatle 1977). We therefore wondered whether this sort of odor familiarity would induce the same alteration in olfactory bulb glomerular activity as did daily odor experience.

Rat pups trained to avoid peppermint odor did not show an enhanced glomerular response to peppermint odor (Coopersmith et al. 1984). Odor-naive day-18 rat pups were subjected to toxicosis or to the control treatment while being exposed either to peppermint odor or clean air. On day 19, pups poisoned in the presence of peppermint showed a robust behavioral aversion to peppermint odor, and did not have the enhanced neural response to the familiar odor. Pups exposed to peppermint for one single time in the absence of aversive conditions had a significant increase in glomerular 2DG response during the subsequent peppermint test. Odor familiarity per se is therefore not sufficient to produce the enhanced neural response.

4 Prospectus

The glomerular layer is diffusely organized at birth and the eventual segregation and differentiation of individual glomeruli may depend on neural activity (Friedman and Price 1984). Intense or repeated olfactory experience may further mold the formation of the glomeruli to reflect this experience. Such changes could facilitate the processing of familiar olfactory information.

It will be important to determine the experiential contingencies and the structural basis underlying this neural change. Such an analysis may be the first to demonstrate naturally occurring, experience-dependent changes in the mammalian brain that form the basis for the variation among individuals in a population.

Acknowledgments. The research presented here is supported in part by grant NS 21484 from NINCDS to M.L., who holds Research Scientist Development Award MH 00317 from NIMH.

References

Altman J (1969) Autoradiographic and histological studies of postnatal neurogenesis. IV. Cell proliferation and migration in the anterior forebrain, with special reference to persisting neurogenesis in the olfactory bulb. J Comp Neurol 137: 433–458

Bayer SA (1983) ^{3}H-Thymidine-radiographic studies of neurogenesis in the rat olfactory bulb. Exp Brain Res 50: 329–340

Coopersmith R, Leon M (1984) Enhanced neural response to familiar olfactory cues. Science 225: 849–851

Coopersmith R, Lee S, Leon M (1984) Potentiated neural response to familiar odors by young rats. Int Soc Dev Psychobiol Abst, Baltimore, p 29

Friedman B, Price JL (1984) Fiber systems in the olfactory bulb and cortex: A study in adult and developing rats, using the Timm method with the light and electron microscope. J Comp Neurol 223: 88–109

Gerling S, Yahr P (1982) Maternal and paternal pheromone in gerbils. Physiol Behav 28: 667–673

Harder W (1949) Zur Morphologie und Physiologie des Blinddarmes der Nagetiere. Verh Dtsch Zool Ges 2: 95–109

Leon M (1974) Maternal pheromone. Physiol Behav 13: 441–453

Leon M (1975) Dietary control of maternal pheromone in the lactating rat. Physiol Behav 14: 311–319

Leon M, Moltz H (1971) Maternal pheromone: Discrimination by pre-weanling albino rats. Physiol Behav 14: 311–319

Leon M, Moltz H (1972) The development of the pheromonal bond in the albino rat. Physiol Behav 8: 683–686

Leon M, Coopersmith R, Ulibarri C, Porter RH, Powers JB (1984) Development of olfactory bulb organization in precocial and altricial rodents. Dev Brain Res 12: 45–53

Mair RG, Gesteland RC (1982) Response properties of mitral cells in the olfactory bulb of the neonatal rat. Neuroscience 7: 3117–3125

Pedersen PE, Blass EM (1981) Olfactory control over suckling in albino rats. In: Aslin RN, Alberts JR, Petersen MR (eds) The development of perception: Psychobiological perspectives, vol 1. Academic Press, London New York

Pedersen PE, Blass EM (1982) Prenatal and postnatal determinants of the first suckling episode in albino rats. Dev Psychobiol 15: 349–356

242

Porter RH, Doane HM (1976) Maternal pheromone in the spiny mouse (Acomys cahirinus). Physiol Behav 16: 75-78

Porter RH, Doane HM (1977) Dietary-dependent cross-species similarities in maternal chemical cues. Physiol Behav 19: 129-131

Porter RH, Etscorn F (1976) A sensitive period for the development of olfactory preference in Acomys cahirinus. Physiol Behav 17: 127-130

Rudy JW, Cheatle MD (1977) Odor-aversion learning in neonatal rats. Science 198: 845-846

Sokoloff L, Reivich M, Kennedy C, Des Rosiers MH, Patlak CS, Pettigrew KD, Sakurada O, Shinohara M (1977) The ^{14}C-deoxyglucose method for the measurement of local cerebral glucose utilization: Theory, procedure, and normal values in the conscious and anesthetized albino rat. J Neurochem 28: 897-916

Stickrod G, Kimble DP, Smotherman WP (1982) In utero taste/odor aversion conditioning in the rat. Physiol Behav 28: 5-7

Yahr P, Anderson-Mitchell K (1983) Attraction of gerbil pups to maternal nest odors: duration, specificity and ovarian control. Physiol Behav 31: 241-247

Olfactory Guidance of Nipple-Search Behaviour in Newborn Rabbits

R Hudson and H Distel

Institut für Medizinische Psychologie, Universität München, München, West Germany.

1 Introduction

As rabbits live in burrows and are most active above ground at dawn and dusk when lighting conditions are poor, they are very dependent on their sense of smell in processing the world around them. The ability to communicate with other rabbits, for example, is largely based on the transmission of subtle olfactory cues. Rabbits rely on odours contained in secretions from specialized skin glands and in the urine and feces for such important tasks as territorial marking and identifying the age, rank and sexual status of individuals (Goodrich and Mykytowycz 1972; Mykytowycz 1965,1972; Schalken 1976; Bell 1980; Hesterman and Mykytowycz 1982a, b; Hesterman et al. 1984). Given their well-developed olfactory system and established position as laboratory animals, rabbits would therefore seem to be good subjects for developmental studies of olfaction. Not only are they notoriously easy to breed and grow rapidly but, as will be argued here, the newborn pups possess a number of behavioural specializations which make them particularly suitable for such investigations. Chief of these is their total dependence on olfactory cues to locate the mother's nipples and to suckle (Schley 1977,1981; Hudson and Distel 1983). The reason for this, and the research opportunities it provides, can be best understood in the context of the rabbit's unusual maternal behaviour and the demands this places on the young.

2 Maternal Behaviour

Rabbits give their young remarkably little care. Before giving birth, the doe digs a nursery burrow separate from the main warren, in which she builds a nest made of grass and fur pulled from her own body. (Niethammer 1937; Deutsch 1957; Zarrow et al. 1963). After giving birth to the pups in this nest, she leaves them and returns to nurse only once approximately every 24 h (Venge 1963; Zarrow et al. 1965). This nursing visit is extremely short, lasting only about 3 to 4 min (Deutsch 1957; Lincoln 1974; Kraft 1979). On entering the nest, the doe simply positions herself over the litter, remaining almost motionless during nursing, and not giving the pups any direct behavioural assistance to suckle (Deutsch 1957; Lincoln 1974; Hudson and Distel 1982). Nursing ends abruptly with the doe jumping out of the nest and leaving the pups alone until the following day. She does not brood them, cleans them little, if at all, and does not even retrieve pups which stray from the nest (Deutsch 1957; Ross et al, 1959).

Fig. 1 Size and developmental state of a newborn rabbit pup in comparison with a 14-day-old animal

This unusual pattern of maternal behaviour may be explained by the fact that rabbits are fugitive animals and heavily preyed upon. Their only protection is to flee into the warren, which, with its many entrances and exits, offers some chance of escaping even those predators able to pursue them underground. In such a situation, the newborn pups would obviously be unable to escape, and it is probably for this reason that the doe constructs a separate nursery burrow with only one entrance, which she carefully closes after every visit (Niethammer 1937; Deutsch 1957; Kraft 1979). However, to reduce the risk of predators trapping her and the pups in this, it is clearly important that the time spent with the young be kept to a minimum (Zarow et al. 1965).

3 Behaviour of the Pups in the Nest

This strategy places great demands on the pups, particularly as they are rather immature at birth. They are born naked, with sealed eyes and outer ears and poor motor coordination (Fig. 1). By day 7 they are capable of limited orienting responses to auditory stimuli (Gottlieb 1971), and may also perceive light changes (Ripisardi et al. 1975), although they only begin to open their eyes on day 9 or 10. Pups start to leave the nest when 13 to 18 days old (Zarrow et al. 1965; Mykytowycz and Dudzinsky 1972; Kraft 1979), by which time they are able to maintain a stable body temperature (Wisham et al. 1979) and have much improved motor coordination (Fig. 1). When one considers that the difference between the two pups in Fig. 1 is only 13 brief nursing episodes, the importance of each of these for the survival and proper growth of the young becomes clear. Such rapid nursing and consequent reduction in the time the burrow must remain open are only possible because of several behavioural specializations.

One of these is the apparent ability of pups to anticipate and prepare themselves for the daily arrival of their mother (Hudson and Distel 1982). While the pups spend most of the time between feeds lying quietly together under an insulating cover of nest material, 1 or 2 h before nursing they become more active and gradually emerge from this cover. At this time they are particularly sensitive to tactile and vibratory stimuli, and react to even slight disturbances with increased activity, rearing movements and vocalization. They respond to the doe's entering the nest by rearing their heads up and pushing their muzzles deep into her belly fur to start the rapid search for nipples. Despite the heat loss presumably incurred by exposure from the nest material (cf. D Hull 1965), anticipatory uncovering appears important, as it enables pups to reach the doe's belly unhindered, and thus reduces nursing time. Experimentally covering young pups with nest material just before the doe's arrival greatly reduces milk intake, despite the doe's spending longer in the nest and trying to uncover them.

When the doe jumps out of the nest at the end of nursing, the pups drop immediately from the nipples. They then urinate simultaneously, becoming very wet, and vigorously burrow back under the nest material and disperse throughout the nest. This activity lasts about 10 to 15 min, during which time the pups become dry and the nest material fluffed up again. The pups then gradually reassemble in the warmest part of the nest (cf. J Hull and D Hull 1982), where they remain covered until the next nursing visit. It is not only important that pups stay together to keep warm, but also because those away from the group at nursing time are ignored by the doe and have little chance of reaching her in time to suckle (Hudson and Distel 1982).

Contrary to what one might expect, the anticipatory uncovering does not appear to be a simple consequence of absence of food in the gut. Litters isolated from their mother and deprived of one nursing uncover from the nest material as usual on the first day of separation but, when the doe does not come, gradually become covered again. The following day, approximately 47 h after last being nursed, the pups again uncover and are able to suckle normally (Hudson and Distel 1982). Interestingly, very similar daily cycles of nursing behaviour have also been reported for other lagomorphs (Rongstad and Tester 1971; Broekhuizen and Maaskamp 1980). Young hares, which are born above ground and with open eyes, gather together at dusk just before the doe's arrival, then disperse after nursing, to spend the following day hidden in the vicinity. Unlike hares, rabbit pups cannot use environmental cues to predict their mother's arrival, and thus their anticipatory behaviour may well represent a circadian rhythm, even at such an early age (Hudson and Distel 1982). However, when and in what way this daily cycle develops and becomes synchronized with that of the mother remains to be investigated.

4 Nipple-Search Behaviour

Uncovering may help pups reach the doe's belly, but how are they then able to find nipples so quickly? Observing nursing through the floor of a glass-bottomed nest box showed that pups take on average only about 6 s to attach to a nipple from the start of searching (Hudson and Distel 1983). This search behaviour is highly stereotyped and is shown in response to any nursing doe. By making very rapid probing movements (5 to 8 s^{-1}) deep into the fur, pups move zig-zag across the doe's belly until a nipple is reached. Surprisingly, they do not remain on the one nipple but change them frequently, repeating the whole search sequence several times within one nursing session. Although nipple-switching reduces the time pups actually spend on nipples to an average of only about 110 s, they are able to drink up to 25% of their own weight in this time (Lincoln 1974; Hudson and Distel 1983).

Fig. 2 The test arena viewed from above with a 5-day-old pup at a nipple of an anaesthetized test doe

To investigate the cues governing this highly effective orienting behaviour, pups were tested on does restrained on their backs in a U-shaped trough. The ends of the trough were closed by dividers which fitted the doe's body so as to form an arena enclosing the six rear nipples (Fig. 2). The response of pups placed on the doe's belly in the arena is similar to that observed during normal nursing, but the arena has the advantage of allowing more rigorous analysis of behavioural components, repeated testing and more accurate measurement of performance parameters (Fig. 3a).

Testing pups under various conditions in the arena showed them to be completely dependent on odour cues on the doe's skin to locate nipples and suckle (Hudson and Distel 1983). Neither shaving the doe's belly nor creating a negative thermal gradient by cooling the nipple areas had any significant effect on the pups' behaviour. However, when the shaved belly was covered with adhesive tape but the nipples left bare, pups did not search, and only attached to nipples when they chanced to bring their noses next to them. Pups also failed to show search behaviour when the doe's belly was covered with a fine nylon net, but attempted to grasp nipples through it when their noses came directly over them. By covering over the area around nipples and/or the nipples themselves in various ways, it could also be shown that the short-ranging odour cues releasing and sustaining the search behaviour increase in strength towards nipples, and thus help guide pups to them. At nipples, high concentrations, or possibly a second odour, then stimulate grasping.

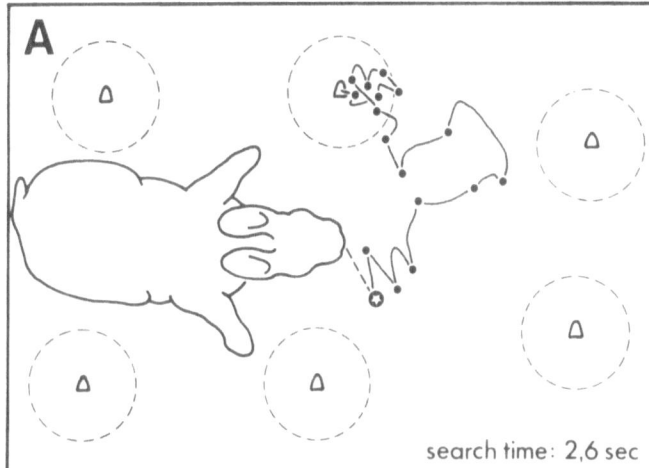

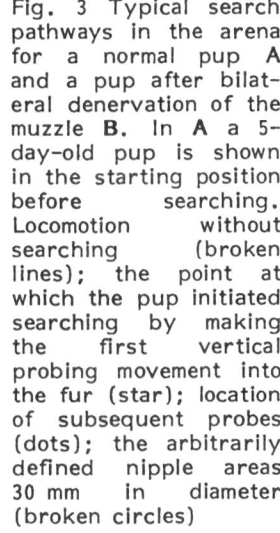

Fig. 3 Typical search pathways in the arena for a normal pup **A** and a pup after bilateral denervation of the muzzle **B**. In **A** a 5-day-old pup is shown in the starting position before searching. Locomotion without searching (broken lines); the point at which the pup initiated searching by making the first vertical probing movement into the fur (star); location of subsequent probes (dots); the arbitrarily defined nipple areas 30 mm in diameter (broken circles)

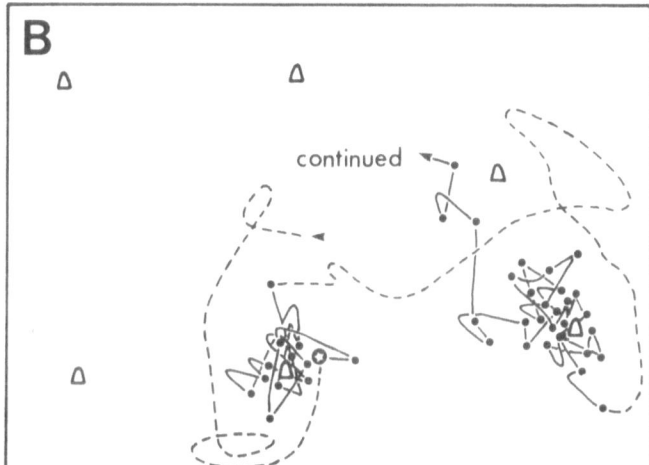

This dependence on highly specific olfactory cues explains why newborn rabbits are so difficult to raise by hand (Appel et al. 1971) and, when rendered anosmic by removing the olfactory bulbs or irrigating the nasal mucosa with zinc sulphate, are completely unable to suckle from the mother (von Gudden 1870; Schley 1977, 1981; Distel and Hudson in preparation). As these treatments also damage the accessory olfactory system, it is possible that the vomeronasal organ mediates the response. However, as removing the vomeronasal organ through the roof of the mouth immediately after birth has little, if any, effect on the nipple-search behaviour, the accessory olfactory system does not appear important for the performance of this task (own unpublished observations).

Despite their dependence on odour cues, pups cannot attach to nipples and suckle using olfaction alone. Bilateral transections of the infraorbital branches of the trigeminal nerve innervating the muzzle demonstrate somatosensory input to be essential also for efficient search behaviour and for nipple attachment (Distel and Hudson in preparation). This operation disrupts suckling as totally as removal of the olfactory bulbs. When tested in the arena, pups with bilaterally denervated muzzles search vigorously, and can locate the nipple areas (Fig. 3b). However, as they do not show the repeated mouth-opening normally performed during searching, they are unable to grasp nipples, even though they are perfectly able to open their mouths, when yawning for example. Despite the importance of somatosensory input, tactile stimuli alone are insufficient to elicit nipple attachment, as anosmic pups cannot attach even when held directly on nipples.

Unilateral denervation of the muzzle also severely handicaps pups, resulting in a lateralization of head movements during searching and nipple attachment. Not surprisingly, nipples are only grasped when contacted with the intact side of the muzzle, and pups turn towards this side when releasing them. In contrast, unilaterally bulbectomized pups show no such lateralization of head movements, and locate nipples almost as quickly as normal pups (Distel and Hudson in preparation). While this suggests that differential input to the two nares is unimportant in guiding the pups' search behaviour, the question remains as to how the rapid probing movements enable pups to sample the odor cues on the doe's belly so as to perceive the gradient so quickly.

5 The Nipple-Seach Pheromone

The unusually reliabe and stereotyped nature of the pups' response to these odour cues would seem to qualify them as a true releasing pheromone, and so they will now be referred to as the nipple-search pheromone. Few of the chemical signals so far identified in mammals have such a reliable and specific effect on behaviour (Beauchamp et al. 1976) as this pheromone. Although the nature and actual site of production is still unclear, the pheromone is not contained in the mother's saliva or urine (Hudson and Distel 1983), but is present in the milk. This is fortunate, as it cannot easily be collected from the doe's skin. Pups presented with a cotton bud soaked in fresh rabbit milk readily grasp it, but do not respond to the milk when more than 10 to 15 min old (Müller 1978; own observations). Thus, the analysis of the pheromonal content of fractionated milk, together with more detailed knowledge of the circumstances under which it is emitted, should facilitate its collection and perhaps even its isolation and chemical characterization.

Interestingly, production of the pheromone is not confined to lactating does. By using the response of newborn pups to regularly test for its presence, it can be shown that emission is under hormonal control and that all mature does produce it under favourable light conditions (Hudson and Distel 1984).

Emission is very low or absent in winter, but increases in early spring to reach a peak at about the time of the summer solstice. The close relationship between pheromone emission and photoperiod can be demonstrated by the experimental reversal of seasonal light conditions. While artificial short days suppress emission, long days stimulate it within 1 or 2 weeks.

Further, the stimulating effect of long day conditions is correlated with an increased readiness of does to mate, and with the production of larger litters.

Only sexually mature does produce the pheromone, and at no time of the year do pups search for or attach to the nipples of animals under 3 months of age.

Pregnancy and lactation have the strongest influence on pheromone emission, overriding the effect of daylength. Pregnancy stimulates emission even in winter, so that by parturition and during early lactation test pups are able to locate and attach to nipples within a few seconds. However, early removal of pups from their mother results in a reduction in pheromone emission within a few days. Strong evidence for the role of the sex steroids in production comes from the finding that emission is suppressed following ovariectomy, but can be stimulated within a few days by the administration of oestradiol to the sterilized does. Exactly how oestradiol stimulates emission is not clear and, as suggested by the effects of early weaning, a synergistic action with prolactin or other hormones seems likely.

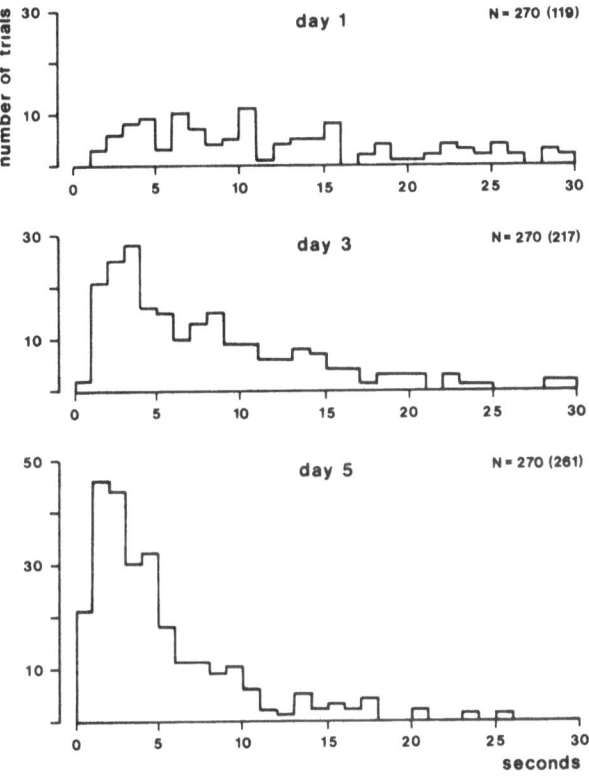

Fig. 4 Improvement in the time taken by pups to locate nipples in the arena within the 30-s time limit. The improvement is also reflected by the increase in the number of successful trials indicated in brackets

6 Development of Reactivity

As the reaction of most mammals to pheromones or pheromone-like substances is at least partially dependent on experience (Beauchamp et al. 1976), it is interesting that rabbit pups are able to respond appropriately to their mother from the very first nursing. This obviously raises the question as to how and when they come to respond to the nipple-search pheromone in such a specific way. More particularly, to what extent is the pups' reaction present at birth and to what extent acquired through postnatal experience?

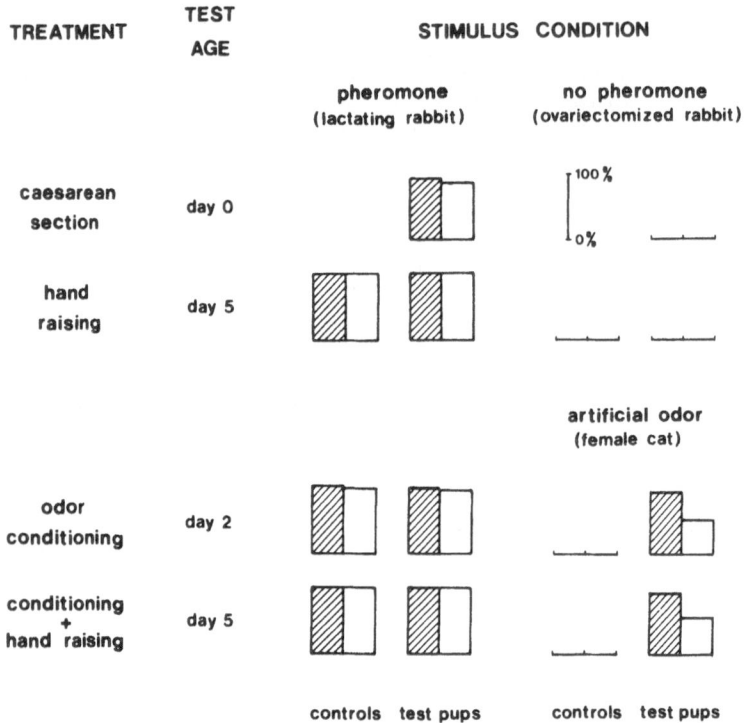

Fig. 5 Summary of the main findings of experiments investigating inborn and acquired responsiveness to odour stimuli releasing nipple-search behaviour. Results are expressed as the percentage of trials in which pups searched for (shaded bars) and attached to nipples (empty bars). Odour conditioning was carried out during the first nursing on day 1

The short exposure of rabbit pups to their mother, and the fact that first-born pups may try to suckle while parturition is still in progress (Hudson and Distel 1982) suggest the postnatal learning of cues associated with nipple location and attachment to be unimportant. This is confirmed by the finding that pups delivered by caesarian section 1 day before term respond with normal search behaviour to a lactating doe (Fig. 5), and take no longer to attach to nipples than normally delivered 1-day-old pups (Hudson in press). Thus knowledge of the pheromone appears to be inborn. However, this does not exclude the possibility that the response is dependent on prenatal experience of, for example, chemical characteristics of the uterine environment (Stickrod et al. 1982).

Suckling experience is also unimportant for the subsequent development of responsiveness to the pheromone. During the first 5 days of life, pups show a considerable improvement in the speed with which they locate nipples both during normal nursing and in the arena (Hudson and Distel 1983; Distel and Hudson 1984; Fig. 4). This appears mainly due to an improvement in reactivity to the pheromone and consequent reduction in the time taken to initiate searching rather than to better motor coordination. Whereas newborn pups crawl around the arena and first start to search near nipples where pheromone concentration is strongest, 5-day-old pups are able to start searching as soon as they are placed in the centre of the doe's belly where the odour cues are weaker (Distel and Hudson 1984). Pups hand-raised to day 5 on an artificial milk formula, and thus without postnatal experience of the pheromone, attach as rapidly as normally nursed pups when held to nipples (Schley 1981), and locate them as quickly when placed on the doe's belly (Hudson in press; Fig. 5). This suggests that the improved reactivity to the pheromone may be due to maturational processes rather than to learning. While the nature of these processes remains unclear, maturation of the olfactory system may well be a significant factor, as it undergoes considerable development postnatally in the rabbit (Schwartze 1967; Meisami personal communication).

A further developmental change in the pattern of reactivity to the pheromone occurs after the first week. Following the general improvement in responsiveness shown during the first few days, the time taken by pups to initiate searching increases again when they are tested between feeds but remains minimal before nursing. This pattern may reflect the development of motivational states, particularly the hunger system, and the increasing control of internal stimuli over suckling behaviour (Distel and Hudson 1984).

7 Odour Conditioning

Although rabbit pups do not learn postnatally the odour cues governing nipple-search behaviour, they can nevertheless rapidly learn to associate novel odours with suckling (Ivanitskii 1962; Hudson in press). Pups nursed by a doe whose ventrum has been scented with a substance such as the perfume Chanel No.5, citral or camphor, come to repond to the scent as to the pheromone itself (Hudson in press). After only one such pairing, pups show the full sequence of search behaviour, nipple attachment and vigorous sucking when placed on the belly of a non-pheromone-producing doe or even on a female cat scented with the odour. In contrast, control pups which have not experienced the scent during nursing respond to the test females by crawling around or resting, and cannot be induced to grasp niples. (Fig. 5).

There are other interesting similarities between the inborn and learned response to odour cues. Just as pups raised without experience of the pheromone respond to it when tested on day 5, so pups conditioned on day 1 but then bottle-fed still show vigorous nipple-search behaviour in response to the conditioned odour when tested on day 5 (Fig. 5). Thus, as in the case of the pheromone, responsiveness to the conditioned odour is retained and perhaps even improves without further experience or practice. It therefore seems possible that the learned odours gain direct access to the neural substrate governing the stereotyped nipple-search sequence, and so come to have a releasing function similar to the pheromone itself. It has been suggested that both inborn responsiveness to odour stimuli and responsiveness acquired through early imprinting may depend on input to the accessory olfactory system (Meredith 1983). It is therefore interesting that disrupting accessory olfactory function by removing the vomeronasal organ in newborn pups impairs neither their inborn responsiveness to the pheromone nor their ability to acquire the conditioned response (own unpublished observations).

The earlier suggestion that odour cues may actually guide pups to nipples is also supported by findings from conditioning experiments. When the conditioned odour is distributed on the test female as an inverse gradient increasing in strength away from nipples, pups concentrate their searching where the odour is strongest and have difficulty finding nipples, if at all (own unpublished observations). The possibility this provides for testing pups on artificial odour gradients should therefore make it possible to investigate the processing of concentration differences and the following of odour trails more precisely. Such rapid odour conditioning also provides a promising tool for investigating the development of olfactory function. Not only does it present a means of investigating the range and sensitivity of early odour perception, but the comparison morphologically and physiologically of the olfactory systems of conditioned and unconditioned pups should help our understanding of the early processing of olfactory stimuli.

Acknowledgments. Our work was supported by the SPP Verhaltensontogenie of the Deutsche Forschungsgemeinschaft, Di 212/2 and 3, and would not have been possible without the facilities and encouragement provided by Prof. Ernst Pöppel.

References

Appel K, Busse H, Schulz K, Wilk W (1971) Beitrag zur Handaufzucht von gnotobiotischen und SPF-Kaninchen. Z Versuchstierkd 13: 282-209

Beauchamp GK, Doty RL, Moulton DG, Mugford RA (1976) The pheromone concept in mammalian chemical communication: A critique. In: Doty RL (ed) Mammalian olfaction, reproductive processes and behavior. Academic Press, London New York, pp 143-160

Bell DJ (1980) Social olfaction in lagomorphs. In: Stoddart DM (ed) Olfaction in mammals. Academic Press, London New York, pp 141-164

Broekhuizen S, Maaskamp F (1980) Behaviour in does and leverets of the European hare (Lepus europaeus) whilst nursing. J Zool (London) 191: 487-501

Deutsch JA (1957) Nest building behaviour of domestic rabbits under semi-natural conditions. Anim Behav 5: 53-54

Distel H, Hudson R (1984) Nipple-search performance by rabbit pups: Changes with age and time of day. Anim Behav 32: 501-507

Distel H, Hudson R (1985) The contribution of the olfactory and tactile modalities to the performance of the nipple-search behaviour of newborn rabbits (in preparation).

Goodrich BS, Mykytowycz R (1972) Individual and sex differences in the chemical composition of pheromone-like substances from the skin glands of the rabbit, Oryctolagus cuniculus. J Mammal 53: 540-548

Gottlieb G (1971) Ontogenesis of sensory function in birds and mammals. In: Tobach E, Aronson LR, Shaw E (eds) The biopsychology of development. Academic Press, London New York, pp 67-128

Gudden von B (1870) Experimentaluntersuchungen über das peripherische und zentrale Nervensystem. Arch Psychiatr 2: 693-723

Hesterman ER, Mykytowycz R (1982a) Misidentification by wild rabbits, Oryctolagus cuniculus, of group members carrying the odor of foreign inguinal gland secretion. I. Experiments with all-male groups. J Chem Ecol 8: 419-427

Hesterman ER, Mykytowycz R (1982b) Misidentification by wild rabbits, Oryctolagus cuniculus, of group members carrying the odor of foreign inguinal gland secretion. II. Experiments with all-female groups. J Chem Ecol 8: 723-729

Hesterman ER, Malafant K, Mykytowycz R (1984) Misidentification by wild rabbits, Oryctolagus cuniculus, of group members carrying the odor of foreign inguinal gland secretion. III. Experiments with mixed sex groups and analysis of further data from all-male and all-female groups. J Chem Ecol 10: 403-419

Hudson R (1986) Do newborn rabbits learn the odor stimuli releasing nipple-search behavior? Dev Psychobiol (in press)

Hudson R, Distel H (1982) The pattern behaviour of rabbit pups in the nest. Behaviour 79: 255-271

Hudson R, Distel H (1983) Nipple location by newborn rabbits: Behavioural evidence for pheromonal guidance. Behaviour 85: 260-275

Hudson R, Distel H (1984) Nipple search pheromone in rabbits: Dependence on season and reproductive state. J Comp Physiol A 155: 13-17

Hull D (1965) Oxygen consumption and body temperature of newborn rabbits and kittens exposed to cold. J Physiol (London 177: 192-202

Hull J, Hull D (1982) Behavioral thermoregulation in newborn rabbits. J Comp Physiol Psychol 96: 143-147

Ivanitskii AM (1962) The morphophysiological investigation of development of conditioned alimentary reaction in rabbits during ontogenesis. In: Works of the institute of higher nervous activity, physiological series, vol IV. Isr Program Sci Transl

Kraft R (1979) Vergleichende Verhaltensstudien an Wild- und Hauskaninchen. I. Das Verhaltensinventar von Wild- und Hauskaninchen. Z Tierzücht Züchtungsbiol 95: 140-162

Lincoln DW (1974) Suckling: A time-constant in the nursing behaviour of the rabbit. Physiol Behav 13: 711-714

Meredith M (1983) Sensory physiology of pheromone communication. In: Vandenbergh JG (ed) Pheromones and reproduction in mammals. Academic Press, London New York, pp 199-252

Müller K (1978) Zum Saugverhalten von Kaninchen unter besonderer Berücksichtigung des Geruchsvermögens. Dissertation, Gießen

Mykytowycz R (1965) Further observations on the territorial function and histology of the submandibular cutaneous (chin) glands in the rabbit, Oryctolagus cuniculus (L). Anim Behav 13: 400-412

Mykytowycz R (1972) The behavioural role of the mammalian skin glands. Naturwissenschaften 59: 133-139

Mykytowycz R, Dudzinski ML (1972) Aggressive and protective behaviour of adult rabbits Oryctolagus cuniculus (L) towards juveniles. Behaviour 43: 97-120

Niethammer G (1937) Ergebnisse von Markierungsversuchen an Wildkaninchen. Z Morphol Oekol Tiere 33: 297-312

Ripisardi SC, Chow KL, Mathers LH (1975) Ontogenesis of receptive field characteristics in the dorsal lateral geniculate nucleus of the rabbit. Exp Brain Res 22: 295-305

Rongstad OJ, Tester JR (1971) Behavior and maternal relations of young snowshoe hares. J Wildlife Manage 35: 338-346

Ross S, Denenberg VH, Frommer GP, Sawin PB (1959) Genetic, physiological and behavioral background of reproduction in the rabbit. V. Nonretrieving of neonates. J Mammal 40: 91-96

Schalken APM (1976) Three types of pheromones in the domestic rabbit, Oryctolagus cuniculus (L). Chem Sens Flavor 2: 139-155

Schley P (1977) Die Ausschaltung des Geruchsvermögens und sein Einfluß auf das Saugverhalten von Jungkaninchen. Berl Münch Tierärztl Wochenschr 90: 382-385

Schley P (1981) Geruchsinn und Saugverhalten bei Jungkaninchen. Kleintierpraxis 26: 261-263

Schwartze P (1967) Die postnatale Entwicklung der Bursttätigkeit im Bulbus olfactorius des Kaninchens. Experientia 23: 725-726

Stickrod G, Kimble DP, Smotherman WP (1982) In utero taste/odor aversion conditioning in the rat. Physiol Behav 28: 5-7

Venge O (1963) The influence of nursing behaviour and milk production on early growth in rabbits. Anim Behav 11: 500-506

Wisham IQ, Flannigan KP, Barnsley RH (1979) Development of tonic immobility in the rabbit: Relation to body temperature. Dev Psychobiol 12: 595-605

Zarrow MX, Farooq A, Denenberg VH, Sawin PB, Ross S (1963) Maternal behaviour in the rabbit: Endocrine control of maternal-nest-building. J Reprod Fertil 6: 375-383

Zarrow MX, Denenberg VH, Anderson CO (1965) Rabbit: Frequency of suckling in the pup. Science 150: 1835-1836

Salivary Pheromones in the Pig and Human in Relation to Sexual Status and Age

DB Gower and WD Booth*

Department of Biochemistry, United Medical and Dental Schools of Guy's and St. Thomas's Hospital, London, SEI 9RT, UK

*AFRC Institute of Animal Physiology, Animal Research Station, 307 Huntingdon Road, Cambridge CB3 OJQ, UK

1 Introduction

Abbreviations used:

An-α(ß)	5α-androst-16-en-3α(ß)-ol
5α-A	5α-androst-16-en-3-one
4,16-androstadienone	4,16-androstadien-3-one
5α-DHT	5α-dihydrotestosterone
RIA	radioimmunoassay
SIM	selective ion monitoring
SDS-PAGE	sodium dodecyl sulfate-polyacrylamide electrophoresis
M_r	relative molecular mass

There is increasing evidence that the sense of olfaction plays a vital role in the survival of neonatal animals. Newly born rabbit pups have their eyes and outer ears sealed, and have poor motor co-ordination. Thus, it is vital for their survival that they are able to rapidly find and grasp the nipples of the mother during her brief (3-4 min) visit to the nest in a 24-h period. This important "nipple-search" behaviour is dominated by olfactory cues; knowledge of the pheromone produced by the lactating doe being inborn in the pups (Hudson and Distel this Vol). Olfaction is again the dominant sense in neonatal rats and the pups use odorous cues to find maternal nipples. In this case, the pheromone is caecotroph, a semi-solid material produced in the caecum and excreted with the faeces. Significantly the emission of the attractant, which starts from 14 days post-partum and continues until the end of the fourth post-partum week, is synchronized with the development of the pups' responsiveness to it (Coopersmith and Leon this Vol).

Of particular relevance to the present review is the possible involvement of odour cues in the teat-order that is taken up by piglets. In a study of behaviour of piglets reared on an artificial sow, Jeppeson (1982a) found that the animals quickly took up fixed positions on the row of teats on the model udder. Odour cues are probably important in this phenomenon because deodorizing this udder resulted in a variable teat-order. Furthermore, "alien" teats (i.e., those which has been used by other piglets) were avoided. Jeppesen (1982b) noted that teats became wet with saliva when they were nuzzled by the animals. The nature of any olfactory cues, which may be present in the saliva of neonatal pigs, is unknown at present. However, the odorous 16-androstenes are present in the boar from foetal life onwards, and it is tempting to speculate that these may play a role in teat-order behaviour. Two of the 16-androstenes are known for their pheromonal role in puberty acceleration and facilitation of the lordosis response (see below) and the remainder of this review is concerned with their significance in post-pubertal animals therefore.

2 Overview of Current Knowledge

2.1 16-Androstenes in the Pig

Prelog and Ruzicka (1944) first showed that musk-smelling steroids are present in boar testis, and this finding has been amply confirmed by other workers (Booth 1970; Claus 1970). The odorous steroids concerned were shown to be 16-unsaturated C_{19} steroids (16-androstenes) and that, one of these, 5α-A was urine-smelling whereas another, An-α possessed a more pleasant musk-like odour (Prelog and Ruzicka 1944; Prelog et al. 1945). Subsequent studies (Gower 1972, 1984) revealed that 16-androstenes are formed readily in boar testis, the quantities being greater than those of the androgens, such as testosterone (Hurden et al. 1984).

Later work, using analytical (Booth 1975; Booth and Polge 1976) and biosynthetic (Katkov et al. 1972) techniques, confirmed the presence of some 16-androstenes in the submaxillary glands of boars and of intersex pigs, and recent research (Booth 1984a) has revealed especially high levels (2 μmol/g) in the submaxillary gland of the Göttingen mini-boar. Although neither the parotid nor the submaxillary glands possess the enzymes necessary for 16-androstene synthesis from the usual testicular steroid precursors, pregnenolone or progesterone (Gower 1984), they do possess active dehydrogenases capable of converting 5α-A or 4,16-androstadienone into An-α and An-ß, the reductive activity of the submaxillary tissue being much greater than that of parotid tissue (Katkov et al. 1972). The relative concentrations of 16-androstenes present in boar submaxillary glands are An-α > 5α-A > An-ß > 5,16-androstadien-3ß-ol; only traces of these steroids are present in the parotid (Katkov et al. 1972). The work of Perry et al. (1980) is in keeping with the suggestion that the submaxillary gland in the pig is the main storage salivary gland for 16-androstenes, and recent confirmation of this has been provided when negligble quantities of the steroids were found in the saliva of a sialectomized boar (Booth et al. unpublished).

Analytical data (Booth 1975) revealed that, in general, the concentrations of An-α, An-ß and 5α-A in the submaxillary gland increase with the age of the animal, especially after the age of 18 weeks. The concentration af An-α always exceeds that of An-ß, irrespective of age (Booth 1975). This is in contrast to the situation in porcine testis where An-α > An-ß in the immature pigs (Booth 1975; Kwan et al. 1985) but where An-ß > An-α in the mature animal (Ahmad and Gower 1968; Booth 1975; Hurden et al. 1984).

In 1967, Sink proposed that some of the 16-androstenes might be involved in porcine communication systems, an idea that was substantiated by field work (Melrose et al. 1971; Reed et al. 1974) in which it was shown that the odour of 5α-A facilitated the lordosis response; some of the other 16-androstenes were less effective. The submaxillary gland was confirmed as being an important source of these pheromones (Perry et al. 1980), and recent evidence indicates that An-α may also possess primer pheromonal activity, since the attainment of puberty is accelerated in gilts exposed to an artificial source of the steroid (Kirkwood et al. 1983; Booth 1984b). Both An-α and 5α-A occur in boar saliva (Gower 1972; Booth

1980), with ratios of alcohol to ketone being approximately 15:1. Levels of both steroids increase when boars are aroused in the presence of alien boars (Booth and Baldwin 1980). The storage and release of 16-androstenes appears to be dependent on testicular hormones, especially androgens and, possible oestrogens. The existence of a biochemical sexual dimorphism in the porcine submaxillary gland with respect to 16-androstenes is also true for testosterone and 5α-DHT, since these can only be isolated from submaxillary glands of boars (Booth 1972). Further results (Booth et al. 1973; Booth 1984a; Parrott and Booth 1984) support the indication that the submaxillary gland is a target organ for androgens because hypertrophy of the serous cells was found in the mature boar, but not in immature or castrate males, or in female pigs. Also consistent with these findings was the demonstration of several oxido-reductases, involved in androgen metabolism (Katkov et al. 1972; Booth 1972, 1977, 1982a; Booth et al. 1973).

A specific binding protein for 5α-A has been isolated from the submaxillary glands of the domestic boar (Booth 1982b,c) and the Göttingen miniboar (Booth 1984a). This protein, called pheromaxein, has a low M_r for a steroid-binding protein (15K) as demonstrated by SDS-PAGE (Fig. 1a), and is present in particularly large amounts in the submaxillary gland of the Göttingen miniboar, complementing the high levels of 16-androstene present there (Booth 1984a). It is probable that the ability of the submaxillary gland to store and, subsequently, to transport the lipophilic 16-androstenes in saliva is primarily dependent on pheromaxein.

2.2 16-Androstenes in the Human

Currently, there is great interest in pheromones, particularly in relation to the possibility of their being involved in human social interactions. Undoubtedly, this interest has arisen, in part, from the finding of odorous substances, including steroids, in human secretions including axillary sweat and saliva. The human neonate can identify the breast odours of its mother (Russell 1976) and this mother-child relationship continues as the age of the child increases. Mothers can also correctly identify their own children (age range 6 months to 17 years) by the odour of a worn T-shirt and preferred these odours to those of strange children. Fully-functional apocrine glands are necessary to make discrimination possible between men and women (Schleidt and Hold 1982). Thus, women cannot discriminate between prepubertal boys and girls, but American, German, Italian and Japanese adults can recognize their own odours (Schleidt and Hold 1982).

Of special interest, so far as human axillary odour is concerned, is the fact that the urinous 5α-A occurs in axillary secretions (Claus and Alsing 1976; Bird and Gower 1981, 1982; Gower et al. 1985) and in saliva (Bird and Gower 1983), with a marked sex difference. The steroid is probably a product of bacterial action since it is present in only minute quantities, if at all, in fresh apocrine sweat (Bird and Gower 1982; Labows et al. 1979) and the typical apocrine odour only develops on incubation with coryneform bacteria (Shehadeh and Kligman 1963; Leyden et al. 1981). The typical axillary odour is associated with a coryneform-dominated axillary microflora (Jackman and Noble 1983) and these bacteria have the ability to perform various transformations of steroid substrates (Nixon et al. 1984).

As well as 5α-A, An-α may also be present in the axillary secretions of men (Brooksbank et al. 1974) and both steroids seem to exhibit pheromonal effects in humans (for review, see Gower 1981; Doty 1981).

3 Salivary Pheromones: Influence of Age and Sexual Status.

3.1 Pig Salivary Pheromones

An interesting finding is the occurrence of significant amounts of the pheromonal 16-androstenes in boars before puberty, even in the foetus at 84 days of gestation (Booth 1975). Extraction of pooled submaxillary glands from 12 domestic boar piglets at birth provided 250pmol/g of An-α and 14pmol/g of 5α-A. More recently 1600pmol/g of An-α and 500pmol/g of 5α-A were determined in pooled submaxillary gland tissue from 4 Göttingen miniboar piglets at birth (Booth unpublished). The much higher concentration of pheromones in the glands of the miniboar piglets compared with that in domestic boar piglets, relates to similar quantitative differences between the adult animals (Booth 1975, 1984a). Incubation of aliquots of porcine submaxillary gland cytosols or saliva with 3H-labelled 5α-A for 1 h at 4°C, and then application of the samples to non-denaturing slab PAGE as described by Booth (1984a) is the technique used to demonstrate the presence of pheromaxein. Submaxillary gland cytosols from mature domestic boars, castrated boars and female pigs and the corresponding immature animals, gave profiles of radioactivity indicative of the presence of pheromaxein with the greatest degree of binding in the intact boar samples (Fig. 1). When samples were subjected to SDS-PAGE, amounts of pheromaxein were greatest in the submaxillary gland cytosols of mature intact boars (Fig. 3.1).

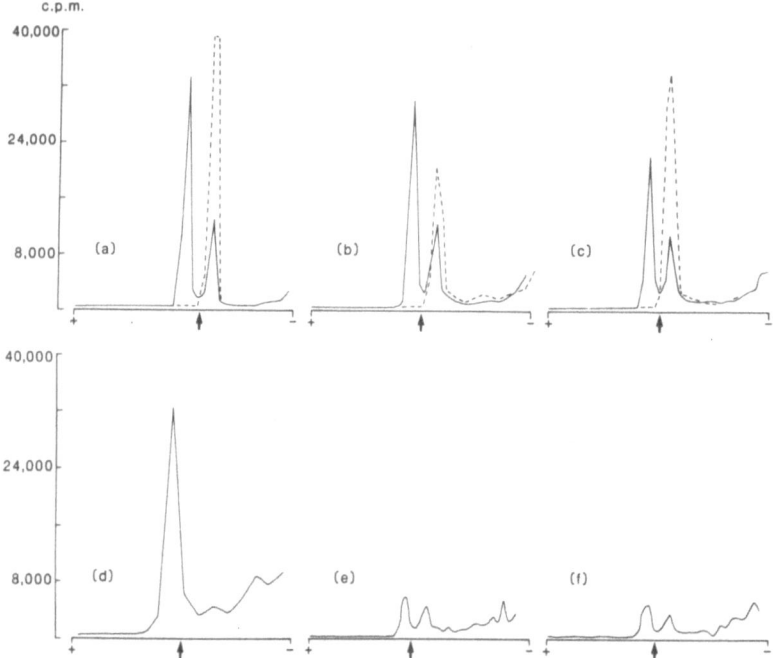

Fig. 1 Binding of ^{3}H-5α-A to pheromaxein in submaxillary gland cytosols of the domestic pig as demonstrated by non-denaturing PAGE **a** 2 mature boars, **b** 2 mature castrated boars, **c** 2 mature females (note in each case **a-c**, one animal has two binding peaks, the other a single binding peak), **d** immature boar, **e** immature castrated boar, **f** immature female

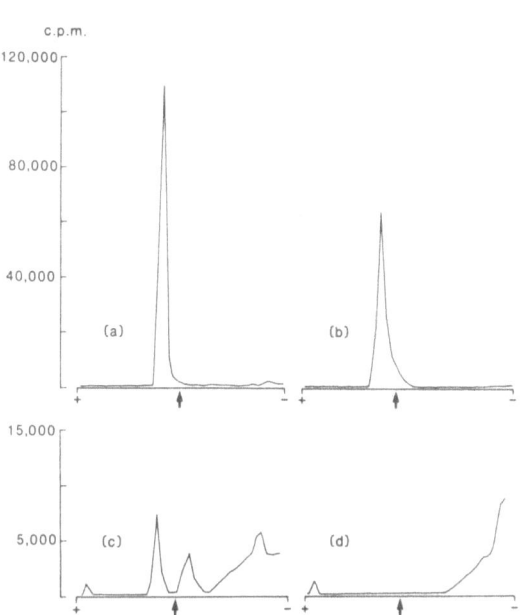

Fig. 2 Binding of ^{3}H-5α-A to pheromaxein in submaxillary gland cytosols of the Göttingen miniature pig as demonstrated by non-denaturing PAGE. **a** mature boar, **b** mature female, **c** male foetuses, **d** female foetuses (no binding). The foetuses were 67 days of gestation

In the mature Göttingen minipig, sexual dimorphism with particular reference to pheromaxein, is reflected in the difference in 16-androstene binding profiles between mature male and female pigs (Fig. 2a,b), and similarly the visible amounts of pheromaxein on SDS-PAGE (Booth 1984a). In the immature minipig, the opportunity arose to investigate the occurrence of pheromaxein in pooled submaxillary glands from seven male foetuses and six female foetuses (exceptionally from a single litter with 100% conception rate) at 67 days of gestation. The submaxillary gland cytosol of the foetal boars showed a binding profile for pheromaxein on non-denaturing PAGE, but there was no binding in the cytosol relating to the female foetuses (Fig. 2c,d). Furthermore, the male submaxillary glands were heavier (mean 114 mg/foetus) than those of the females (mean 88 mg/foetus). Together, these findings indicate a possible effect of testicular steroids on the submaxillary gland early in development and support the occurrence of 16-androstenes in boar submaxillary glands at birth.

The binding profiles for pheromaxein on non-denaturing PAGE have demonstrated the presence of charge isomers (Fig. 3.2) with profiles conforming to one of three types (a) pre-albumin binding (Fig. 1d), (b) post-albumin binding (Fig. 1a, b, c) and (c) both pre- and post-albumin binding with the pre-albumin binding predominating with a ratio about 2:1 (Fig. 1a, b, c). The underlying genetical control of the synthesis of the charge isomers and their possible physiological significance, remains to be determined. When submaxillary gland cytosols are incubated with ^{3}H-labelled 5α-A or ^{3}H-labelled An-α for 1 h at 4°C prior to non-denaturing PAGE, the 3H-labelled 5α-A is metabolized in greater part (68%) to An-α, whereas ^{3}H-labelled An-α is metabolized to much less (11%) 5α-A (Booth unpublished); this finding is in keeping with the known high activity of 3α-hydroxysteroid dehydrogenase in porcine submaxillary glands (Katkov et al. 1972; Booth 1977, 1982a). Binding of 16-androstenes under these conditions therefore,

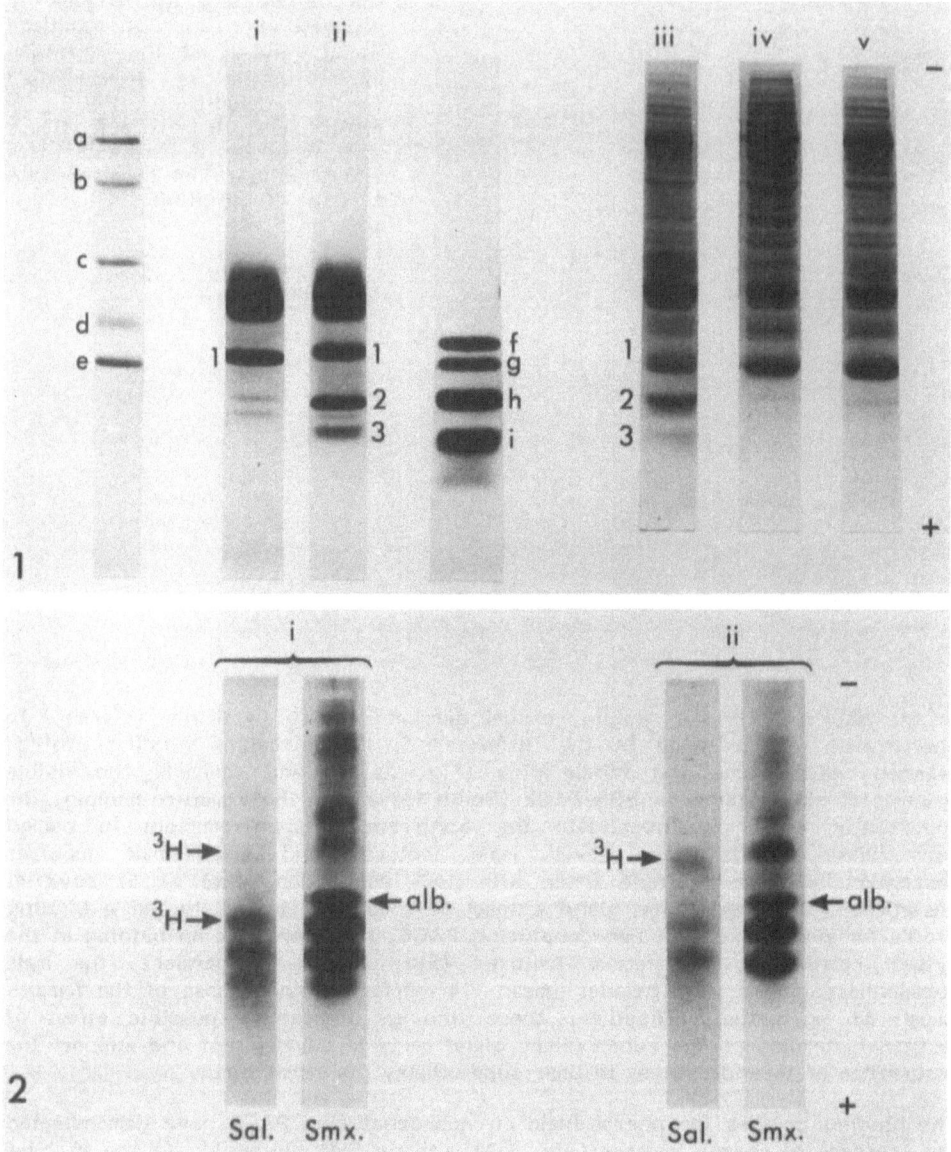

Fig. 3. 1 Presence of pheromaxein in porcine submaxillary glands as demonstrated by SDS-PAGE. i miniboar submaxillary gland ultrafiltrate not treated with thiol reducing agent; ii as i, but treated with thiol reducing agent; iii domestic boar; iv domestic castrated boar; v domestic female (note iii - v treated with thiol reducing agent). 1 pheromaxein; 2 and 3 pheromaxein fragments. Molecular weight standards a bovine serum albumin 66 K; b ovalbumin 45 K; c trypsinogen 24 K; d ß-lactoglobulin 18.4 K; e lysozyme 14.3 K; f myoglobin 16.9 K; g myoglobin I and II 14.4 K; h myoglobin I 8.2 K/myoglobin II 6.2 K (not separated); i myoglobin III 2.5 K. Gel stained with Coomassie blue 2 Position of pheromaxein on non-denaturing PAGE. i two peak binding in saliva and submaxillary gland cytosol of domestic boar; ii single peak binding in saliva and submaxillary gland cytosol of another domestic boar (see also Text - Fig. 1 a). Note visible appearance of pheromaxein in ii. Gel stained with Coomassie blue

has to be interpreted as a mixture and not that of the original radioactive ligand. In incubations with saliva which lacks secreted 3α-hydroxysteroid dehydrogenase (Katkov et al. 1972), metabolism of 16-androstenes only occurs after several days at 21°C, due it seems, to oral bacteria, with An-α primarily converted to 5α-A and 5α-A to An-ß (Booth unpublished). Pheromaxein is relatively stable since its integrity in saliva is maintained for at least a week at 4°C, but it is degraded at 21°C and 37°C within 3 days (Booth unpublished). A micropartition system (Amicon Corporation, USA) has shown that at 4°C about 10% An-α and 5α-A are in the free form in saliva, whereas at 21°C and 37°C as much as 30% of the steroids are unbound (Booth unpublished). These temperature-related phenomena could be important to the ecology of pheromone dispersal, particularly in wild or feral pigs.

3.2 Estimation of 5α-A in Human Saliva

Earlier estimations (Bird and Gower 1983) of 5α-A in human saliva utilized a radioimmunoassay (RIA) method which was time-consuming and possible non-specific; the anti-serum used could have cross-reacted with 4,16-androstadienone, if this steroid had been present in the salivary extract. Preliminary work, using SIM (Kwan et al. 1985), indicates that 5α-A is present in the pooled saliva of men at a level of approximately 160pmol/l, a value much less than that found by RIA (Bird and Gower 1983) of 600-1800 pmol/l for individual men. No 4,16-androstadienone was shown to be present by SIM. This discrepancy between the results of the two methods may indicate the non-specificity of our earlier method (Bird and Gower 1983) and this is currently under investigation.

4 Conclusions

Several studies have demonstrated the importance of the pheromonal 16-androstenes to reproduction in the pig. However, there are still aspects of reproductive and social behaviour in the pig which need further investigation with respect to the role 16-androstenes might play in these behaviours. The presence of these odorous steroids in the saliva of boar piglets may aid the establishment of a teat order, in so far as the boar piglets may preferentially occupy those teats marked with the pheromones. In older animals, there is the possibility that 16-androstenes might be involved in dominance relationships in both heterosexual and homosexual (boars) groups of pigs.

In contrast with some other species, information is lacking in the pig for a role of steroid hormones in the development of the olfactory system at the neural level.

Further work on the significance of 16-androstenes in human saliva and in other external secretions such as axillary sweat, is needed to either support or refute current thinking among many people that these musk-smelling steroids play a role in human olfactory communication.

Summary

The important role of olfaction in neonatal animals, especially with regard to nipple-search behaviour, is highlighted. The significance of the odorous 16-androstenes in porcine salivary glands and porcine and human saliva is discussed in relation to the pheromonal activity of these steroids. Evidence is presented for the presence of a specific binding protein (pheromaxein) for 16-androstenes in the submaxillary salivary glands of the domestic boar and, especially, the Göttingen miniboar. Sexual dimorphism with respect to pheromaxein was noted and it was found in the submaxillary glands of male, but not female, foetuses. The importance of the binding protein is emphasized because it may be that the porcine submaxillary gland can store, and then subsequently, transport the lipophilic pheromonal 16-androstenes in saliva.

Acknowledgments. D.B.G. is grateful to the MRC (Grant No. G8217853SB); the AFRC (Grant nos. AG35/76 and 26), The British Council and The Herbert Dunhill Trust for financial support. Drs. B. Ruparalia, G. Taylor and D. Watson kindly performed the salivary 5α-androstenone estimations. The authors wish to thank Miss Carol White for her able technical assistence.

References

Ahmad N, Gower DB (1968) The biosynthesis of some androst-16-enes from C_{21} and C_{19} steroids in boar testicular and adrenal tissue. Biochem J 108: 223-241

Bird S, Gower DB (1981) The validation and use of radio-immunoassay for 5α-androst-16-en-3-one in human axillary collections. J Steroid Biochem 14: 213-219

Bird S, Gower DB (1982) Axillary 5α-androst-16-en-3-one, cholesterol and squalene in men; preliminary evidence for 5α-androst-16-en-3-one being a product of bacterial action. J Steroid Biochem 17: 517-522

Bird S, Gower DB (1983) Estimation of the odorous steroid, 5α-androst-16-en-3-one, in human saliva. Experientia 39: 790-793

Booth WD (1970) The occurence of some C_{19} steroids and vitamin A in boar testis. J Reprod Fertil 23: 533-534

Booth WD (1972) The occurence of testosterone and 5α-dihydrotestosterone in the submaxillary salivary gland of the boar. J Endocrinol 55: 119-125

Booth WD (1975) Changes with age in the occurence of C_{19} steroids in the testis and submaxillary gland of the boar. J Reprod Fertil 42: 459-472

Booth WD (1977) Metabolism of androgens in vitro by the submaxillary gland of the mature domestic boar. J Endocrinol 75: 145-154

Booth WD (1980) Endocrine and exocrine factors in the reproductive behaviour of the pig. In: Stoddart DM (ed) Olfaction in mammals. Academic Press, London New York (Symp Zool Soc London) 45: 289-311

Booth WD (1982a) Metabolism of C_{19} steroids in vitro by the submaxillary salivary gland of mature and immature domestic pigs. J Endocrinol 93: 91-97

Booth WD (1982b) Steroid hormone and pheromone binding proteins in the submaxillary gland and saliva of the pig. In: Breipohl W (ed) Olfaction and endocrine regulation. IRL Press, London, pp 353-354

Booth WD (1982c) Testicular steroids and boar taint. In: Cole DJA, Foxcroft GF (eds) Control of pig reproduction. Butterworths, London, pp 25-48

Booth WD (1984a) Sexual dimorphism involving steroidal pheromones and their binding protein in the submaxillary salivary gland of the Göttingen miniature pig. J Endocrinol 100: 195-202

Booth WD (1984b) A note on the significance of boar salivary pheromones to the male effect on puberty attainment in gilts. Anim Prod 39: 149-152

Booth WD, Baldwin BA (1980) Lack of effect on sexual behaviour on the development of testicular function after removal of olfactory bulbs in prepubertal boars. J Reprod Fertil 58: 173-182

Booth WD, Polge C (1976) The occurence of C_{19} steroids in testicular tissue and submaxillary glands of intersex pigs in relation to morphological characteristics. J Reprod Fertil 46: 115-121

Booth WD, Hay MF, Dott HM (1973) Sexual dimorphism in the submaxillary gland of the pig. J Reprod Fertil 33: 163-166

Brooksbank BWL, Brown D, Gustafsson JA (1974) The detection of 5α-androst-16-en-3α-ol in human male axillary sweat. Experientia 30: 864-865

Claus R (1970) Bestimmung von Testosteron und 5α-androst-16-en-3-on, einem Ebergeruchsstoff, bei Schweinen. Thesis, Tech Univ, München-Weihenstephan

Claus R, Alsing W (1976) Occurence of 5α-androst-16-en-3-one, a boar pheromone, in man and its relationship to testosterone. J Endocrinol 68: 483-484

Doty RL (1981) Olfactory communication in humans. Chem Sens 6: 351-376

Gower DB (1972) 16-Unsaturated C_{19} steroids. A review of their chemistry, biochemistry and possible physiological role. J Steroid Biochem 3: 45-103

Gower DB (1981) The biosynthesis and occurence of 16-androstenes in man. In: Fotherby K, Pal SB (eds) Hormones in normal and abnormal human tissues, vol 1. de Gruyter, Berlin New York, pp 1-27

Gower DB (1984) In: Matkin HLJ (ed) Biochemistry of steroid hormones. 2nd edn Blackwell, Oxford, pp 170-206

Gower DB, Bird S, Sharma P, House FR (1985) Axillary 5α-androst-16-en-3-one in men and women: relationships with olfactory acuity to odorous 16-androstenes. Experientia (in press)

Hurden EL, Gower DB, Harrison FA (1984) Comparative rates of formation, in vitro, of 16-androstenes, testosterone and androstenedione in boar testis. J Endocrinol 103: 179-186

Jackman PJH, Noble WC (1983) Norman axillary skin microflora in various populations. Clin Exp Dermatol 8: 259-268

Jeppeson LE (1982a) Teat-order in groups of piglets reared on an artificial sow. 1. Formation of teat-order and influence of milk yield on teat performance. Appl Anim Ethol 8: 335-345

Jeppeson LE (1982b) Teat-order in groups of piglets reared on an artificial sow. II. Maintenance of teat-order with some evidence for the use of odour cues. Appl Anim Ethol 8: 347-355

Katkov T, Booth WD, Gower DB (1972) The metabolism of 16-androstenes on boar salivary glands. Biochim Biophys Acta 170: 546-556

Kirkwood RN, Hughes PE, Booth WD (1983) The influence of boar-related odours on puberty attainment in gilts. Anim Prod 36: 131-136

Kwan TK, Taylor NF, Watson D, Gower DB (1985) Gas chromatographic-mass spectometric study of metabolites of C_{21} and C_{19} steroids in neonatal porcine testicular microsomes. Biochem J 227: 909-916

Kwan TK, Orengo C, Gower DB (1986) Biosynthesis of androgens and pheromonal steroids in neonatal porcine testicular preparations. FEBS Lett (in press)

Labows JN, Preti G, Hoelzle J, Leyden JJ, Kligman A (1979) Steroid analysis of human apocrine secretion. Steroids 34: 249-258

Leyden J, McGinley KJ, Hoelzle E, Labows JN, Kligman A (1981) J Invest Dermatol 77: 413-416

Melrose DR, Reed HCB, Patterson RLS (1971) Androgen steroids associated with boar odour as an aid to the detection of oestrus in pig artificial insemination. Br Vet J 127: 497-501

Nixon A, Mallet AI, Jackman PJH, Gower DB (1984) Production of 5α- and 5ß-dihydrotestosterone by isolated human axillary bacteria. FEMS Microbiol Lett 25: 153-157

Parrot RF, Booth WD (1984) Behavioural and morphological effects of 5α-dihydotestosterone and oestradiol-17ß in the prepubertally castrated boar. J Reprod Fertil 71: 453-461

Perry GC, Patterson RLS, MacFie HJH, Stinson CG (1980) Pig courtship behaviour: pheromonal property of androstene steroids in male submaxillary secretion. Anim Prod 31: 191-199

Prelog V, Ruzicka L (1944) Über zwei moschusartig riechende Steroide aus Schweinetestes-extrakten. Helv Chim Acta 27: 61-66

Prelog V, Ruzicka L, Meister P, Wieland P (1945) Steroide und Sexualhormone. 113 Mitteilung. Untersuchungen über den Zusammenhang zwischen Konstitution und Geruch bei Steroiden. Helv Chim Acta 28: 618-627

Reed HCB, Melrose DR, Patterson RLS (1974) Androgen steroids as an aid to the detection of oestrus in pig artificial insemination. Br Vet J 130: 61-67

Russell M (1976) Human olfactory communication. Nature (London) 260: 520-522

Schleidt M, Hold B (1982) Human odour and identity. In: Breipohl W (ed) Olfaction and endocrine regulation. IRL Press, London, pp 181-194

Shehadeh NH, Kligman A (1963) The bacteria responsible for axillary odour. J Invest Dermatol 41: 3

Sink JD (1967) Theoretical aspects of sex odor in swine. J Theoret Biol 17: 174-180